Patrick Moore's
Yearbook of Astronomy
2012

Patrick Moore's Yearbook of Astronomy 2012

Special 50th Anniversary Edition

EDITED BY

Patrick Moore

AND

John Mason

MACMILLAN

First published 2011 by Macmillan
an imprint of Pan Macmillan, a division of Macmillan Publishers Limited
Pan Macmillan, 20 New Wharf Road, London N1 9RR
Basingstoke and Oxford
Associated companies throughout the world
www.panmacmillan.com

ISBN 978-0-230-75984-8

Some articles from Part II of this book are republished from previous editions:
'The Barwell Meteorite' first published in the *Yearbook of Astronomy 1968*;
'Brighter Than a Million Suns' first published in the *Yearbook of Astronomy 1979*;
'The Night Sky – AD 50,000' first published in the *Yearbook of Astronomy 1984*;
'The Biggest Structures in the Universe' first published in the *Yearbook of Astronomy 1991*;
'A Universe of Darkness' first published in the *Yearbook of Astronomy 2002*.

1 3 5 7 9 8 6 4 2

A CIP catalogue record for this book is available from
the British Library.

Typeset by Ellipsis Digital Limited
Printed and bound by CPI Group (UK) Ltd, Croydon, CR0 4YY

Visit **www.panmacmillan.com** to read more about all our books
and to buy them. You will also find features, author interviews and
news of any author events, and you can sign up for e-newsletters
so that you're always first to hear about our new releases.

Contents

Contents

Part Two
Article Section

Contents

Part Three
Miscellaneous

Editors' Foreword

With the *2012 Yearbook*, we are celebrating fifty years since the first edition appeared in print. The very first *Yearbook of Astronomy* was published in 1962 under the editorship of J. G. Porter, with Patrick Moore as Associate Editor. It contained the same basic mix of astronomical data for the year, general information and contributed articles as the *Yearbook* does today, but there have been a great many refinements over the years. Dr Porter retired as Editor in 1965 and Patrick Moore carried out the task single-handedly until 2001, when he was joined by John Mason as Associate Editor; the two of us became Co-editors in 2005.

Over the past five decades that the *Yearbook* has appeared, we have been fortunate to have contributions from some of the world's most eminent astronomers, both amateur and professional. For this special fiftieth anniversary publication, we decided to look back through past editions and pick out one contributed article from each decade which we thought would be of interest to readers today. In all cases the articles have been updated as necessary. From the 1960s we have Howard Miles's description of the Barwell meteorite fall, still the heaviest recorded meteoritic fall in England and the second heaviest in the British Isles. From the 1970s, we have the late David Allen's article about the remarkable star Eta Carinae. David was one of the *Yearbook*'s most regular and welcome contributors until his untimely death in 1994, a few days before his forty-eighth birthday. From the 1980s, we have included Steve Bell's fascinating look into the night sky of AD 50,000. From the 1990s, there is the late Tony Fairall's article about mapping the biggest structures in the universe. Tony died in a tragic diving accident in 2008. Finally, from 2002, we have Iain Nicolson's important article on dark matter and dark energy. We are indebted to Professor Fred Watson for carrying out the necessary updating of the original articles by David Allen and Tony Fairall.

As well as looking back at some past highlights, this bumper edition of the *Yearbook* follows the long-established pattern. Wil Tirion, who produced our stars maps for the Northern and Southern Hemispheres, has again drawn all of the line diagrams showing the positions and movements of the planets to accompany the Monthly Notes. Martin Mobberley has provided the notes on eclipses, comets and minor planets, and Nick James has provided the data for the phases of the Moon, longitudes of the Sun, Moon and planets, and details of lunar occultations. As always, John Isles and Bob Argyle have provided the information on variable stars and double stars, respectively.

We also have a fine selection of previously unpublished longer articles, both from our regular contributors and from those new to the *Yearbook* this year. As usual, we have done our best to give you a wide range, both of subject and of technical level. In keeping with the main theme of the *2012 Yearbook*, some authors have looked back over the past fifty years of progress in their chosen subject areas. For example, David Harland reviews the many highlights of Solar System exploration by robot probes since 1962. Ninian Boyle looks back briefly at fifty years of solar observing before reviewing the very latest observational techniques. In a similar vein, Fred Watson looks at the remarkable progress made in the technology of the instruments used on telescopes, before bringing the story right up-to-date with applications of fibre optics and astrophotonics. For the amateur observer, Martin Mobberley gives some important advice on choosing a telescope and Bill Leatherbarrow looks at observing the Moon, still a favourite with many observers. We also have a fascinating discussion of stellar mass black holes by Paul Abel, a look at some curious episodes in the history of planetary observation by Richard Baum and, in a year which will see the last transit of Venus for over 105 years, Allan Chapman gives an absorbing account of the eighteenth-century transits of Venus.

Patrick Moore
John Mason
Selsey, August 2011

Preface

New readers will find that all the information in this *Yearbook* is given in diagrammatic or descriptive form; the positions of the planets may easily be found from the specially designed star charts, while the Monthly Notes describe the movements of the planets and give details of other astronomical phenomena visible in both the Northern and Southern hemispheres. Two sets of star charts are provided. The **Northern Charts** (pp. 7–32) are designed for use at latitude 52°N, but may be used without alteration throughout the British Isles, and (except in the case of eclipses and occultations) in other countries of similar northerly latitude. The **Southern Charts** (pp. 33–57) are drawn for latitude 35°S, and are suitable for use in South Africa, Australia and New Zealand, and other locations in approximately the same southerly latitude. The reader who needs more detailed information will find *Norton's Star Atlas* an invaluable guide, while more precise positions of the planets and their satellites, together with predictions of occultations, meteor showers and periodic comets, may be found in the *Handbook of the British Astronomical Association*. Readers will also find details of forthcoming events given in the American monthly magazine *Sky and Telescope* and the British periodicals *The Sky at Night*, *Astronomy Now* and *Astronomy and Space*.

Important note

The times given on the star charts and in the Monthly Notes are generally given as local times, using the 24-hour clock, the day beginning at midnight. All the dates, and the times of a few events (e.g. eclipses) are given in Greenwich Mean Time (GMT), which is related to local time by the formula:

Local Mean Time = GMT − west longitude

In practice, small differences in longitude are ignored, and the observer will use local clock time, which will be the appropriate Standard (or

Zone) Time. As the formula indicates, places in west longitude will have a Standard Time slow on GMT, while places in east longitude will have a Standard Time fast on GMT. As examples we have:

Standard Time in

New Zealand	GMT + 12 hours
Victoria, NSW	GMT + 10 hours
Western Australia	GMT + 8 hours
South Africa	GMT + 2 hours
British Isles	GMT
Eastern ST	GMT − 5 hours
Central ST	GMT − 6 hours, etc.

If Summer Time is in use, the clocks will have been advanced by one hour, and this hour must be subtracted from the clock time to give Standard Time.

Part One

Monthly Charts and Astronomical Phenomena

Notes on the Star Charts

The stars, together with the Sun, Moon and planets, seem to be set on the surface of the celestial sphere, which appears to rotate about the Earth from east to west. Since it is impossible to represent a curved surface accurately on a plane, any kind of star map is bound to contain some form of distortion.

Most of the monthly star charts which appear in the various journals and some national newspapers are drawn in circular form. This is perfectly accurate, but it can make the charts awkward to use. For the star charts in this volume, we have preferred to give two hemispherical maps for each month of the year, one showing the northern aspect of the sky and the other showing the southern aspect. Two sets of monthly charts are provided, one for observers in the Northern Hemisphere and one for those in the Southern Hemisphere.

Unfortunately, the constellations near the overhead point (the zenith) on these hemispherical charts can be rather distorted. This would be a serious drawback for precision charts, but what we have done is give maps which are best suited to star recognition. We have also refrained from putting in too many stars, so that the main patterns stand out clearly. To help observers with any distortions near the zenith, and the lack of overlap between the charts of each pair, we have also included two circular maps, one showing all the constellations in the northern half of the sky, and one showing those in the southern half. Incidentally, there is a curious illusion that stars at an altitude of 60° or more are actually overhead, and beginners may often feel that they are leaning over backwards in trying to see them.

The charts show all stars down to the fourth magnitude, together with a number of fainter stars which are necessary to define the shapes of constellations. There is no standard system for representing the outlines of the constellations, and triangles and other simple figures have been used to give outlines which are easy to trace with the naked eye. The names of the constellations are given, together with the proper names of the brighter stars. The apparent magnitudes of the stars are

3

indicated roughly by using different sizes of dot, the larger dots representing the brighter stars.

The two sets of star charts – one each for Northern and Southern Hemisphere observers – are similar in design. At each opening there is a single circular chart which shows all the constellations in that hemisphere of the sky. (These two charts are centred on the North and South Celestial Poles, respectively.) Then there are twelve double-page spreads, showing the northern and southern aspects for each month of the year for observers in that hemisphere. In the **Northern Charts** (drawn for latitude 52°N) the left-hand chart of each spread shows the northern half of the sky (lettered 1N, 2N, 3N . . . 12N), and the corresponding right-hand chart shows the southern half of the sky (lettered 1S, 2S, 3S . . . 12S). The arrangement and lettering of the charts is exactly the same for the **Southern Charts** (drawn for latitude 35°S).

Because the sidereal day is shorter than the solar day, the stars appear to rise and set about four minutes earlier each day, and this amounts to two hours in a month. Hence the twelve pairs of charts in each set are sufficient to give the appearance of the sky throughout the day at intervals of two hours, or at the same time of night at monthly intervals throughout the year. For example, charts 1N and 1S here are drawn for 23 hours on 6 January. The view will also be the same on 6 October at 05 hours; 6 November at 03 hours; 6 December at 01 hours and 6 February at 21 hours. The actual range of dates and times when the stars on the charts are visible is indicated on each page. Each pair of charts is numbered in bold type, and the number to be used for any given month and time may be found from the following table:

Local Time	18h	20h	22h	0h	2h	4h	6h
January	11	12	1	2	3	4	5
February	12	1	2	3	4	5	6
March	1	2	3	4	5	6	7
April	2	3	4	5	6	7	8
May	3	4	5	6	7	8	9
June	4	5	6	7	8	9	10
July	5	6	7	8	9	10	11
August	6	7	8	9	10	11	12
September	7	8	9	10	11	12	1

Local Time	18h	20h	22h	0h	2h	4h	6h
October	8	9	10	11	12	1	2
November	9	10	11	12	1	2	3
December	10	11	12	1	2	3	4

On these charts, the ecliptic is drawn as a broken line on which longitude is marked every 10°. The positions of the planets are then easily found by reference to the table on p. 64. It will be noticed that on the **Southern Charts** the ecliptic may reach an altitude in excess of 62.5° on the star charts showing the northern aspect (5N to 9N). The continuations of the broken line will be found on the corresponding charts for the southern aspect (5S, 6S, 8S and 9S).

Northern Star Charts

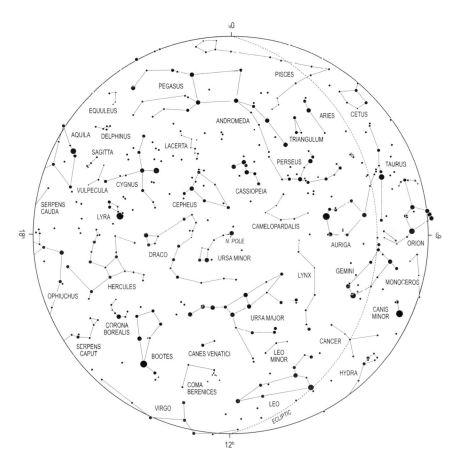

Northern Hemisphere

Note that the markers at 0ʰ, 6ʰ, 12ʰ and 18ʰ
indicate hours of Right Ascension.

1N

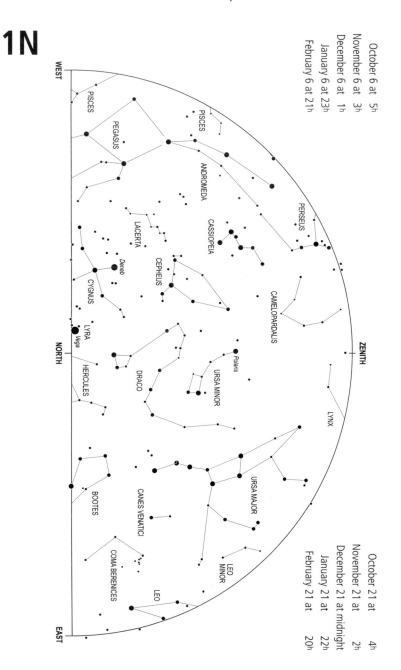

October 6 at 5h
November 6 at 3h
December 6 at 1h
January 6 at 23h
February 6 at 21h

October 21 at 4h
November 21 at 2h
December 21 at midnight
January 21 at 22h
February 21 at 20h

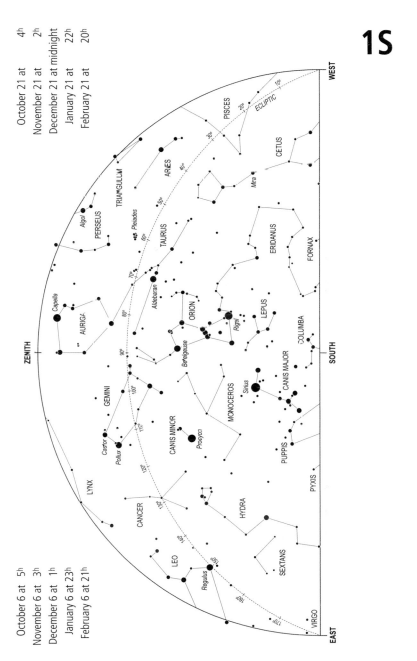

1S

October 21 at 4ʰ
November 21 at 2ʰ
December 21 at midnight
January 21 at 22ʰ
February 21 at 20ʰ

October 6 at 5ʰ
November 6 at 3ʰ
December 6 at 1ʰ
January 6 at 23ʰ
February 6 at 21ʰ

2N

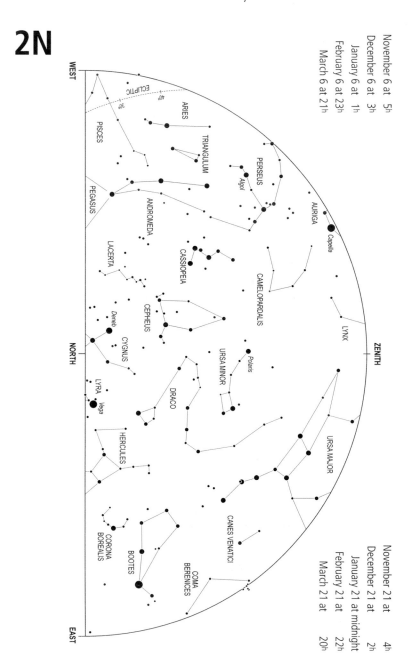

November 21 at 4h
December 21 at 2h
January 21 at midnight
February 21 at 22h
March 21 at 20h

WEST

ECLIPTIC

ARIES

PISCES

TRIANGULUM

PERSEUS

Algol

PEGASUS

ANDROMEDA

AURIGA

Capella

LACERTA

CASSIOPEIA

CAMELOPARDALIS

ZENITH

CEPHEUS

Deneb

LYNX

CYGNUS

URSA MINOR

Polaris

NORTH

LYRA

Vega

DRACO

URSA MAJOR

HERCULES

CANES VENATICI

CORONA
BOREALIS

BOOTES

COMA
BERENICES

EAST

Northern Star Charts

2S

WEST

ZENITH

SOUTH

EAST

CETUS
Mira
PERSEUS
Pleiades
TAURUS
Aldebaran
ERIDANUS
Capella
AURIGA
ORION
Rigel
LEPUS
Betelgeuse
GEMINI
Castor
Pollux
CANIS MINOR
Procyon
MONOCEROS
Sirius
CANIS MAJOR
LYNX
HYDRA
PUPPIS
URSA MAJOR
CANCER
LEO MINOR
LEO
Regulus
SEXTANS
PYXIS
ANTLIA
CRATER
CORVUS
COMA BERENICES
VIRGO
ECLIPTIC

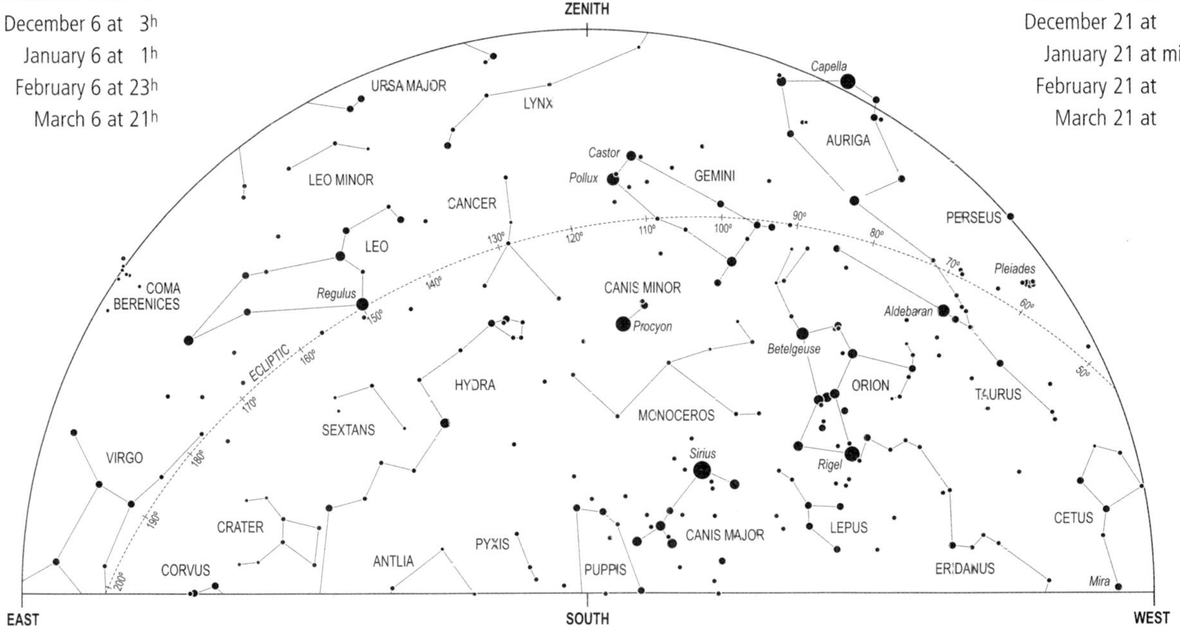

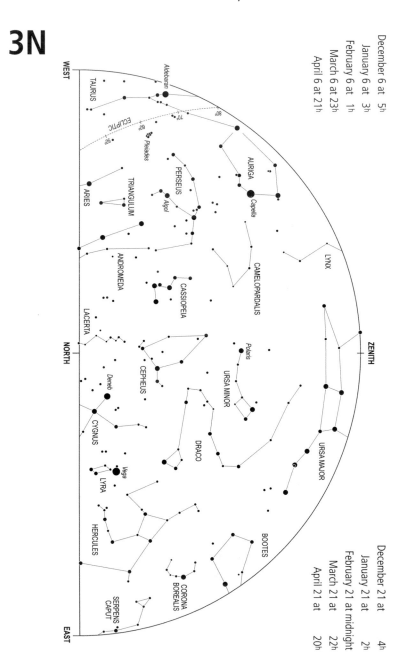

3N

December 6 at 5h
January 6 at 3h
February 6 at 1h
March 6 at 23h
April 6 at 21h

December 21 at 4h
January 21 at 2h
February 21 at midnight
March 21 at 22h
April 21 at 20h

WEST

TAURUS
Aldebaran
ECLIPTIC
Pleiades
ARIES
TRIANGULUM
PERSEUS
Algol
AURIGA
Capella
LYNX
ANDROMEDA
CAMELOPARDALIS
CASSIOPEIA
LACERTA
CEPHEUS
Polaris
URSA MINOR
ZENITH
Deneb
CYGNUS
DRACO
URSA MAJOR
Vega
LYRA
HERCULES
BOOTES
CORONA BOREALIS
SERPENS CAPUT

NORTH

EAST

3S

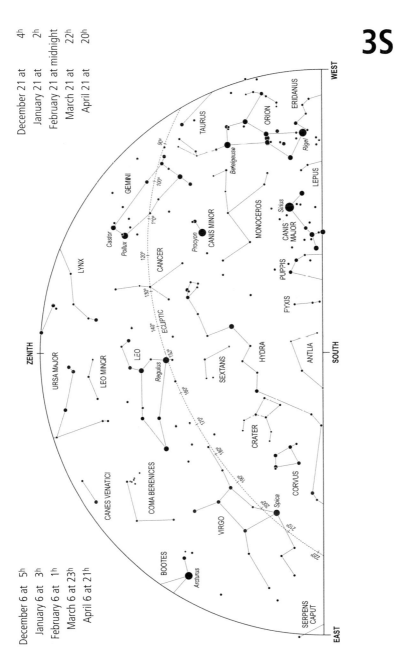

December 21 at 4ʰ
January 21 at 2ʰ
February 21 at midnight
March 21 at 22ʰ
April 21 at 20ʰ

December 6 at 5ʰ
January 6 at 3ʰ
February 6 at 1ʰ
March 6 at 23ʰ
April 6 at 21ʰ

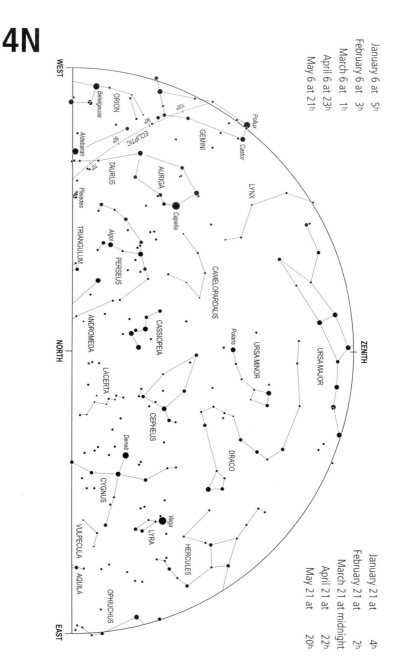

4N

January 6 at 5h
February 6 at 3h
March 6 at 1h
April 6 at 23h
May 6 at 21h

January 21 at 4h
February 21 at 2h
March 21 at midnight
April 21 at 22h
May 21 at 20h

WEST

NORTH

ZENITH

EAST

14

4S

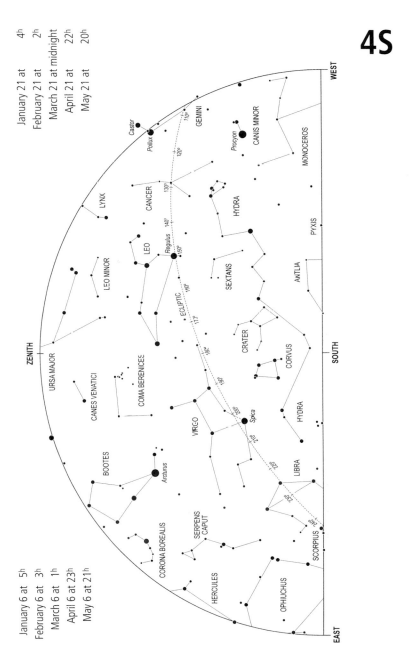

January 21 at 4ʰ
February 21 at 2ʰ
March 21 at midnight
April 21 at 22ʰ
May 21 at 20ʰ

January 6 at 5ʰ
February 6 at 3ʰ
March 6 at 1ʰ
April 6 at 23ʰ
May 6 at 21ʰ

WEST
SOUTH
EAST
ZENITH

GEMINI
Castor
Pollux
CANIS MINOR
Procyon
MONOCEROS
LYNX
CANCER
HYDRA
LEO
LEO MINOR
Regulus
SEXTANS
PYXIS
ANTLIA
URSA MAJOR
ECLIPTIC
CRATER
CANES VENATICI
COMA BERENICES
CORVUS
HYDRA
VIRGO
Spica
BOOTES
Arcturus
LIBRA
SERPENS CAPUT
CORONA BOREALIS
HERCULES
OPHIUCHUS
SCORPIUS

110°
120°
130°
140°
150°
160°
170°
180°
190°
200°
210°
220°
230°
240°

15

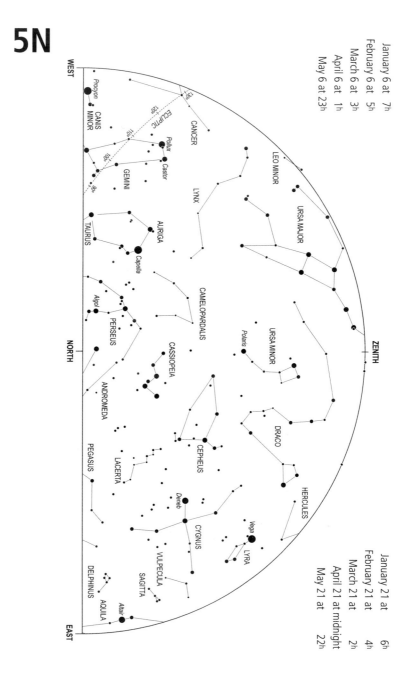

5N

WEST

Procyon
CANIS MINOR
ECLIPTIC
CANCER
Pollux
Castor
GEMINI
LYNX
TAURUS
AURIGA
Capella
Algol
PERSEUS
CAMELOPARDALIS
CASSIOPEIA
ANDROMEDA
PEGASUS
LACERTA
CEPHEUS
Polaris
URSA MINOR
DRACO
Deneb
CYGNUS
VULPECULA
SAGITTA
DELPHINUS
Altair
AQUILA
Vega
LYRA
HERCULES
LEO MINOR
URSA MAJOR
ZENITH

NORTH

EAST

January 6 at 7h
February 6 at 5h
March 6 at 3h
April 6 at 1h
May 6 at 23h

January 21 at 6h
February 21 at 4h
March 21 at 2h
April 21 at midnight
May 21 at 22h

Northern Star Charts

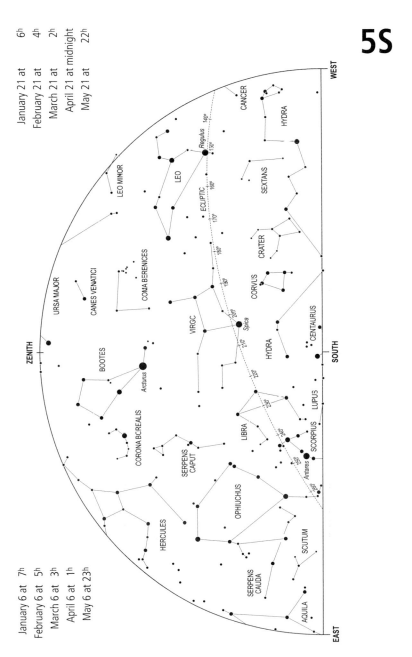

January 21 at 6ʰ
February 21 at 4ʰ
March 21 at 2ʰ
April 21 at midnight
May 21 at 22ʰ

January 6 at 7ʰ
February 6 at 5ʰ
March 6 at 3ʰ
April 6 at 1ʰ
May 6 at 23ʰ

6N

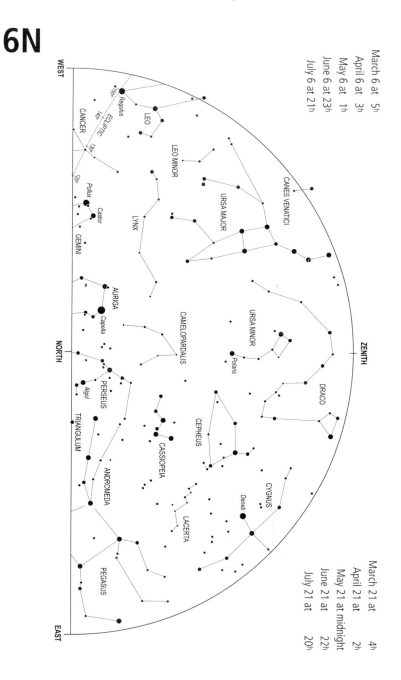

WEST

March 6 at 5h
April 6 at 3h
May 6 at 1h
June 6 at 23h
July 6 at 21h

ZENITH

NORTH

EAST

March 21 at 4h
April 21 at 2h
May 21 at midnight
June 21 at 22h
July 21 at 20h

6S

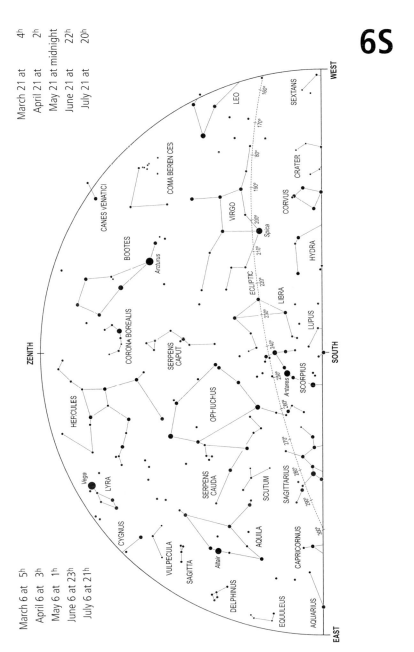

March 21 at 4h
April 21 at 2h
May 21 at midnight
June 21 at 22h
July 21 at 20h

March 6 at 5h
April 6 at 3h
May 6 at 1h
June 6 at 23h
July 6 at 21h

WEST

ZENITH

SOUTH

EAST

LEO
SEXTANS
COMA BERENCES
CANES VENATICI
CRATER
CORVUS
VIRGO
HYDRA
BOOTES
Arcturus
Spica
ECLIPTIC
CORONA BOREALIS
LIBRA
SERPENS CAPUT
LUPUS
HERCULES
OPHIUCHUS
Antares
SCORPIUS
Vega
LYRA
SERPENS CAUDA
CYGNUS
SCUTUM
SAGITTARIUS
VULPECULA
AQUILA
CAPRICORNUS
SAGITTA
Altair
DELPHINUS
EQUULEUS
AQUARIUS

160°
170°
80°
190°
200°
210°
220°
230°
240°
250°
260°
270°
280°
290°
300°

19

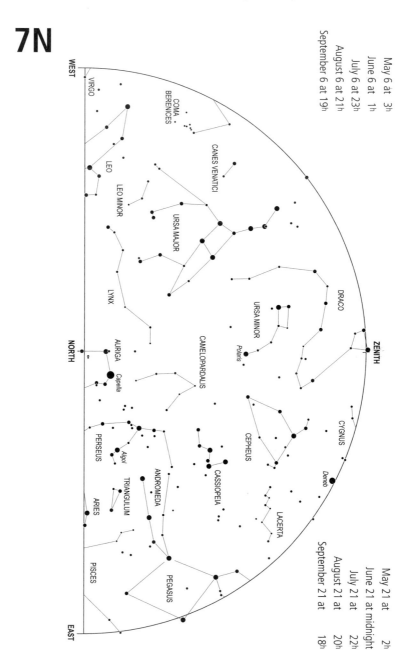

7N

May 6 at 3h
June 6 at 1h
July 6 at 23h
August 6 at 21h
September 6 at 19h

May 21 at 2h
June 21 at midnight
July 21 at 22h
August 21 at 20h
September 21 at 18h

Northern Star Charts

7S

May 21 at 2ʰ
June 21 at midnight
July 21 at 22ʰ
August 21 at 20ʰ
September 21 at 18ʰ

May 6 at 3ʰ
June 6 at 1ʰ
July 6 at 23ʰ
August 6 at 21ʰ
September 6 at 19ʰ

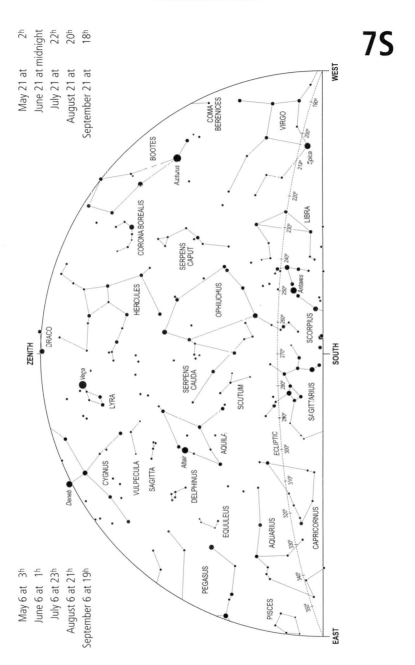

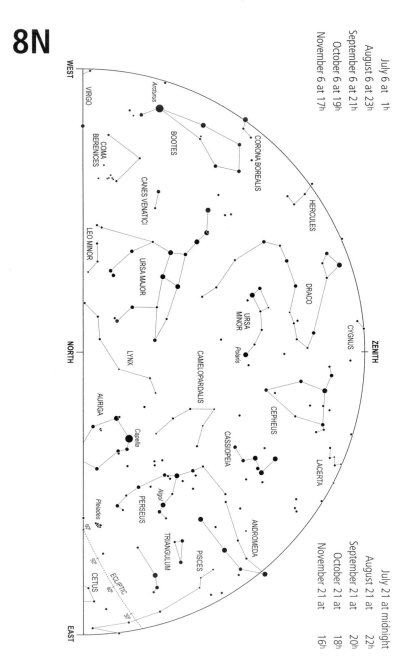

Northern Star Charts

8S

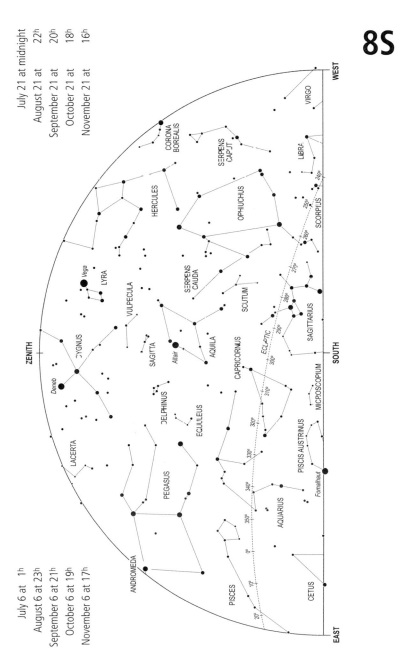

July 21 at midnight
August 21 at 22ʰ
September 21 at 20ʰ
October 21 at 18ʰ
November 21 at 16ʰ

July 6 at 1ʰ
August 6 at 23ʰ
September 6 at 21ʰ
October 6 at 19ʰ
November 6 at 17ʰ

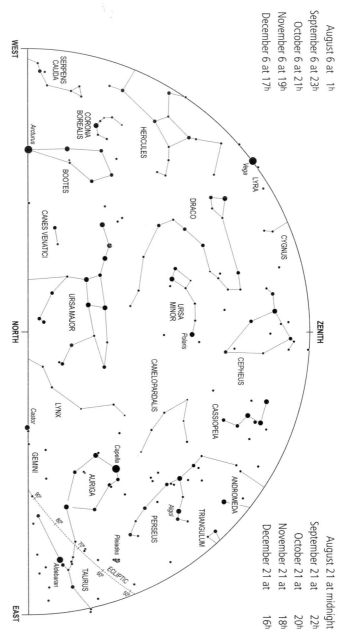

Northern Star Charts

August 21 at midnight
September 21 at 22ʰ
October 21 at 20ʰ
November 21 at 18ʰ
December 21 at 16ʰ

August 6 at 1ʰ
September 6 at 23ʰ
October 6 at 21ʰ
November 6 at 19ʰ
December 6 at 17ʰ

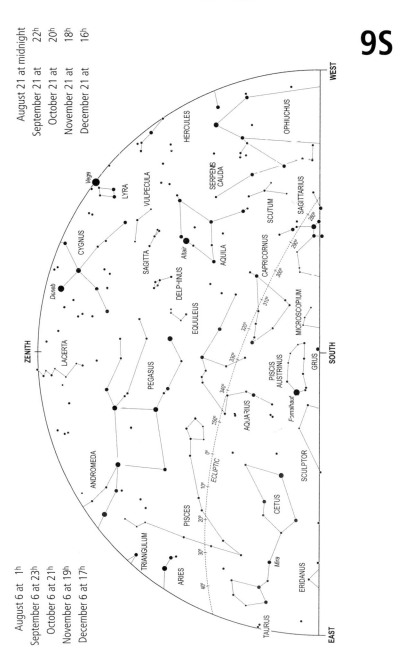

WEST

ZENITH

SOUTH

EAST

HERCULES
OPHIUCHUS
Vega
LYRA
VULPECULA
SERPENS
CAUDA
SCUTUM
SAGITTARIUS
CYGNUS
SAGITTA
DELPHINUS
Altair
AQUILA
CAPRICORNUS
MICROSCOPIUM
Deneb
EQUULEUS
LACERTA
PEGASUS
PISCIS
AUSTRINUS
GRUS
ANDROMEDA
AQUARIUS
Fomalhaut
TRIANGULUM
PISCES
CETUS
SCULPTOR
ARIES
Mira
ERIDANUS
TAURUS
ECLIPTIC

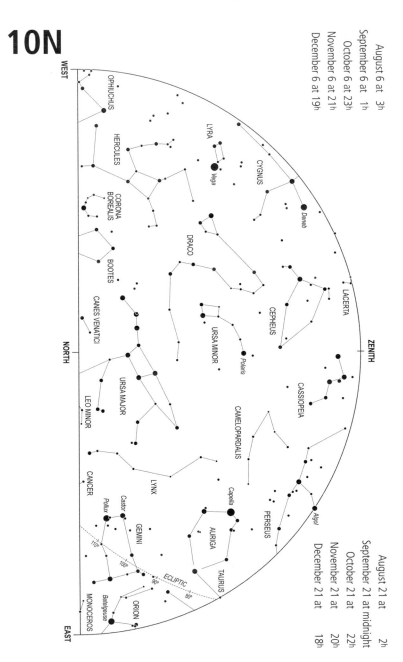

10N

August 6 at 3h
September 6 at 1h
October 6 at 23h
November 6 at 21h
December 6 at 19h

August 21 at 2h
September 21 at midnight
October 21 at 22h
November 21 at 20h
December 21 at 18h

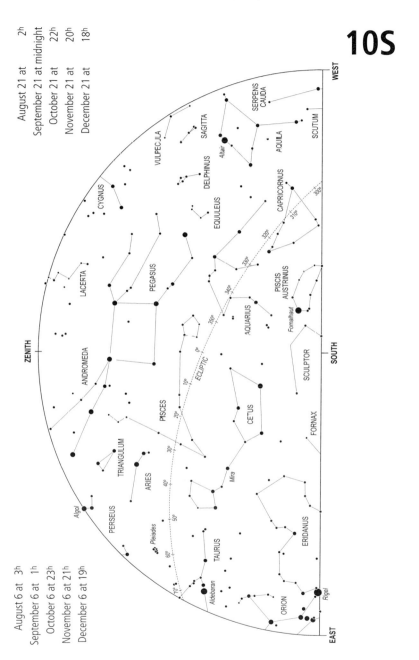

10S

August 6 at 3ʰ
September 6 at 1ʰ
October 6 at 23ʰ
November 6 at 21ʰ
December 6 at 19ʰ

11N

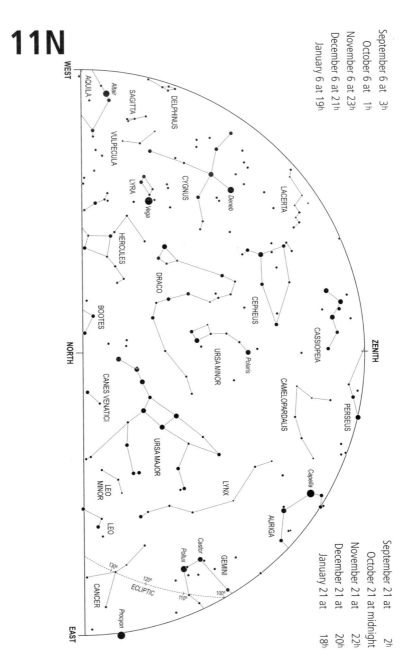

11S

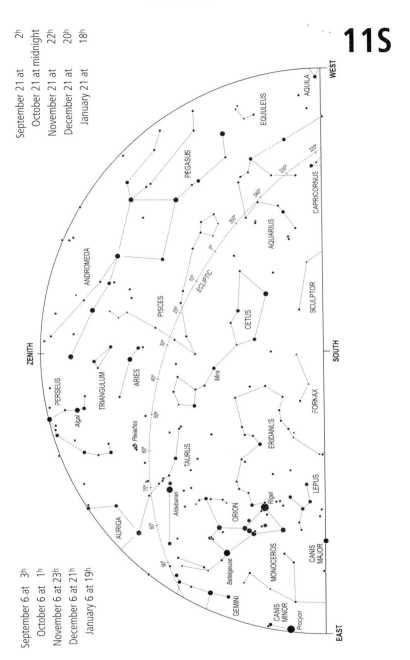

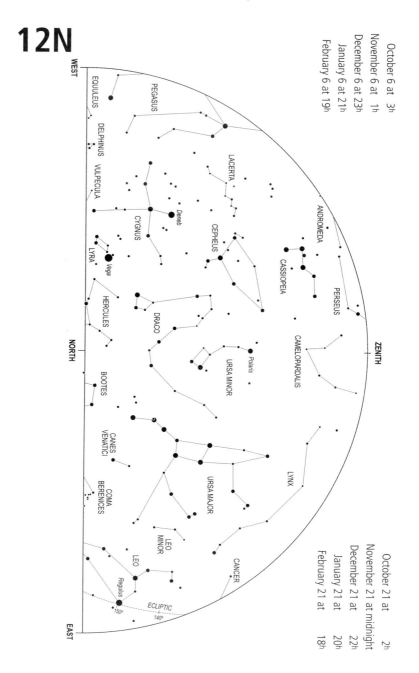

12N

12S

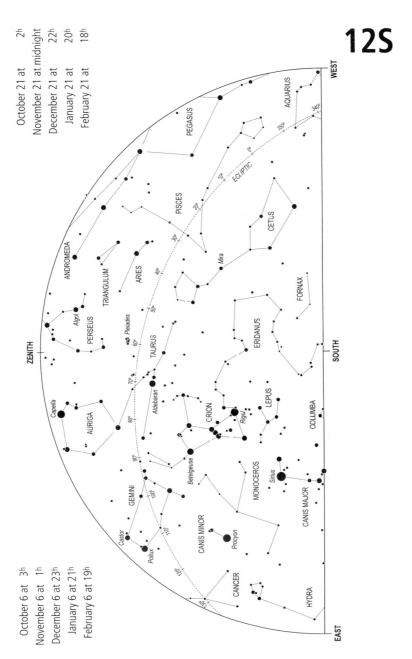

Southern Star Charts

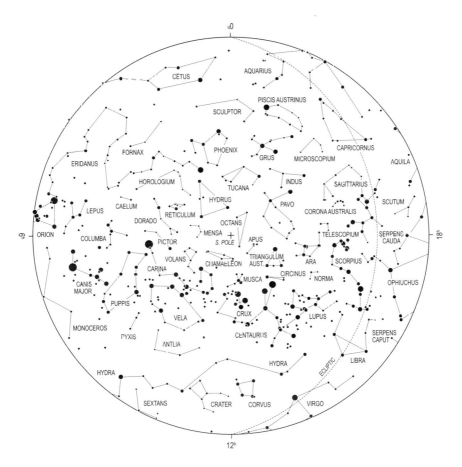

Southern Hemisphere

Note that the markers at 0ʰ, 6ʰ, 12ʰ and 18ʰ
indicate hours of Right Ascension.

33

1N

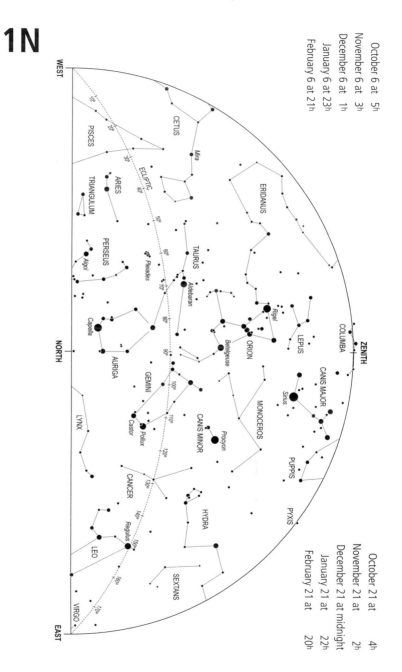

October 6 at 5h
November 6 at 3h
December 6 at 1h
January 6 at 23h
February 6 at 21h

October 21 at 4h
November 21 at 2h
December 21 at midnight
January 21 at 22h
February 21 at 20h

Southern Star Charts

October 21 at 4ʰ
November 21 at 2ʰ
December 21 at midnight
January 21 at 22ʰ
February 21 at 20ʰ

October 6 at 5ʰ
November 6 at 3ʰ
December 6 at 1ʰ
January 6 at 23ʰ
February 6 at 21ʰ

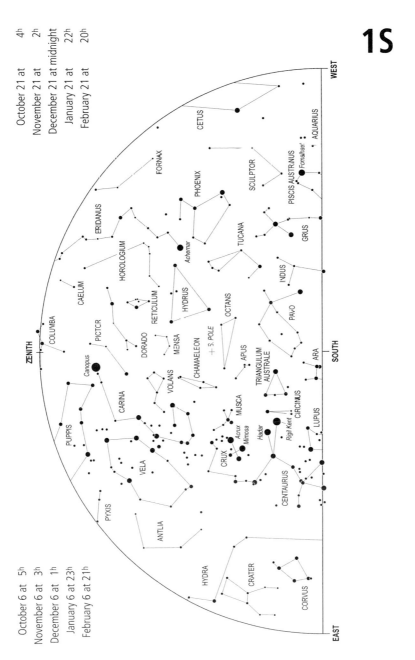

WEST

SOUTH

EAST

ZENITH

CETUS
FORNAX
PHOENIX
SCULPTOR
PISCIS AUSTRINUS
AQUARIUS
Fomalhaut
ERIDANUS
HOROLOGIUM
CAELUM
Achernar
TUCANA
GRUS
INDUS
COLUMBA
PICTOR
RETICULUM
HYDRUS
OCTANS
PAVO
DORADO
MENSA
S. POLE
Canopus
CARINA
VOLANS
CHAMAELEON
APUS
TRIANGULUM AUSTRALE
ARA
PUPPIS
MUSCA
CIRCINUS
Acrux
Mimosa
Hadar
Rigil Kent
LUPUS
VELA
CRUX
CENTAURUS
PYXIS
ANTLIA
HYDRA
CRATER
CORVUS

35

2N

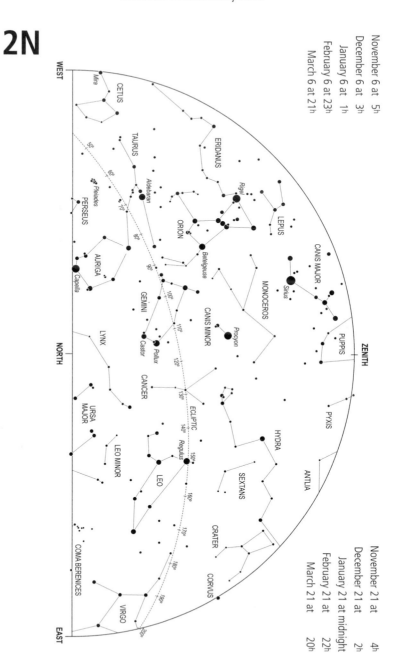

Southern Star Charts

November 21 at 4ʰ
December 21 at 2ʰ
January 21 at midnight
February 21 at 22ʰ
March 21 at 20ʰ

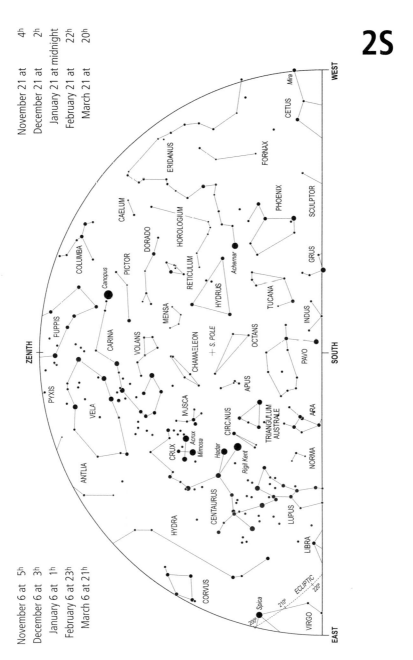

November 6 at 5ʰ
December 6 at 3ʰ
January 6 at 1ʰ
February 6 at 23ʰ
March 6 at 21ʰ

37

3N

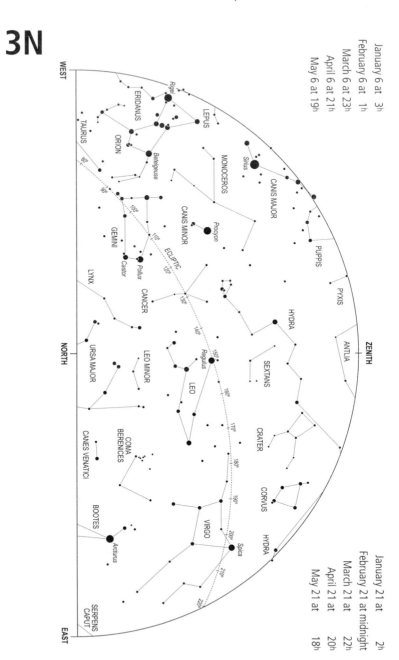

3S

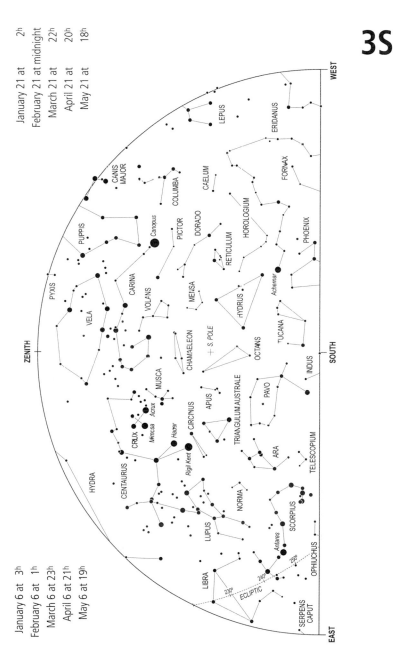

January 21 at 2ʰ
February 21 at midnight
March 21 at 22ʰ
April 21 at 20ʰ
May 21 at 18ʰ

January 6 at 3ʰ
February 6 at 1ʰ
March 6 at 23ʰ
April 6 at 21ʰ
May 6 at 19ʰ

WEST

ZENITH

SOUTH

EAST

LEPUS
ERIDANUS
CANIS MAJOR
COLUMBA
CAELUM
FORNAX
PUPPIS
Canopus
PICTOR
DORADO
HOROLOGIUM
PHOENIX
PYXIS
CARINA
RETICULUM
Achernar
VELA
VOLANS
MENSA
HYDRUS
TUCANA
+ S. POLE
CHAMAELEON
OCTANS
INDUS
MUSCA
Acrux
APUS
TRIANGULUM AUSTRALE
PAVO
Mimosa
Hadar
CIRCINUS
CRUX
Rigil Kent
ARA
TELESCOPIUM
HYDRA
CENTAURUS
NORMA
SCORPIUS
LUPUS
Antares
250°
240°
230°
LIBRA
ECLIPTIC
OPHIUCHUS
SERPENS CAPUT

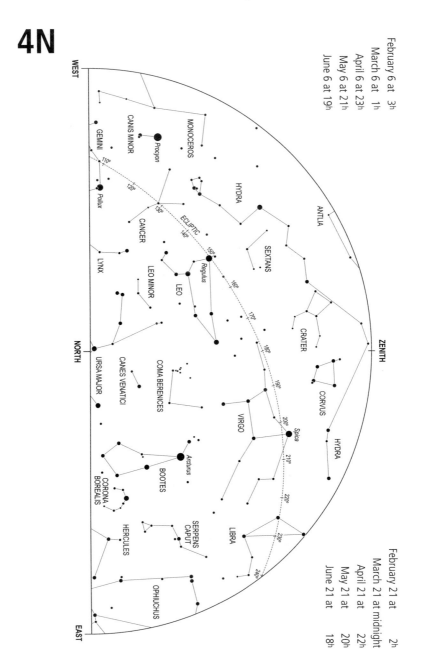

Southern Star Charts

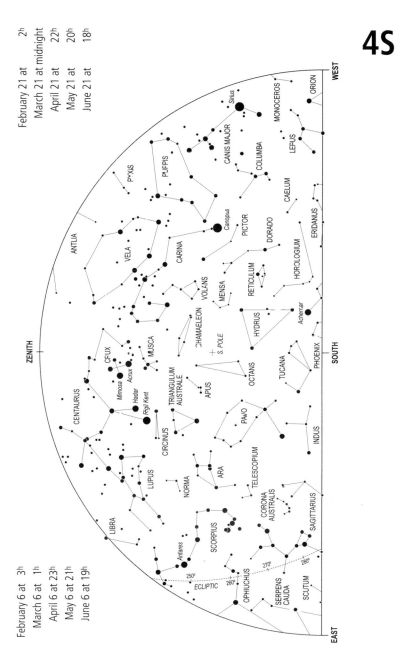

4S

February 21 at 2ʰ
March 21 at midnight
April 21 at 22ʰ
May 21 at 20ʰ
June 21 at 18ʰ

February 6 at 3ʰ
March 6 at 1ʰ
April 6 at 23ʰ
May 6 at 21ʰ
June 6 at 19ʰ

WEST
EAST
SOUTH
ZENITH

ORION
MONOCEROS
LEPUS
Sirius
CANIS MAJOR
COLUMBA
CAELUM
ERIDANUS
PUPPIS
PYXIS
Canopus
PICTOR
DORADO
HOROLOGIUM
CARINA
RETICULUM
VELA
ANTLIA
VOLANS
MENSA
Achernar
HYDRUS
CHAMAELEON
PHOENIX
S. POLE
CRUX
MUSCA
Acrux
Mimosa
Hadar
OCTANS
TUCANA
CENTAURUS
TRIANGULUM
AUSTRALE
APUS
PAVO
Rigil Kent
CIRCINUS
INDUS
LUPUS
NORMA
ARA
TELESCOPIUM
LIBRA
CORONA
AUSTRALIS
SAGITTARIUS
Antares
SCORPIUS
270°
280°
250°
260°
ECLIPTIC
OPHIUCHUS
SERPENS
CAUDA
SCUTUM

41

5N

March 6 at 3h
April 6 at 1h
May 6 at 23h
June 6 at 21h
July 6 at 19h

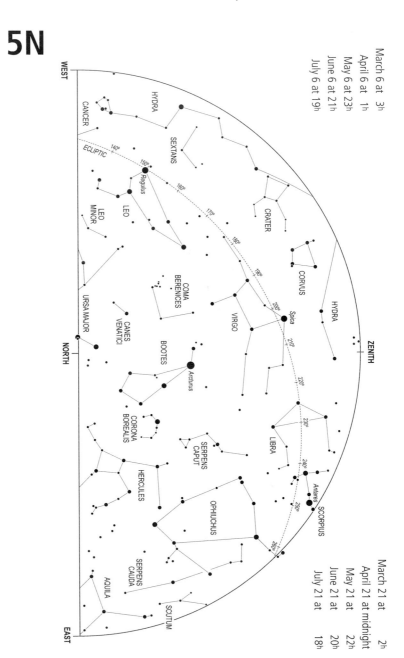

March 21 at 2h
April 21 at midnight
May 21 at 22h
June 21 at 20h
July 21 at 18h

Southern Star Charts

March 21 at 2ʰ
April 21 at midnight
May 21 at 22ʰ
June 21 at 20ʰ
July 21 at 18ʰ

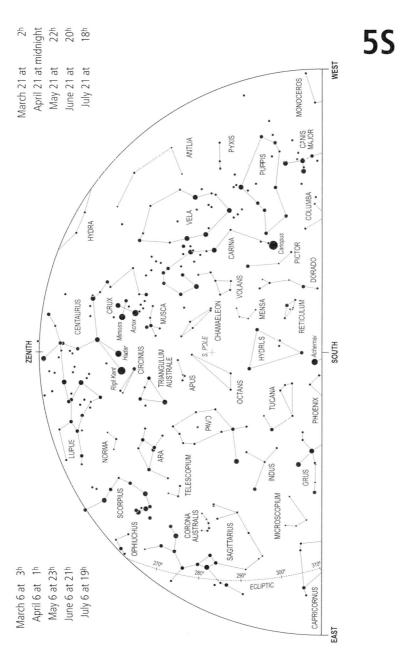

WEST

MONOCEROS
CANIS MAJOR
ANTLIA
PYXIS
PURPIS
COLUMBA
HYDRA
VELA
CARINA
Canopus
PICTOR
DORADO
CENTAURUS
CRUX
MUSCA
VOLANS
MENSA
RETICULUM
ZENITH
Mimosa
Acrux
CHAMAELEON
S. POLE
HYDRUS
Hadar
CIRCINUS
TRIANGULUM AUSTRALE
Achernar
Rigil Kent
APUS
OCTANS
TUCANA
SOUTH
LUPUS
NORMA
ARA
PAVO
INDUS
PHOENIX
TELESCOPIUM
GRUS
SCORPIUS
OPHIUCHUS
CORONA AUSTRALIS
SAGITTARIUS
MICROSCOPIUM
270°
280°
290°
300°
310°
ECLIPTIC
CAPRICORNUS

EAST

March 6 at 3ʰ
April 6 at 1ʰ
May 6 at 23ʰ
June 6 at 21ʰ
July 6 at 19ʰ

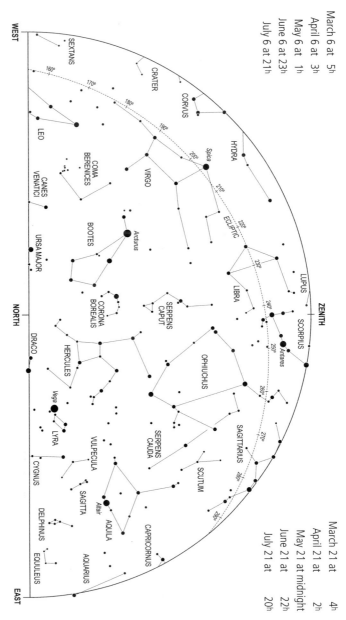

6N

March 6 at 5h
April 6 at 3h
May 6 at 1h
June 6 at 23h
July 6 at 21h

March 21 at 4h
April 21 at 2h
May 21 at midnight
June 21 at 22h
July 21 at 20h

Southern Star Charts

6S

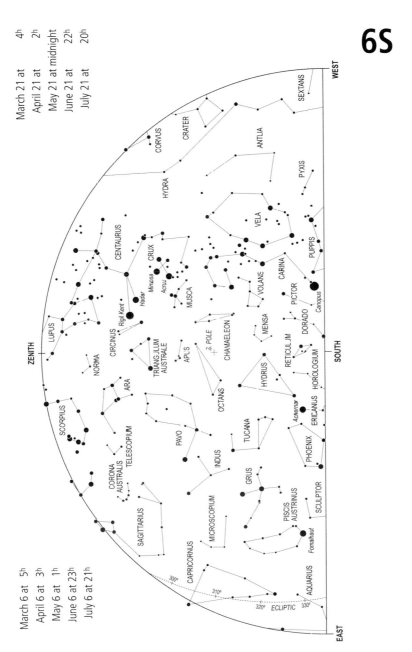

March 6 at 5ʰ
April 6 at 3ʰ
May 6 at 1ʰ
June 6 at 23ʰ
July 6 at 21ʰ

WEST

SEXTANS

CORVUS
CRATER
ANTLIA
PYXIS

HYDRA

VELA

CENTAURUS
CRUX
Mimosa
Acrux
MUSCA
CARINA
VOLANS
PUPPIS

Hadar
Rigil Kent
CIRCINUS
PICTOR
DORADO
Canopus

LUPUS
δ. POLE
CHAMAELEON
MENSA
RETICULUM
HOROLOGIUM
SOUTH

NORMA
TRIANGULUM AUSTRALE
APLS
HYDRUS
ERIDANUS

ARA
OCTANS
Achernar
PHOENIX

SCORPIUS
PAVO
TUCANA

CORONA AUSTRALIS
TELESCOPIUM
INDUS
GRUS
PISCIS AUSTRINUS
SCULPTOR

SAGITTARIUS
MICROSCOPIUM
Fomalhaut

CAPRICORNUS
AQUARIUS
300°
310°
320° ECLIPTIC 330°

ZENITH

EAST

45

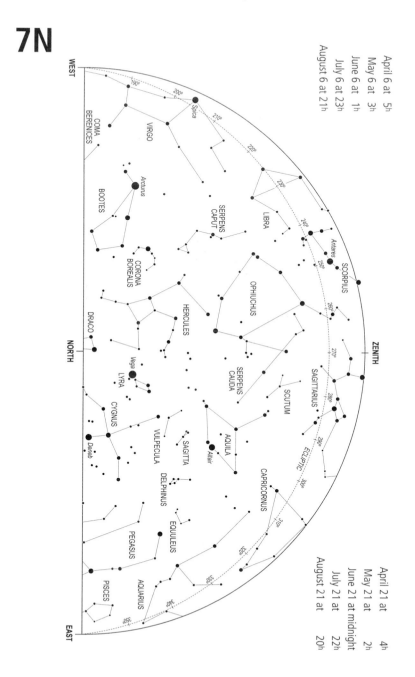

7N

April 6 at 5ʰ
May 6 at 3ʰ
June 6 at 1ʰ
July 6 at 23ʰ
August 6 at 21ʰ

April 21 at 4ʰ
May 21 at 2ʰ
June 21 at midnight
July 21 at 22ʰ
August 21 at 20ʰ

Southern Star Charts

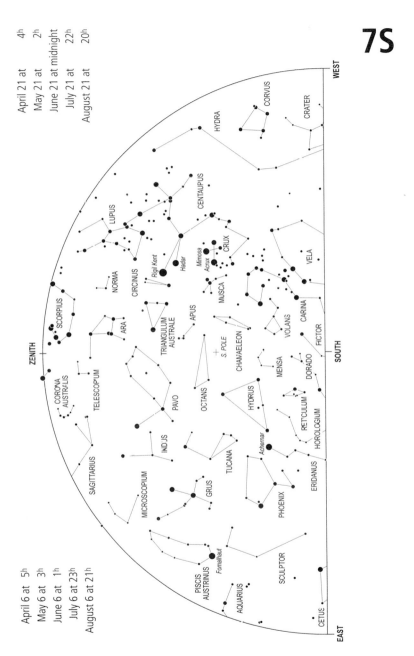

April 21 at 4ʰ
May 21 at 2ʰ
June 21 at midnight
July 21 at 22ʰ
August 21 at 20ʰ

April 6 at 5ʰ
May 6 at 3ʰ
June 6 at 1ʰ
July 6 at 23ʰ
August 6 at 21ʰ

WEST

SOUTH

EAST

ZENITH

CORVUS
CRATER
HYDRA
CENTAUPUS
LUPUS
Rigil Kent
Hadar
Mimosa
Acrux
CRUX
VELA
NORMA
CIRCINUS
MUSCA
SCORPIUS
ARA
TRIANGULUM AUSTRALE
APUS
CARINA
VOLANS
PICTOR
CORONA AUSTRALIS
TELESCOPIUM
PAVO
OCTANS
S. POLE
CHAMAELEON
MENSA
DORADO
SAGITTARIUS
INDJS
HYDRUS
RETICULUM
HOROLOGIUM
MICROSCOPIUM
TUCANA
Achernar
ERIDANUS
GRUS
PHOENIX
PISCIS AUSTRINUS
Fomalhaut
SCULPTOR
AQUARIUS
CETUS

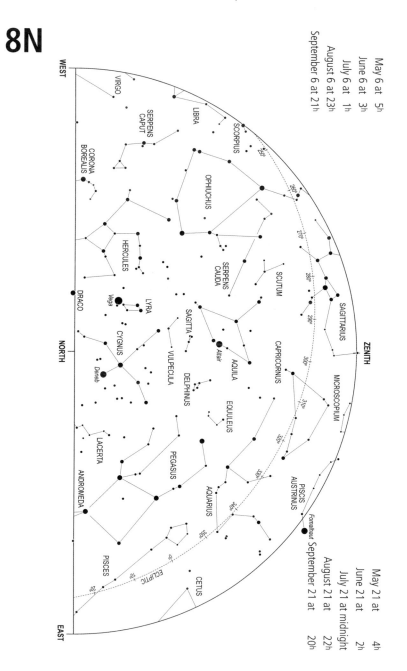

8N

May 6 at 5h
June 6 at 3h
July 6 at 1h
August 6 at 23h
September 6 at 21h

May 21 at 4h
June 21 at 2h
July 21 at midnight
August 21 at 22h
September 21 at 20h

8S

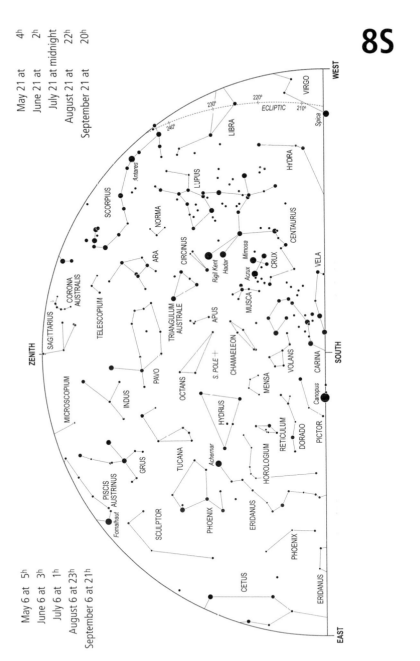

May 21 at 4ʰ
June 21 at 2ʰ
July 21 at midnight
August 21 at 22ʰ
September 21 at 20ʰ

May 6 at 5ʰ
June 6 at 3ʰ
July 6 at 1ʰ
August 6 at 23ʰ
September 6 at 21ʰ

WEST

VIRGO
ECLIPTIC
210°
220°
230°
240°
LIBRA
Spica
HYDRA
Antares
LUPUS
SCORPIUS
NORMA
CENTAURUS
ARA
CIRCINUS
VELA
Rigil Kent
Hadar
Mimosa
Acrux
CRUX
CORONA AUSTRALIS
TRIANGULUM AUSTRALE
APUS
MUSCA
SAGITTARIUS
TELESCOPIUM
S. POLE +
CHAMAELEON
CARINA
ZENITH
OCTANS
VOLANS
Canopus
PAVO
MENSA
SOUTH
MICROSCOPIUM
INDUS
HYDRUS
RETICULUM
DORADO
PICTOR
GRUS
TUCANA
Achernar
HOROLOGIUM
PISCIS AUSTRINUS
Fomalhaut
SCULPTOR
PHOENIX
ERIDANUS
PHOENIX
CETUS
ERIDANUS

EAST

49

9N

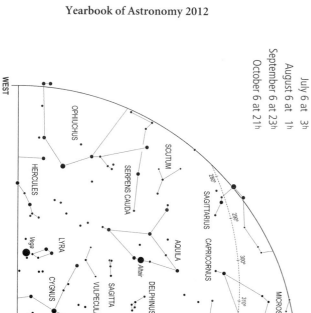
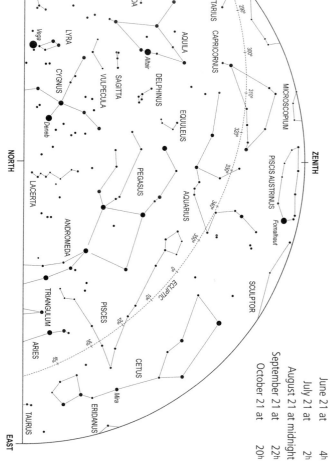

WEST

OPHIUCHUS

HERCULES

SCUTUM

SERPENS CAUDA

Vega

LYRA

CYGNUS

Deneb

VULPECULA

SAGITTA

Altair

AQUILA

DELPHINUS

SAGITTARIUS

CAPRICORNUS

EQUULEUS

280°

290°

300°

310°

320°

330°

340°

350°

MICROSCOPIUM

ZENITH

PISCIS AUSTRINUS

Fomalhaut

NORTH

LACERTA

ANDROMEDA

PEGASUS

AQUARIUS

SCULPTOR

TRIANGULUM

PISCES

ECLIPTIC

0°

10°

20°

30°

40°

ARIES

CETUS

Mira

ERIDANUS

TAURUS

EAST

Southern Star Charts

June 21 at 4h
July 21 at 2h
August 21 at midnight
September 21 at 22h
October 21 at 20h

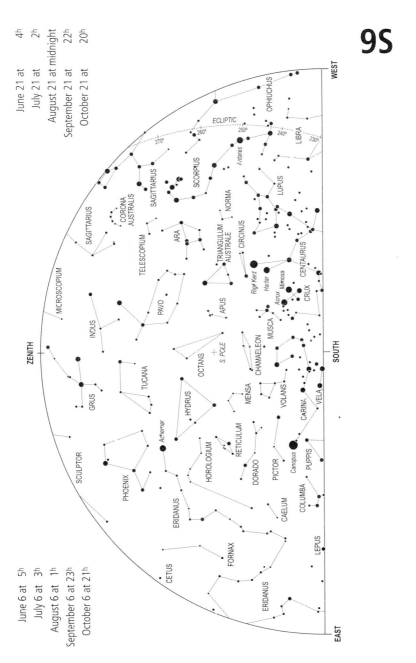

June 6 at 5h
July 6 at 3h
August 6 at 1h
September 6 at 23h
October 6 at 21h

WEST

OPHIUCHUS

ECLIPTIC
270° 260° 250° 240° 230°
Antares
LIBRA

SCORPIUS
LUPUS

CORONA AUSTRALIS
SAGITTARIUS
NORMA
CIRCINUS
CENTAURUS

SAGITTARIUS
TELESCOPIUM
ARA
TRIANGULUM AUSTRALE
Rigil Kent
Hadar
Acrux Mimosa
CRUX

MICROSCOPIUM
PAVO
APUS
MUSCA

ZENITH

INDUS
OCTANS
S. POLE
CHAMAELEON
MENSA
VOLANS

GRUS
TUCANA
HYDRUS
MUSCA

SCULPTOR
Achernar
RETICULUM
DORADO
CARINA
VELA
SOUTH

PHOENIX
HOROLOGIUM
PICTOR
Canopus
PUPPIS

ERIDANUS
CAELUM
COLUMBA

CETUS
FORNAX
LEPUS

ERIDANUS

EAST

51

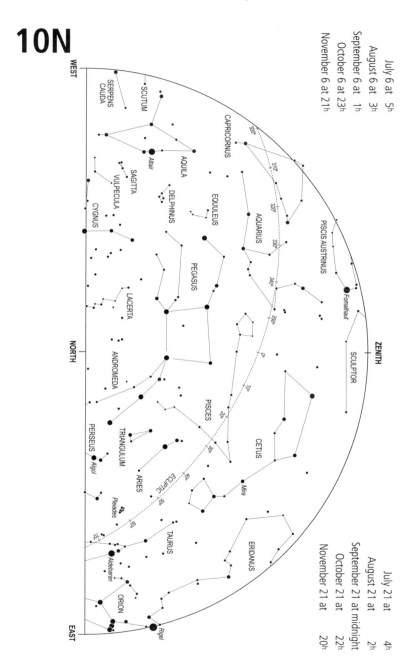

Southern Star Charts

July 21 at 4ʰ
August 21 at 2ʰ
September 21 at midnight
October 21 at 22ʰ
November 21 at 20ʰ

July 6 at 5ʰ
August 6 at 3ʰ
September 6 at 1ʰ
October 6 at 23ʰ
November 6 at 21ʰ

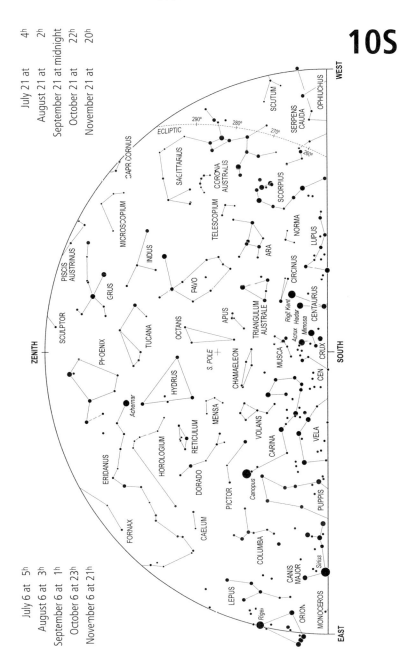

WEST

SOUTH

EAST

ZENITH

OPHIUCHUS
SCUTUM
SERPENS CAUDA
ECLIPTIC
290° 280° 270° 260°
CAPR CORNUS
SAGITTARIUS
CCROÑA AUSTRALIS
SCORPIUS
TELESCOPIUM
NORMA
ARA
CIRCINUS
LUPUS
MICROSCOPIUM
INDUS
PAVO
CENTAURUS
Rigil Kent
Hadar
TRIANGULUM AUSTRALE
Acrux
Mimosa
PISCIS AUSTRINUS
GRUS
APUS
CRUX
SCULPTOR
TUCANA
OCTANS
MUSCA
CEN
PH·OENIX
S. POLE
CHAMAELEON
HYDRUS
MENSA
VOLANS
VELA
Achernar
RETICULUM
CARINA
HOROLOGIUM
ERIDANUS
DORADO
PICTOR
PUPPIS
Canopus
FORNAX
CAELUM
COLUMBA
Sirius
CANIS MAJOR
LEPUS
MONOCEROS
ORION
Rigel

53

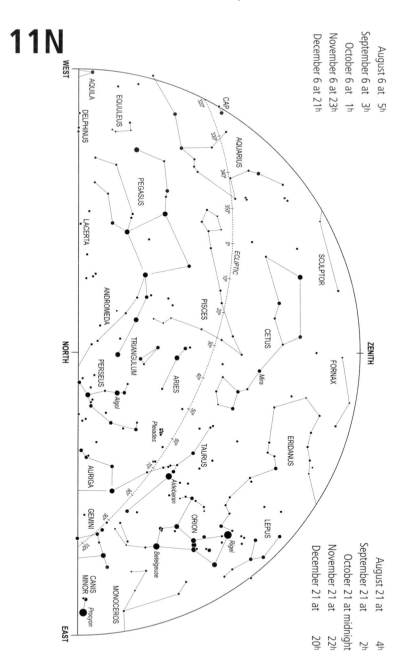

11N

August 6 at 5h
September 6 at 3h
October 6 at 1h
November 6 at 23h
December 6 at 21h

August 21 at 4h
September 21 at 2h
October 21 at midnight
November 21 at 22h
December 21 at 20h

11S

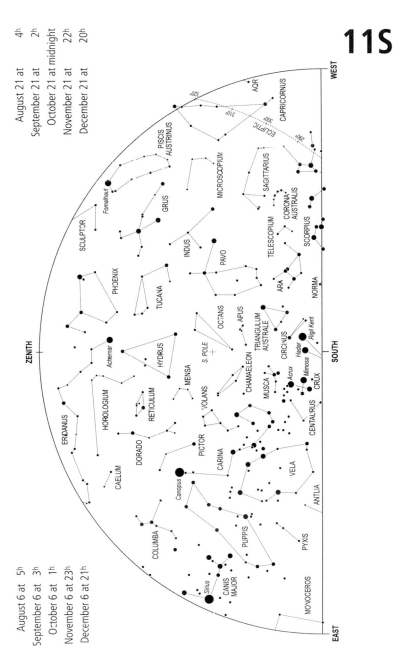

August 21 at 4ʰ
September 21 at 2ʰ
October 21 at midnight
November 21 at 22ʰ
December 21 at 20ʰ

August 6 at 5ʰ
September 6 at 3ʰ
October 6 at 1ʰ
November 6 at 23ʰ
December 6 at 21ʰ

ZENITH

WEST

SOUTH

EAST

AQR
CAPRICORNUS
320°
310°
ECLIPTIC
300°
290°
PISCIS AUSTRINUS
Fomalhaut
SCULPTOR
GRUS
MICROSCOPIUM
SAGITTARIUS
CORONA AUSTRALIS
TELESCOPIUM
SCORPIUS
INDUS
PAVO
ARA
NORMA
PHOENIX
TUCANA
OCTANS
APUS
TRIANGULUM AUSTRALE
CIRCINUS
Rigil Kent
Hadar
Achernar
S. POLE
HYDRUS
CHAMAELEON
Acrux
Mimosa
CRUX
MENSA
MUSCA
CENTAURUS
HOROLOGIUM
RETICULUM
VOLANS
ERIDANUS
DORADO
PICTOR
CARINA
VELA
CAELUM
COLUMBA
PUPPIS
ANTLIA
PYXIS
Canopus
Sirius
CANIS MAJOR
MONOCEROS

55

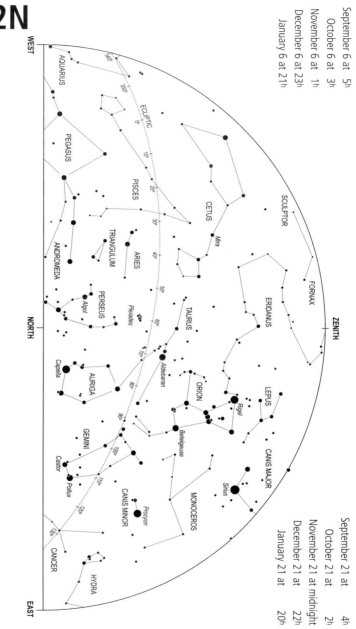

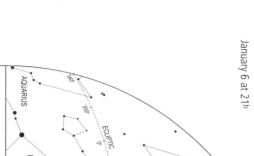

12N

September 6 at 5h
October 6 at 3h
November 6 at 1h
December 6 at 23h
January 6 at 21h

September 21 at 4h
October 21 at 2h
November 21 at midnight
December 21 at 22h
January 21 at 20h

12S

September 21 at 4h
October 21 at 2h
November 21 at midnight
December 21 at 22h
January 21 at 20h

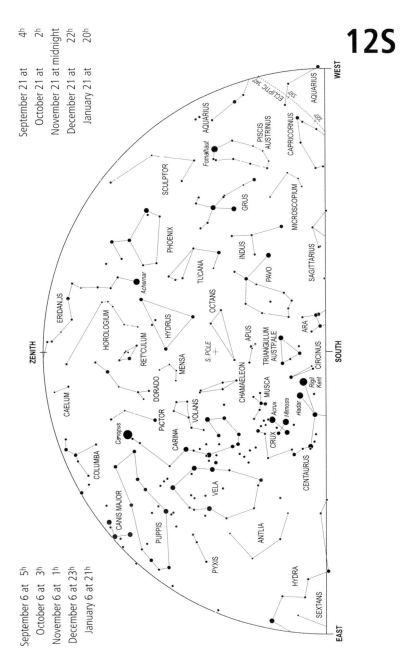

September 6 at 5h
October 6 at 3h
November 6 at 1h
December 6 at 23h
January 6 at 21h

The Planets and the Ecliptic

The paths of the planets about the Sun all lie close to the plane of the ecliptic, which is marked for us in the sky by the apparent path of the Sun among the stars, and is shown on the star charts by a broken line. The Moon and naked-eye planets will always be found close to this line, never departing from it by more than about 7°. Thus the planets are most favourably placed for observation when the ecliptic is well displayed, and this means that it should be as high in the sky as possible. This avoids the difficulty of finding a clear horizon, and also overcomes the problem of atmospheric absorption, which greatly reduces the light of the stars. Thus a star at an altitude of 10° suffers a loss of 60 per cent of its light, which corresponds to a whole magnitude; at an altitude of only 4°, the loss may amount to two magnitudes.

The position of the ecliptic in the sky is therefore of great importance, and since it is tilted at about 23.5° to the Equator, it is only at certain times of the day or year that it is displayed to best advantage. It will be realized that the Sun (and therefore the ecliptic) is at its highest in the sky at noon in midsummer, and at its lowest at noon in midwinter. Allowing for the daily motion of the sky, it follows that the ecliptic is highest at midnight in winter, at sunset in the spring, at noon in summer and at sunrise in the autumn. Hence these are the best times to see the planets. Thus, if Venus is an evening object in the western sky after sunset, it will be seen to best advantage if this occurs in the spring, when the ecliptic is high in the sky and slopes down steeply to the horizon. This means that the planet is not only higher in the sky, but also will remain for a much longer period above the horizon. For similar reasons, a morning object will be seen at its best on autumn mornings before sunrise, when the ecliptic is high in the east. The outer planets, which can come to opposition (i.e. opposite the Sun), are best seen when opposition occurs in the winter months, when the ecliptic is high in the sky at midnight.

The seasons are reversed in the Southern Hemisphere, spring beginning at the September Equinox, when the Sun crosses the Equator on its way south, summer beginning at the December Solstice, when the

Sun is highest in the southern sky, and so on. Thus, the times when the ecliptic is highest in the sky, and therefore best placed for observing the planets, may be summarized as follows:

	Midnight	**Sunrise**	**Noon**	**Sunset**
Northern latitudes	December	September	June	March
Southern latitudes	June	March	December	September

In addition to the daily rotation of the celestial sphere from east to west, the planets have a motion of their own among the stars. The apparent movement is generally *direct*, i.e. to the east, in the direction of increasing longitude, but for a certain period (which depends on the distance of the planet) this apparent motion is reversed. With the outer planets this *retrograde* motion occurs about the time of opposition. Owing to the different inclination of the orbits of these planets, the actual effect is to cause the apparent path to form a loop, or sometimes an S-shaped curve. The same effect is present in the motion of the inferior planets, Mercury and Venus, but it is not so obvious, since it always occurs at the time of inferior conjunction.

The *inferior planets*, Mercury and Venus, move in smaller orbits than that of the Earth, and so are always seen near the Sun. They are most obvious at the times of greatest angular distance from the Sun (greatest elongation), which may reach 28° for Mercury, and 47° for Venus. They are seen as evening objects in the western sky after sunset (at eastern elongations) or as morning objects in the eastern sky before sunrise (at western elongations). The succession of phenomena, conjunctions and elongations, always follows the same order, but the intervals between them are not equal. Thus, if either planet is moving round the far side of its orbit its motion will be to the east, in the same direction in which the Sun appears to be moving. It therefore takes much longer for the planet to overtake the Sun – that is, to come to superior conjunction – than it does when moving round to inferior conjunction, between Sun and Earth. The intervals given in the table below are average values; they remain fairly constant in the case of Venus, which travels in an almost circular orbit. In the case of Mercury, however, conditions vary widely because of the great eccentricity and inclination of the planet's orbit.

		Mercury	Venus
Inferior Conjunction	to Elongation West	22 days	72 days
Elongation West	to Superior Conjunction	36 days	220 days
Superior Conjunction	to Elongation East	35 days	220 days
Elongation East	to Inferior Conjunction	22 days	72 days

The greatest brilliancy of Venus always occurs about thirty-six days before or after inferior conjunction. This will be about a month after greatest eastern elongation (as an evening object), or a month before greatest western elongation (as a morning object). No such rule can be given for Mercury, because its distances from the Earth and the Sun can vary over a wide range.

Mercury is not likely to be seen unless a clear horizon is available. It is seldom as much as 10° above the horizon in the twilight sky in northern temperate latitudes, but this figure is often exceeded in the Southern Hemisphere. This favourable condition arises because the maximum elongation of 28° can occur only when the planet is at aphelion (furthest from the Sun), and it then lies well south of the Equator. Northern observers must be content with smaller elongations, which may be as little as 18° at perihelion. In general, it may be said that the most favourable times for seeing Mercury as an evening object will be in spring, some days before greatest eastern elongation; in autumn, it may be seen as a morning object some days after greatest western elongation.

Venus is the brightest of the planets and may be seen on occasions in broad daylight. Like Mercury, it is alternately a morning and an evening object, and it will be highest in the sky when it is a morning object in autumn, or an evening object in spring. Venus is to be seen at its best as an evening object in northern latitudes when eastern elongation occurs in June. The planet is then well north of the Sun in the preceding spring months, and is a brilliant object in the evening sky over a long period. In the Southern Hemisphere a November elongation is best. For similar reasons, Venus gives a prolonged display as a morning object in the months following western elongation in October (in northern latitudes) or in June (in the Southern Hemisphere).

The *superior planets*, which travel in orbits larger than that of the Earth, differ from Mercury and Venus in that they can be seen opposite the Sun in the sky. The superior planets are morning objects after conjunction with the Sun, rising earlier each day until they come to opposition. They will then be nearest to the Earth (and therefore at their brightest), and will be on the meridian at midnight, due south in northern latitudes, but due north in the Southern Hemisphere. After opposition they are evening objects, setting earlier each evening until they set in the west with the Sun at the next conjunction. The difference in brightness from one opposition to another is most noticeable in the case of Mars, whose distance from Earth can vary considerably and rapidly. The other superior planets are at such great distances that there is very little change in brightness from one opposition to the next. The effect of altitude is, however, of some importance, for at a December opposition in northern latitudes the planets will be among the stars of Taurus or Gemini, and can then be at an altitude of more than 60° in southern England. At a summer opposition, when Mars is in Sagittarius, it may only rise to about 15° above the southern horizon, and so makes a less impressive appearance. In the Southern Hemi sphere the reverse conditions apply, a June opposition being the best, with the planet in Sagittarius at an altitude which can reach 80° above the northern horizon for observers in South Africa.

Mars, whose orbit is appreciably eccentric, comes nearest to the Earth at oppositions at the end of August. It may then be brighter even than Jupiter, but rather low in the sky in Aquarius for northern observers, though very well placed for those in southern latitudes. These favourable oppositions occur every fifteen or seventeen years (e.g. in 1988, 2003 and 2018). In the Northern Hemisphere the planet is probably better seen at oppositions in the autumn or winter months, when it is higher in the sky – such as in 2005 when opposition was in early November. Oppositions of Mars occur at an average interval of 780 days, and during this time the planet makes a complete circuit of the sky.

Jupiter is always a bright planet, and comes to opposition a month later each year, having moved, roughly speaking, from one zodiacal constellation to the next.

Saturn moves much more slowly than Jupiter, and may remain in the same constellation for several years. The brightness of Saturn depends on the aspects of its rings, as well as on the distance from Earth and Sun. The Earth passed through the plane of Saturn's rings in 1995 and 1996, when they appeared edge-on; we saw them at maximum opening, and Saturn at its brightest, in 2002. The rings last appeared edge-on in 2009, and they are now opening once again.

Uranus and *Neptune* are both visible with binoculars or a small telescope, but you will need a finder chart to help you locate them (such as those reproduced in this *Yearbook* on pages 128 and 120). *Pluto* (now officially classified as a 'dwarf planet') is hardly likely to attract the attention of observers without adequate telescopes.

Phases of the Moon in 2012

NICK JAMES

New Moon			First Quarter			Full Moon			Last Quarter		
d	h	m	d	h	m	d	h	m	d	h	m
			Jan	1	06 15	Jan	9	07 30	an	16	09 08
Jan	23	07 39	Jan	31	04 10	Feb	7	21 54	Feb	14	17 04
Feb	21	22 35	Mar	1	01 22	Mar	8	09 40	Mar	15	01 25
Mar	22	14 37	Mar	30	19 41	Apr	6	19 19	Apr	13	10 50
Apr	21	07 18	Apr	29	09 58	May	6	03 35	May	12	21 47
May	20	23 47	May	28	20 16	June	4	11 12	June	11	10 41
June	19	15 02	June	27	03 30	July	3	18 52	July	11	01 48
July	19	04 24	July	26	08 56	Aug	2	03 28	Aug	9	18 55
Aug	17	15 54	Aug	24	13 54	Aug	31	13 58	Sept	8	13 15
Sept	16	02 11	Sept	22	19 41	Sept	30	03 19	Oct	8	07 33
Oct	15	12 03	Oct	22	03 32	Oct	29	19 50	Nov	7	00 36
Nov	13	22 08	Nov	20	14 31	Nov	28	14 46	Dec	6	15 31
Dec	13	08 42	Dec	20	05 19	Dec	28	10 21	Jan	5	03 58

All times are UTC (GMT)

Longitudes of the Sun, Moon and Planets in 2012

NICK JAMES

Date		Sun °	Moon °	Venus °	Mars °	Jupiter °	Saturn °	Uranus °	Neptune °
Jan	6	285	67	320	171	31	209	1	329
	21	300	271	338	173	32	209	1	330
Feb	6	317	113	357	172	33	210	2	330
	21	332	321	15	168	35	209	3	331
Mar	6	346	134	31	163	38	209	3	331
	21	1	343	47	157	41	208	4	332
Apr	6	17	185	62	154	44	207	5	332
	21	31	28	74	154	48	206	6	333
May	6	46	224	82	156	51	205	7	333
	21	60	60	83	161	55	204	7	333
June	6	76	277	76	167	59	203	8	333
	21	90	106	68	174	62	203	8	333
July	6	104	314	69	181	65	203	9	333
	21	119	140	76	190	68	203	9	333
Aug	6	134	2	89	199	71	204	8	332
	21	148	191	103	208	73	205	8	332
Sept	6	164	46	119	219	75	207	7	331
	21	178	244	136	229	76	208	7	331
Oct	6	193	78	153	239	76	210	6	331
	21	208	283	171	250	76	212	6	330
Nov	6	224	122	190	262	75	214	5	330
	21	239	334	209	273	73	215	5	330
Dec	6	254	156	227	284	71	217	5	331
	21	270	8	246	296	69	219	5	331

Moon: Longitude of the ascending node: Jan 1: 253°   Dec 31: 234°

Mercury moves so quickly among the stars that it is not possible to indicate its position on the star charts at convenient intervals. The monthly notes should be consulted for the best times at which the planet may be seen.

The positions of the Sun, Moon and planets other than Mercury are given in the table on page 64. These objects move along paths which remain close to the ecliptic and this list shows the apparent ecliptic longitude for each object on dates which correspond to those of the star charts. This information can be used to plot the position of the desired object on the selected chart.

EXAMPLES

Two bright planets are seen close together low in the western sky at around 8 p.m. in early March. What are they?

The northern star chart 2N shows the northern sky for 6 March at 21h. The range of ecliptic longitude visible low in the west ranges from 20° to 40°. Reference to the table of longitudes on page 64 for 6 March shows that two planets are in this range: Venus is at longitude 31° and Jupiter is at 38°. The two planets are therefore Venus and Jupiter on the Aries/Pisces border. From the table it can be seen that by 21 March Venus has overtaken Jupiter and, in fact, the two planets come to within 3° of each other on the evening of 13 March.

The positions of the Sun and Moon can be plotted on the star charts in the same way as the planets. This is straightforward for the Sun, since it always lies on the ecliptic and it moves on average only 1° per day. The Moon is more difficult since it moves more rapidly, at an average of 13° per day, and it moves up to 5° north or south of the ecliptic during the month. An indication of the Moon's position relative to the ecliptic may be obtained by considering its longitude relative to that of the ascending node. The latter changes only slowly during the year as will be seen from the values given on page 64. If d is the difference in longitude between the Moon and its ascending node, then the Moon is on the ecliptic when $d = 0°$, $180°$ or $360°$. The Moon is 5° north of the ecliptic if $d = 90°$ and the Moon is 5° south of the ecliptic if $d = 270°$.

As an example, the Moon is full at 21h on 7 February. The table shows that the Moon's longitude is 113º at 0h on 6 February. Extrapolating at 13º per day the Moon's longitude at 21h on the 7th is around 137º. At this time the longitude of the node is found by interpolation to be around 251°. Thus $d = -114°$ (or +246º) and the Moon is about 4° south of the ecliptic. Its position may be plotted on northern star charts 1S, 2S, 3S, 4S and southern star charts 1N, 2N, 3N, 4N.

Some Events in 2012

Jan	2	Moon at Apogee (404,580 km)
	5	*Earth* at Perihelion
	9	Full Moon
	17	Moon at Perigee (369,880 km)
	23	New Moon
	30	Moon at Apogee (404,325 km)

Feb	7	Full Moon
	7	*Mercury* in Superior Conjunction
	11	Moon at Perigee (367,920 km)
	19	*Neptune* in Conjunction with Sun
	21	New Moon
	27	Moon at Apogee (404,860 km)

Mar	3	*Mars* at Opposition in Leo
	5	*Mercury* at Greatest Eastern Elongation (18°)
	5	*Mars* closest to *Earth*
	8	Full Moon
	10	Moon at Perigee (362,400 km)
	20	Equinox (Spring Equinox in Northern Hemisphere)
	21	*Mercury* in Inferior Conjunction
	22	New Moon
	24	*Uranus* in Conjunction with Sun
	25	Summer Time Begins in the UK
	26	Moon at Apogee (405,780 km)
	27	*Venus* at Greatest Eastern Elongation (46°)

Apr	6	Full Moon
	7	Moon at Perigee (358,315 km)
	15	*Saturn* at Opposition in Virgo
	18	*Mercury* at Greatest Western Elongation (28°)
	21	New Moon
	22	Moon at Apogee (406,420 km)

May	6	Full Moon
	6	Moon at Perigee (356,955 km)
	13	*Jupiter* in Conjunction with Sun
	19	Moon at Apogee (406,450 km)
	20	New Moon
	20	Annular Eclipse of the Sun
	27	*Mercury* in Superior Conjunction
June	3	Moon at Perigee (358,480 km)
	4	Full Moon
	4	Partial Eclipse of the Moon
	5–6	*Venus* in Inferior Conjunction
	5–6	Transit of *Venus* across face of Sun
	16	Moon at Apogee (405,790 km)
	19	New Moon
	20	Solstice (Summer Solstice in Northern Hemisphere)
	29	*Pluto* at Opposition in Sagittarius
July	1	Moon at Perigee (362,360 km)
	1	*Mercury* at Greatest Eastern Elongation (26°)
	3	Full Moon
	5	*Earth* at Aphelion
	13	Moon at Apogee (404,780 km)
	19	New Moon
	28	*Mercury* in Inferior Conjunction
	29	Moon at Perigee (367,315 km)
Aug	2	Full Moon
	10	Moon at Apogee (404,125 km)
	15	*Venus* at Greatest Western Elongation (46°)
	16	*Mercury* at Greatest Western Elongation (19°)
	17	New Moon
	23	Moon at Perigee (369,730 km)
	24	*Neptune* at Opposition in Aquarius
	31	Full Moon
Sept	7	Moon at Apogee (404,295 km)
	10	*Mercury* in Superior Conjunction
	16	New Moon

	19	Moon at Perigee (365,750 km)
	22	Equinox (Autumnal Equinox in Northern Hemisphere)
	29	*Uranus* at Opposition in Pisces
	30	Full Moon
Oct	5	Moon at Apogee (405,160 km)
	15	New Moon
	17	Moon at Perigee (360,670 km)
	25	*Saturn* in Conjunction with Sun
	26	*Mercury* at Greatest Eastern Elongation (24°)
	28	Summer Time Ends in the UK
	29	Full Moon
Nov	1	Moon at Apogee (406,050 km)
	13	New Moon
	13	Total Eclipse of the Sun
	14	Moon at Perigee (357,360 km)
	17	*Mercury* in Inferior Conjunction
	28	Full Moon
	28	Penumbral Eclipse of the Moon
	28	Moon at Apogee (406,365 km)
Dec	3	*Jupiter* at Opposition in Taurus
	4	*Mercury* at Greatest Western Elongation (21°)
	12	Moon at Perigee (357,075 km)
	13	New Moon
	21	Solstice (Winter Solstice in Northern Hemisphere)
	25	Moon at Apogee (406,100 km)
	28	Full Moon

Monthly Notes 2012

January

Full Moon: 9 January   *New Moon:* 23 January

EARTH is at perihelion (nearest to the Sun) on 5 January at a distance of 147 million kilometres (91.3 million miles).

MERCURY was at greatest western elongation (22°) on 23 December last year. From northern temperate latitudes the planet may be glimpsed low above the south-eastern horizon at the beginning of morning civil twilight, at the beginning of January. From equatorial and more southerly latitudes the planet should be visible in the east-south-east, and the period of visibility extends to the middle of the month. The planet's brightness remains fairly constant at magnitude −0.4 during this time.

VENUS is a brilliant object in the south-western sky at dusk. From northern temperate latitudes the planet will be setting more than three hours after the Sun by the middle of the month; from the tropics and the Southern Hemisphere the planet sets about two hours after the Sun. The planet is in Capricornus at the beginning of the year, moving into Aquarius during January, and brightening very slightly from magnitude −4.0 to −4.1. During the month the phase of the planet decreases from 83 per cent to 74 per cent. On 26 January the waxing crescent Moon will pass close to Venus and the pair will make a pleasing spectacle in the evening twilight sky.

MARS rises in the mid-evening by the end of January, brightening noticeably from magnitude +0.2 to −0.5 during the month. The planet begins the year moving direct in Leo, crosses into neighbouring Virgo mid-month, and reaches its first stationary point on 24 January; thereafter it begins to move retrograde, back towards the border with Leo.

JUPITER is a splendid object visible in the southern sky as soon as darkness falls and does not set until the early morning hours. It begins

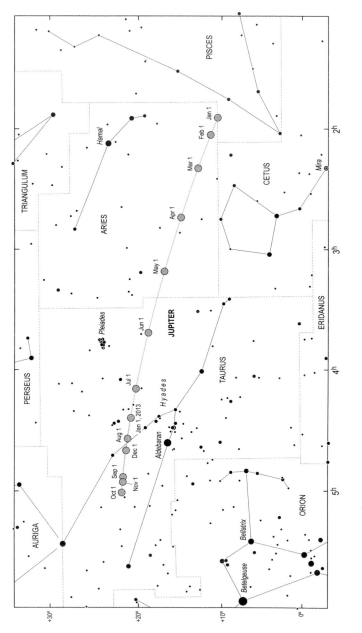

Figure 1. The path of Jupiter as it moves against the background stars of Pisces (briefly in early January), Aries and Taurus during 2012.

the month in Pisces but its direct motion carries it across the border into neighbouring Aries. Figure 1 shows the path of Jupiter against the background stars during 2012. Jupiter fades very slightly from magnitude −2.6 to −2.4 during January. The waxing first-quarter Moon will appear close to Jupiter in the sky on 2 and 30 January and the two objects will be a nice sight in the evening sky.

SATURN rises in the early morning hours, brightening slightly from magnitude +0.7 to +0.6 during the month. The planet is moving direct in Virgo. Following the ring-plane crossing in September 2009, the rings have now opened nicely again and are displayed at an angle of 14.2° as viewed from the Earth at the beginning of January, so the planet is once again a fine sight even in a small telescope.

Red Sirius. The brilliant star Sirius, in the constellation of Canis Major, is well placed in the south-eastern sky in the mid-evening throughout January. Sirius, or Alpha Canis Majoris, at magnitude −1.4, is much the brightest star in the night sky, more than twice as bright as its nearest rival, Canopus (magnitude −0.7). Yet appearances can be deceptive. Sirius is an A-type main sequence star, a mere 26 times as luminous as our Sun and 8.6 light years away, so that on the cosmic scale it is a very near neighbour of our Sun. Canopus, the brightest star in Carina, the keel of the great ship Argo Navis, is considerably further away and is a rare F-type yellow-white giant star. Recent determinations (from Hipparcos satellite data) give the distance of Canopus as about 310 light years with a five per cent uncertainty, and a luminosity 13,300 times that of the Sun. Canopus is not visible from latitudes north of 37°N, which excludes virtually all of Europe, and much of North America although, from southern states of the USA, Sirius and Canopus make a fine sight in the winter months, as they do during the summertime in the Southern Hemisphere.

Sirius is described in all the old records; it was called Sothis by the ancient Egyptians who based their calendar on its 'heliacal rising', that is to say the day when it first became visible just before sunrise in pre-dawn twilight sky, after a period of invisibility. This occurred in early July about three thousand years ago (now about three weeks later due to precession), and at that time it immediately preceded the annual flooding of the River Nile, upon which the whole economy of the country depended, and marked the start of the Egyptian new year.

Of course, the star was described by Ptolemy, last of the great Greek astronomers (c. AD 150), but there is a definite mystery here. We find it in his books VII and VIII of the *Almagest*, where he includes it in his list of bright naked-eye red stars, together with Betelgeux (also spelt Betelgeuse), Aldebaran, Antares, Arcturus and Pollux. The others are certainly red, or at least orange, but Sirius is a glittering white. Has it changed colour since the second century AD?

There have been endless arguments about this, so let us examine the possibilities:

1. Sirius has genuinely changed from a red star to a white one.
2. The change is linked with Sirius's faint binary companion, Sirius B, discovered by Alvan Graham Clark in January 1862 – exactly 150 years ago this month.
3. Sirius was reddened by an interstellar cloud passing between the star and ourselves.
4. Ptolemy simply made a mistake.
5. What Ptolemy wrote was miscopied or misinterpreted by other writers, and it is these versions of his work that have come down to us.
6. The 'red' description comes from the undeniable fact that when Sirius is seen low down in the sky (as it is from the British Isles) and its light has to pass through thicker layers of the Earth's unsteady atmosphere, its great brilliance makes it flash various colours of the rainbow, including red.

The first of these alternatives can be discarded at once. Sirius is simply not the kind of star to show significant colour changes on a timescale of a few millenia; it is on the main sequence, classified as spectral type A1 and about twice as massive as the Sun.

The second possibility seems initially more promising, and leads us on to a description of Sirius B. Since Sirius itself is generally called 'the Dog Star', the companion has been nicknamed 'the Pup'. In 1844, the German astronomer Friedrich Bessel, who had studied the proper motion of Sirius, deduced that it must have an unseen companion. Much later, on 31 January 1862, Alvan Graham Clark, while busy testing the new 18.5-inch refractor at the Dearborn Observatory, discovered 'the Pup' precisely where Bessel had expected it to be. The two stars actually orbit each other in a period of just over fifty years, but the orbit is quite eccentric, taking them from 1,210 million kilometres apart (almost the distance of Saturn from our Sun) to 4,710 million kilometres (the distance of Neptune) and back again. The two stars were furthest apart in 1969, at their closest in 1994 and will be farthest apart again in 2019.

The magnitude of Sirius B is 8.4 and normally it would be an easy object to see were it not so drowned by the glare of the primary, Sirius A, which is visually nearly ten thousand times brighter than its companion (Figure 2). Initially, Sirius B was assumed to be cool and red.

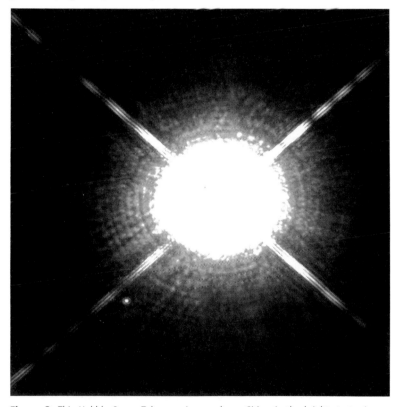

Figure 2. This Hubble Space Telescope image shows Sirius A, the brightest star in our night-time sky, along with its faint, tiny stellar companion, Sirius B. Astronomers overexposed the image of Sirius A (at centre) so that the very dim Sirius B (tiny dot at lower left) could be seen. The cross-shaped diffraction spikes and concentric rings around Sirius A, and the small ring around Sirius B, are artefacts produced within the telescope's imaging system. (Image courtesy of NASA, ESA, H. Bond (STScI) and M. Barstow (University of Leicester, UK.)

Not so! In 1915, W. S. Adams, using the 60-inch reflector at Mount Wilson, examined its spectrum and found that the Pup is actually the hotter of the two stars, being blue-white with a surface temperature of around 25,000°. The only way the companion star could be both hot and dim – its total luminosity (including the ultraviolet light) is just 2.4 per cent of that of the Sun – is if it is small and amazingly dense. With a diameter of just 12,000 kilometres (smaller than the Earth), but with the mass of the Sun, it has an astonishing average density of 1.7 tons per cubic centimetre. Such a star is known as a white dwarf.

We can trace the story of Sirius B. White dwarfs are the end products in the lives of ordinary stars like our Sun. Sirius B used to be a bright binary companion of Sirius A, but it was slightly more massive and luminous, evolved more quickly, ran out of nuclear fuel in its core, and bloated to become a red giant. That its mass is now lower than that of Sirius A is proof that stars lose considerable amounts of mass as they die. About 120 million years ago Sirius B collapsed into its present state as a white dwarf, a ball of carbon and oxygen whose only fate is to cool down very gradually until it eventually becomes a cold, dark globe. By then, Sirius A will have passed through the same evolutionary stages.

So, if the Pup was once a red giant star could this be the answer to our problem? The answer is a resounding 'no', because the timescale is hopelessly wrong. We must think again.

A passing interstellar dust cloud is also most unlikely. It would have manifested itself in many ways, including significantly dimming the brightness of Sirius (and there is no evidence of that), and it would have reddened and dimmed other nearby stars, such as Procyon in Canis Minor.

I would tend to discount the possibility that Ptolemy himself made mistake; he was much too good an observer for that. It is possible that we have been left garbled versions of what he actually wrote, since even the earliest surviving text of the *Almagest* was written many hundreds of years after the original, but the inclusion of Sirius in the same section of the book as the five other red or orange stars would tend not to support this hypothesis.

Lastly, the effects of Earth's moving atmosphere would be the obvious answer but for the fact that Ptolemy observed from Mediterranean latitudes, where Sirius rises much higher in the sky than it does from Britain, and so flashes less.

It was Sherlock Holmes who (in *The Sign of Four*) said: 'When you have eliminated all which is impossible, then whatever remains, however improbable, must be the truth.' Following Holmes, I will opt for the last of my alternatives. On the next clear night, go out and look for yourself. You will see that Sirius sparkles with many different colours – including red.

Seasonal Variations. The Earth's orbit round the Sun is not perfectly circular; it is an ellipse of very slight eccentricity, with the Sun located at one focus of the ellipse while the other focus is empty. Consequently, the orbit is not centred on the Sun, so that Earth's distance from it varies during the course of the year. On 5 January this year, the Earth is at perihelion (that is to say, its closest point to the Sun), and the distance is reduced to about 147.1 million kilometres. At aphelion (when furthest from the Sun) on 5 July, the distance is 152.1 million kilometres. So the Earth receives just 6.9 per cent more solar radiation at perihelion than at aphelion.

It may seem curious that Earth's perihelion occurs during winter in the Northern Hemisphere, but the seasons actually have very little to do with Earth's changing distance from the Sun. They are caused by the fact that the axis of rotation is tilted at about 23.5°; in northern winter, the North Pole is tilted away from the Sun, whereas in northern summer, the North Pole is tilted towards the Sun, thus receiving the solar rays much more directly. The slight effects of the varying distance are masked by geographical peculiarities of the Earth, inasmuch as there is a great deal more ocean in the Southern Hemisphere.

Conditions on the planet Mars, which comes to opposition in early March this year, are roughly analogous. The axial tilt is almost 24°, much the same as that of the Earth, and so the seasons are of the same general type, though they are much longer (the Martian 'year' is 687 Earth-days, though the axial rotation is a little over half-an-hour longer than ours). There is, however, one important difference. The orbit of Mars is considerably more eccentric than that of the Earth, so that its varying distance plays a fairly major rôle. When Mars is at perihelion, its distance from the Sun is 206.7 million kilometres, whereas at aphelion, the distance is 249.1 million kilometres. So Mars receives 45 per cent more solar radiation at perihelion than at aphelion. Consequently, the southern winters, when Mars is near aphelion, are significantly longer and colder than those of the northern hemisphere. This is shown by the behaviour of the polar caps. The southern cap may become larger than the maximum extent of its northern counterpart – but at Martian midsummer it may disappear completely, whereas the northern cap never does so (Figure 3).

A Tireless Observer of Double Stars. The Italian astronomer Ercole (Hercules) Dembowski was born in Milan, two hundred years ago this month, on 12 January 1812, and spent most of his life there. He specialized in observing double stars and he made tens of thousands of micrometer measurements of them. His careful measurements were of great value. He remeasured a great many double stars from Friedrich von Struve's Dorpat Catalogue, demonstrating how some of them had changed position over the years due to their orbital motion. Dembowski was awarded the Gold Medal of the Royal Astronomical Society in 1878. The crater Dembowski on the Moon is named after him, as well as the minor planet 349 Dembowska. He died on 19 January 1881, shortly after his sixty-ninth birthday.

Figure 3. The images opposite show the polar ice cap at the North Pole of Mars at two different times. The upper picture was taken during springtime, shortly after the winter season. The ice cap grew very large during the winter. It is still quite large in this picture, though warmer spring temperatures have begun to erode some of the ice. The lower picture is from summertime on Mars. There is still some ice, but a lot less than before. Some of the ice in the polar cap is made from water ice, which stays frozen all year round on Mars. The ice in the lower picture is water ice. In the winter, a different kind of ice forms when the temperatures get really cold. Carbon dioxide gas from the atmosphere freezes and forms dry ice. The dry ice only forms a thin layer (a couple of metres deep), but it covers a very large area. That is why the polar ice cap is so much larger in the upper picture. In the summer the dry ice sublimates (turns back into a gas in the atmosphere) as the temperature gets warmer. The ice cap at the South Pole of Mars also shrinks and grows as the seasons change. (Image courtesy of NASA/JPL/ Malin Space Science Systems.)

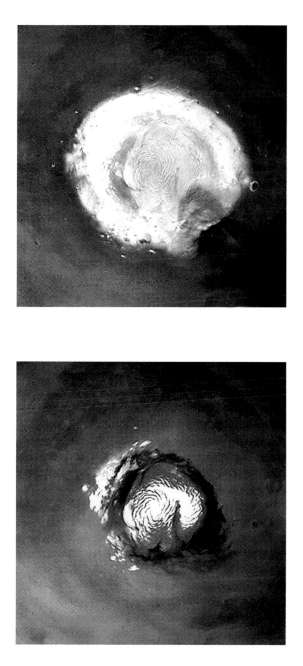

February

Full Moon: 7 February   *New Moon:* 23 February

MERCURY is in superior conjunction on 7 February and so will not be visible for the first two weeks of the month. However, by the end of the third week of February, the planet will have moved far enough east of the Sun to become visible in the western sky in the evenings for observers in tropical and northern latitudes. For Northern Hemisphere observers this is the most favourable evening apparition of the year. Figure 4 shows, for observers in latitude 52°N, the changes in azimuth (true bearing from the north through east, south and west) and altitude of Mercury on successive evenings when the Sun is 6° below the horizon. This is at the end of evening civil twilight and in this latitude and at this time of year occurs about thirty-five minutes after sunset.

The changes in the brightness of the planet are indicated on the diagram by the relative sizes of the white circles marking Mercury's

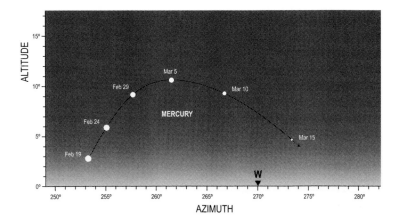

Figure 4. Evening apparition of Mercury from latitude 52°N. The planet reaches greatest eastern elongation on 5 March. It will be at its brightest in late February, before elongation.

position at five-day intervals: Mercury is at its brightest before it reaches greatest eastern elongation on 5 March. Between 19 and 29 February, Mercury fades from magnitude −1.3 to −0.8. The diagram gives positions for a time at the end of evening civil twilight on the Greenwich meridian on the stated date. Observers in different longitudes should note that the actual positions of Mercury in azimuth and altitude will differ slightly from those given in the diagram due to the motion of the planet.

VENUS is a brilliant object in the south-western sky at dusk and throughout the early evening. From northern temperate latitudes the planet will be setting nearly four hours after the Sun by the end of the month, but from tropical and more southerly latitudes the planet sets only about two hours after the Sun. The planet is in Aquarius at the beginning of the month, but moves into neighbouring Pisces during February, and brightens very slightly from magnitude –4.1 to - 4.2. The planet appears gibbous, the phase decreasing from 74 per cent to 64 per cent during the month. On 25 February, the waxing crescent Moon will appear close to Venus and the following night will be roughly midway between Venus and Jupiter. These events will present interesting opportunities for digital photographers against the backdrop of the evening twilight sky.

MARS rises in the mid-evening, brightening obviously from magnitude −0.6 to − 1.2 during the month. The planet's retrograde motion carries it from Virgo back into neighbouring Leo early in February. Mars passes aphelion on 15 February, when it will lie 1.66598 AU (249.2 million kilometres), its greatest distance, from the Sun.

JUPITER will be seen in the southern sky as soon as darkness falls, but it now sets before midnight. The planet is moving direct in Aries and fades very slightly from magnitude −2.4 to −2.2 during February. The waxing first-quarter Moon will appear quite close to Jupiter on 26 February and the two objects will be a pleasing sight in the evening sky.

SATURN now rises before midnight, brightening slightly from magnitude +0.6 to +0.4 during the month. The planet reaches its first stationary point on 8 February and thereafter its motion is retrograde. The planet remains in Virgo.

The Search for Life. To the pioneer observers of SETI (the Search for Extra-Terrestrial Intelligence) two stars were particularly attractive: Tau Ceti and Epsilon Eridani, both of which are easy naked-eye stars on view in the early evening during February. At a distance of less than a dozen light years they were the nearest solar-type (Sun-like) stars, though they were of slightly later spectral type, orange-yellow stars, and somewhat less massive and less luminous than the Sun; there seemed no reason why they should not have systems of planets.

Although in itself a very stable star, Tau Ceti proved to be a disappointment. It is metal-deficient and so less likely to have any rocky planets, and is surrounded by an extensive débris disk containing a tremendous amount of dust and rubble, so any planets would suffer a heavy bombardment. However, thus far there has been absolutely no sign of a planet, so the main attention has shifted to Epsilon Eridani.

Epsilon Eridani does actually have an old proper name, Sadira. It is seldom used, and this seems a pity, so let us use it here. The star is quite easy to find, not so very far from Rigel in Orion; its magnitude is 3.7, and it makes a pair with Delta Eridani (Rana) magnitude 3.5. Close to the pair is the much brighter Gamma Eridani (Zaurak), magnitude 3.0.

Associated with Sadira there are certainly two asteroid-type rocky belts, one at around 3 astronomical units from the star and the other at 20 astronomical units (one astronomical unit (AU) is the average distance between the Earth and the Sun – 150 million kilometres). There is evidence for a gas giant planet (known as Epsilon Eridani b) moving at a distance of 3.4 astronomical units (500 million kilometres) in a period of approximately 7 years, and it is possible that there is a second planet associated with the outermost of the two asteroid belts. In addition, Epsilon Eridani has an extensive outer disk of débris left over from the system's formation. Smaller planets similar to the Earth could possibly exist – so can there be life?

In 1960, a SETI team headed by the American astronomer Frank Drake used the big 85-foot (26-metre) Howard E. Tatel radio 'dish' antenna at the Green Bank Observatory in West Virginia (Figure 5), to listen out to Sadira and Tau Ceti, tuning the receiver to a wavelength of about 21cm (1420 MHz). This is the wavelength of radiation emitted by the clouds of interstellar hydrogen spread around the galaxy, so that any alien operators might well pay special attention to it. Alas, 150 hours of observation over a 4-month period picked up no

Figure 5. The Howard E. Tatel Radio Telescope, built in 1958, was the first major radio telescope at the National Radio Astronomy Observatory (NRAO) at Green Bank, West Virginia. It is also known as 85-1, and was the first of three 85-foot telescopes of similar design built at NRAO. Built by the Blaw-Knox Corporation of Pittsburgh, it began regular observations in February 1959 and did much of the pioneering radio astronomy at NRAO. In 1964 when the NRAO decided to build an interferometer, the Tatel Telescope became the fixed element in the NRAO 3-element interferometer system. (Image courtesy of NRAO, Green Bank, West Virginia.)

recognizable signals – a spurious signal detected on 8 April 1960 probably came from a high-flying aircraft – and the programme was soon given up. It was always the longest of long shots, but it was worth trying. It was known as 'Project Ozma', after Princess Ozma, the ruler of L. Frank Baum's fictional land of Oz, but it was commonly known as 'Project Little Green Men!'

If intelligent beings do exist in planets moving around Sadira, then the only reasonable way to try to contact them seems to be by radio, but later attempts have been no more successful than the first, which means that either no intelligent beings exist there, that they have not picked up our signals, or that they have no wish to communicate – perhaps they know too much about us! But attempts will go on.

It was also Frank Drake who made the first really serious speculations about the numbers of civilizations possibly able to communicate with us if they want to. His figure was about ten thousand civilizations in our Galaxy, although this estimate is subject to very large uncertainties. Of course, there is always the chance that they will contact us before we contact them, and there is one point here that I think is worth making. There have been suggestions that to communicate with other civilizations might be dangerous, because we might attract the attention of undesirable beings. To me this makes very little sense, because any civilizations able to conquer interstellar flight will be far more advanced than we are, and will have had to put war far behind them. We must simply wait and see what happens.

Eridanus and the Last in the River. Eridanus, the River, is one of the ancient constellations, and one of the largest in the entire sky. It begins near Rigel in Orion with the star Beta Eridani (Kursa) and sprawls southwards, ending with Achernar (magnitude +0.5, declination 57°S) in the far southern sky, which is not visible north of latitude 33°N. Of the top ten apparent brightest stars (excluding our Sun), Achernar is the hottest and bluest, lying at a distance of about 144 light years.

Achernar's name means 'The Last in the River' or 'River's End', but it has been suggested that the star seen by Hipparchus and later Ptolemy (and considered by them to mark the end of the river) was not Achernar, but Theta Eridani, also known as Acamar. Although only a third-magnitude star, Theta Eridani has a declination of 40°S, and is thus visible from locations further north. Whether or not this is correct, Theta Eridani is a lovely double star, whose components are magnitudes 3.4 and 4.5; the separation is over 8 seconds of arc, so that the pair is well seen with a small telescope. Both components are white. Of course it is always very low down as seen from southern Britain, but from my home in Selsey, I have just managed to get a glimpse of it as it peers above the horizon. See if you can locate it!

Architect of the Modern Gregorian Calendar. Cristoph Clavius, the Jesuit mathematician and astronomer, was born at Bamberg in Germany, on 25 March 1538. He became one of the most respected authors in the field of mathematics and astronomy during the late sixteenth and early seventeenth centuries, and his textbooks were used for education over many decades, not only in Europe but also much further afield. In 1579, Clavius was asked to work out the basis for a new calendar that would stop the Church's holidays shifting in relation to the seasons of the year. He proposed a reformed calendar that was adopted in 1582 by Catholic countries on the orders of Pope Gregory XIII, and which is now the modern Gregorian calendar used worldwide.

Clavius corresponded with Galileo, who visited him in 1611, and they discussed the new observations and discoveries made possible by the invention of the telescope. But Clavius could never bring himself to accept the theory that the Earth goes around the Sun and that it is not the central body in the universe. Clavius died in Rome on 6 February 1612, four hundred years ago this month. Later, the third largest crater on the visible nearside of the Moon was named in his honour (Figure 6 overleaf).

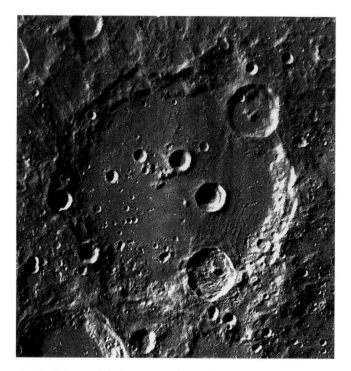

Figure 6. Mosaic image of the lunar crater Clavius obtained by the Lunar Reconnaissance Orbiter Camera (LROC) Wide Angle Camera (WAC). The floor of the crater forms a convex plain that is marked by some interesting smaller impacts. The most notable of these is a curving chain of craters that begins in the south then arcs across the floor in an anticlockwise direction, forming a sequence of ever decreasing size. This series of diminishing craters is useful for any amateur astronomers who wish to test the resolution of a small telescope. (Image courtesy of NASA Goddard/Arizona State University.)

March

Full Moon: 8 March   *New Moon:* 22 March

Equinox: 20 March

Summer Time in the United Kingdom commences on 25 March.

MERCURY reaches greatest eastern elongation (18°) on 5 March. It will be visible in the western sky in the evenings, at the end of civil twilight, during the first two weeks of the month, for observers in tropical and northern latitudes. For Northern Hemisphere observers this is the most favourable evening apparition of the year. Figure 4, given with the notes for February, shows observers in latitude 52°N the changes in azimuth (true bearing from the north through east, south and west) and altitude of Mercury on successive evenings when the Sun is 6° below the horizon, about 35 minutes after sunset in this latitude. The changes in the brightness of the planet are indicated on the diagram by the relative sizes of the white circles marking Mercury's position at five-day intervals: Mercury is at its brightest before it reaches elongation. During its period of visibility this month, before and after greatest eastern elongation, Mercury fades from magnitude −0.8 to +1.7. In mid-March, Mercury's elongation from the Sun rapidly decreases as the planet draws in towards inferior conjunction on 21 March.

VENUS reaches greatest eastern elongation (46°) on 27 March. It is a spectacular object in the evening sky, brightening from magnitude −4.2 to −4.4 during March. Over the same period the phase decreases from 63 per cent to 49 per cent. The planet moves rapidly from Pisces, through Aries and into Taurus during the month, its northerly declination meaning that it sets over four hours after the Sun throughout the month from northern temperate latitudes, but is visible for less than half this time from the tropics and more southerly locations. The planet's easterly motion through Aries takes it past Jupiter between 12 and 15 March and the two bright planets will make a lovely pairing in

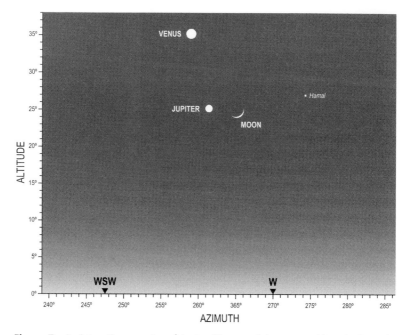

Figure 7. An interesting grouping of Jupiter, Venus and the crescent Moon in the early evening sky on 25 March 2012 as viewed from latitude 52°N. The angular diameters of the planets and the Moon are not to scale.

the evening twilight sky during this period. On 25 and 26 March the waxing crescent Moon will join the party, presenting another interesting opportunity for digital photographers. Figure 7 shows the relative positions of the crescent Moon, Jupiter and Venus on the evening of 25 March from latitude 52°N.

MARS is at opposition in Leo on 3 March, when its magnitude will attain −1.2, and it is visible all night long. Mars is at its closest to the Earth two days later, on 5 March, when it will be 0.67368 AU (100.8 million kilometres) distant. Given that this is an aphelic opposition of the Red Planet, the apparent diameter of the disk of Mars reaches only 13.9 arc seconds at opposition this year – about half the apparent disk diameter enjoyed at a perihelic opposition when it can attain 25.1 arc seconds, as in August 2003. Opposition in 2012 occurs close to the northern hemisphere summer/southern hemisphere winter solstice on

Mars, which takes place on 30 March, so that the north pole of the planet will be tilted towards the Earth. Mars will fade quite quickly after opposition, and the apparent size of the disk will decrease, as its distance from the Earth increases; its magnitude will be -0.7 and the disk diameter 12.6 arc seconds by the end of the month.

JUPITER is moving direct through Aries, and sets well before midnight; its magnitude decreases from -2.2 to -2.0 during March. The close approach of the brilliant Venus to Jupiter between 12 and 15 March and the interesting configuration with the crescent Moon on 25 March have already been mentioned above, and these will be interesting events to watch.

SATURN may be seen rising in the eastern sky in the mid-evening. The planet is moving retrograde in Virgo. The planet is at opposition next month and it brightens slightly from magnitude $+0.4$ to $+0.3$ during March. Figure 8 overleaf shows the path of Saturn against the background stars during the year.

Mars: A Look into the Past. Mars comes to opposition this month, on 3 March, and this means that even small telescopes will show the main dark markings on the disk. It is only about a hundred years ago that Mars was widely believed to be inhabited, but we know so much more about the planet now, and it is certainly not the kind of world where we can exist under natural conditions.

But what did astronomers think about it a hundred years ago? Here is a quotation from a book by the French astronomer Guillemin, dated 1912:

'It is generally told that the reddish and bright spots of Mars are the solid parts of the surface on the continents, whilst the dark bluish spots form the liquid parts on the seas. This distinction is founded on the unequal reflection of the light by the land and the water. According to Mr Lockyer, if we admit that the darkest spots indicate water, the darkest among them are those portions which are land-locked. Whence comes the reddish colouring, which characterizes the bright parts of the disk? If Mars were self-luminous, this tint would doubtless be attributed to the very nature of its light; but it only reflects to us the white light of the Sun; it is evident, therefore, that the colour is imparted by

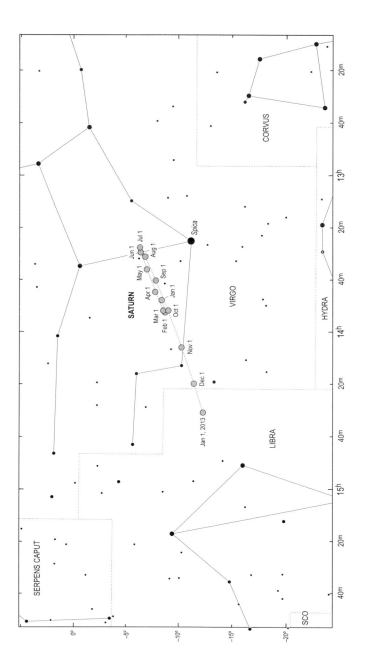

Figure 8. The path of Saturn against the background stars of Virgo and Libra during 2012.

the planet or its atmosphere. Several hypotheses have been suggested on this subject. Some have attributed sandstone. Others, among them Lambert, have thought that the colour of the vegetation, instead of being green, as it is on our Earth, is red on Mars. This explanation is not an impossible one; but, if it is true, there should be variations in the intensity of the tint on each of the hemispheres of the planet corresponding to the seasons; the tint should diminish during the winter, to reappear in the spring, and to attain its maximum in the summer.'

We know much better now. There can be no doubt that Mars had extensive oceans in the remote past, but there is no water left on the surface – although we cannot rule out the possibility of underground pools or even lakes. Neither can we rule out the chance of very primitive life forms which have survived since early times. If we do locate them, it will show that life may appear anywhere where conditions are suited to it, and this would be a pointer to the fact that life is likely to be common in the universe.

Mars is in Leo at opposition this year, which means that from Britain it is high in the sky. Against this, the apparent diameter is always below fourteen seconds of arc, and the magnitude is only −1.2 at opposition, so that Mars is outshone by the brightest star, Sirius. Its minimum distance from the Earth, on 5 March, will be 101 million kilometres, but the next few oppositions will be closer than this, and in 2018 the planet's apparent diameter will exceed twenty-four seconds of arc, but Mars will be much lower down then, in the constellation of Capricornus.

For two or three weeks before and after opposition this year the main markings upon Mars will not be hard to see with a telescope of fair size. Whether there will be any extensive Martian clouds remains to be seen.

Occultations of Jupiter by Venus. The planet Venus is at greatest elongation east on 27 March, when the planet will lie in Aries, close to the border with neighbouring Taurus, and well north of the celestial equator. It then sets over four hours after the Sun, and is extremely well placed for observers in northern Europe and North America. Earlier in the month, between 12 and 15 March, Venus and Jupiter are quite close together in the evening sky, and the superior brilliance of Venus is very obvious; a good chance here for astrophotographers. Close planetary conjunctions such as this are beautiful to look at, but the occultation of one planet by another is an exceptionally rare event, and of these, occultations of Jupiter by Venus are the most spectacular. The last such occultation occurred on 3 January 1818, when Venus passed in front of Jupiter as seen from the Far East. The next occasion will be on 22 November 2065, when Venus will again pass in front of Jupiter, but the elongation from the Sun will be only 8°, so the event will be essentially unobservable.

A Great French Astronomer. Nicolas-Louis de Lacaille was born at Rumigny, Ardennes, France, on 15 March 1713. He was one of the greatest observers of the eighteenth century and the pioneer of mapping the southern sky. In August 1751, he set up an observatory at the Cape of Good Hope, beneath the slopes of Table Mountain, and in two years he measured the positions of over 9,800 stars in the far southern sky. Although he used a very small telescope, his measurements were amazingly accurate, obtaining accurate positions for 240 principal stars. He also catalogued forty-two nebulae and clusters in the Southern Hemisphere of the sky, and this list was published in 1755, sixteen years before the first instalment of Charles Messier's catalogue of nebulae and clusters. Lacaille introduced fifteen new constellations, fourteen of which are still in use today, breaking up the original large constellation Argo Navis into three parts, Carina, Puppis and Vela. He also made careful measurements of the Moon, Mars and Venus, and in 1761 he observed the transit of Venus. He died in Paris on 21 March 1762, two hundred and fifty years ago this month. The crater La Caille on the Moon and the asteroid 9135 Lacaille, discovered on 17 October 1960 at the Palomar Observatory, are named after him.

April

Full Moon: 6 April   *New Moon:* 21 April

MERCURY reaches greatest western elongation (28°) on 18 April and the planet is visible in the morning for observers in tropical and southern latitudes. Unfortunately for observers in the latitudes of the British Isles, the planet remains unsuitably placed for observation throughout the month. For Southern Hemisphere observers this is the most favourable morning apparition of the year. Figure 9 shows, for observers in latitude 35°S, the changes in azimuth (true bearing from the north through east, south and west) and altitude of Mercury on successive mornings when the Sun is 6° below the horizon. This is at the beginning of morning civil twilight, which in this latitude and at this time of year occurs about twenty-five minutes before sunrise.

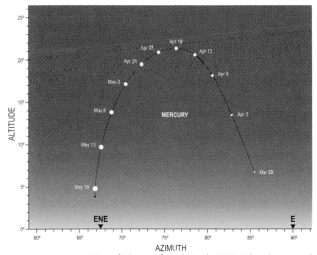

Figure 9. Morning apparition of Mercury from latitude 35°S. The planet reaches greatest western elongation on 18 April. It will be at its brightest in mid-May, after elongation.

During its long period of visibility, which runs throughout April and until the middle of May, Mercury brightens from magnitude +1.6 to −1.2. The changes in the brightness of the planet are indicated on the diagram by the relative sizes of the white circles marking Mercury's position at five-day intervals. It should be noted that Mercury is at its brightest in mid-May, some four weeks after it reaches greatest western elongation. The diagram gives positions for a time at the beginning of morning civil twilight on the Greenwich meridian on the stated date. Observers in different longitudes should note that the actual positions of Mercury in azimuth and altitude will differ slightly from those given in the diagram due to the motion of the planet. The waning crescent Moon will appear fairly close to Mercury in the sky on the morning of 18 April.

VENUS remains a spectacular object in the evening sky, brightening very slightly from magnitude −4.4 to −4.5 during April. From northern temperate latitudes the planet sets over four-and-a-half hours after the Sun, although from latitude 35°S this length of time is reduced to just two hours. The phase of the planet decreases from 48 per cent to 27 per cent during the month. Venus is moving against the background stars of Taurus, the Bull, and will appear to pass very close to the beautiful open star cluster, the Pleiades or Seven Sisters, on 3 April, presenting an interesting photographic opportunity for observers.

MARS, now past opposition, fades from magnitude −0.7 to 0.0 during the month. The planet begins the month moving retrograde in Leo, but reaches its second stationary point on 14 April and thereafter begins to move direct once more. The apparent diameter of the planet's disk has dropped below 10 arc seconds by the end of the month.

JUPITER, magnitude −2.0, is in conjunction with the Sun in May, and will be lost in the glare of the evening twilight sky by the end of the month. Earlier in April, the planet may be spotted low in the twilight sky just north of west about thirty to forty minutes after sunset. On 22 April, at this time of the evening, Jupiter will lie very close to the incredibly thin crescent Moon, about 6° high in the west-north-western sky.

SATURN, magnitude +0.3, is at opposition in Virgo on 15 April. The planet becomes visible in the east-south-eastern sky as soon as darkness falls and is observable all night long. The planet is a lovely sight in even a small telescope, with the rings displayed at an angle of 13.7° as viewed from the Earth.

Phœbe. Saturn is at opposition this month, and still fairly well placed for observers in the Northern Hemisphere. The rings are opening, and Saturn has started to regain its beauty, but do not forget its disk and its family of satellites.

Figure 10. Saturn's irregular satellite Phœbe is revealed in this mosaic of two images acquired from a distance of about 32,500 kilometres during the flyby of the Cassini spacecraft on 11 June 2004. Phœbe appears to be an ice-rich body coated with a thin layer of dark material. Small bright craters in the image are probably fairly young features. When impacting bodies slammed into the surface of Phœbe, the collisions excavated fresh, bright material – probably ice – underlying the surface layer. Further evidence for this can be seen on some crater walls where the darker material appears to have slid downwards, exposing light-coloured material. (Image courtesy of NASA/JPL/Space Science Institute, Boulder, Colorado.)

Figure 11. This artist's impression simulates an infrared view of the giant, nearly invisible ring around Saturn – the largest of the giant planet's many rings – that was discovered by NASA's Spitzer Space Telescope. The ring is probably associated with Saturn's irregular, retrograde satellite Phœbe (Figure 10). Saturn appears as just a small dot from outside the band of ice and dust. The giant ring, stars and wispy clouds are part of the artist's representation. The bulk of the ring material starts about six million kilometres away from the planet and extends outward roughly another 12 million kilometres. The inset shows an enlarged image of Saturn, as seen by the W.M. Keck Observatory at Mauna Kea, Hawaii, in infrared light. (Image courtesy of NASA/JPL-Caltech/Keck.)

There have been interesting developments on the disk – a huge equatorial storm, large enough to be seen with a very small telescope. Whether it will have abated by the time you read these words I know not; go out and have a look!

Eight satellites visible with small or modest telescopes (Mimas to Iapetus) were known before 1898. In that year, W. H. Pickering discovered the ninth satellite, Phœbe, the first discovery to be made photographically. At magnitude 16.5 it is well within the range of modern amateur equipment. All the remaining outer satellites, discovered since 2000, are very small indeed. The smallest, Fenrir, a satellite with a retrograde orbit, and the sixty-first Saturnian satellite to be discovered, can be no more than four kilometres in diameter.

All the small outer moons are presumably asteroidal, but Phœbe (Figure 10), the largest of the retrograde satellites, diameter 220 kilometres, is rather different and is much more likely to be a captured Centaur object, an icy planetoid that has escaped from the Kuiper Belt. Its surface temperature is −198°C. Phœbe is more or less spherical with an extremely dark, heavily cratered surface; one crater, Jason, is eighty kilometres across with walls sixteen kilometres high. The orbital period of Phœbe is 550 days, but the rotation period is only 9 hours 17 minutes. The globe seems to be about fifty per cent rock, mixed with a substantial amount of water ice; the dark surface coating is probably very thin.

An important discovery made in 2009 by NASA's Spitzer Space Telescope has solved one mystery. Phœbe is associated with a very tenuous outer ring, that is tilted by 27° to Saturn's equatorial plane and the other rings (Figure 11). Particles knocked off this ring migrate inwards and strike the leading hemisphere of the surface of the next inner moon, Iapetus, darkening it. This is why Iapetus has its 'yin-yang' surface, one part as bright as snow and the other part as black as a blackboard.

Two Astronomers to Remember. Two astronomers should be specially remembered this month: W. W. Campbell and C. V. L. Charlier.

William Wallace Campbell was born in Ohio on 11 April 1862, 150 years ago this month. He graduated from the University of Michigan, and began to take a serious interest in astronomical spectroscopy. His abilities were noted by Edward Holden, then Director of the Lick Observatory in California, and in 1891 Campbell joined the observatory staff. He quickly became the most successful astronomical spectroscopist in the world, and in 1901 he was appointed Director at Lick, a post he retained for thirty years. He published catalogues of stellar radial velocities, and published papers dealing with many topics such as the discoveries of spectroscopic binary systems. Campbell also took part in seven total solar eclipse expeditions, and led a team to Wallal, Australia in 1922, where the observations confirmed the deflection of starlight during totality, providing further evidence supporting Einstein's general theory of relativity.

Campbell was happily married with three sons, and was universally popular and respected. He received many honours, including the Lalande Medal of the French Academy of Sciences (1903) and the Gold Medal of the Royal Astronomical Society (1906). He died in San Francisco on 14 June 1938. Craters on the Moon and on Mars and the asteroid 2751 Campbell are named in his honour.

Carl Vilhelm Ludwig Charlier was born at Östermund (Sweden) on 1 April 1862, 150 years ago this month. He graduated from Uppsala in 1887, and ten years later became professor of astronomy at Lund University; it was here that he made his main contributions to astronomy, mainly in statistical studies of the stars in our Galaxy and their positions and motions.

In the early twentieth century, it was still not certain whether objects such as the Andromeda 'nebula', M31, were independent galaxies or were simply minor features of our own Galaxy. One argument advanced in favour of the second of these theories was that few of the objects were found near the Milky Way in the sky – there is a distinct 'Zone of Avoidance'. Charlier explained this neatly, and correctly, as being due to obscuring clouds of matter near the main plane of our Galaxy.

In 1897, Charlier married Siri Leissner from Stockholm. He retired to Lund and died there on 5 November 1934. Craters on the Moon and on Mars and the asteroid 8677 Charlier are named in his honour.

One minor fact might be worth noting here. Charlier might have been regarded as the co-discoverer of asteroid 433 Eros, the first asteroid known to come within the orbit of Mars. He observed it in 1898 slightly earlier than Witt, on the same night – but Witt reduced and published his results first!

May

Full Moon: 6 May   *New Moon:* 20 May

MERCURY is still on view before dawn, in the east-north-eastern twilight sky, for observers in the tropics and Southern Hemisphere during the first two weeks of May. From more northerly latitudes the planet will not be visible this month. Figure 9, given with the notes for April, shows, for observers in latitude 35°S, the changes in azimuth (true bearing from the north through east, south and west) and altitude of Mercury on successive mornings when the Sun is 6° below the horizon, about 25 minutes before sunrise in this latitude. The changes in the brightness of the planet are indicated on the diagram by the relative sizes of the white circles marking Mercury's position at five-day intervals: Mercury is at its brightest after it reaches elongation. During its period of visibility this month, Mercury brightens from magnitude 0.0 to -1.2. After mid-May, Mercury's elongation from the Sun rapidly decreases as the planet draws in towards superior conjunction on 27 May.

VENUS begins the month as a prominent object in the south-western sky in the evening, shining at magnitude −4.5, and setting four hours after the Sun in northern temperate latitudes, although only half that time in the tropics and more southerly climes. However, the planet draws very rapidly in towards the Sun as the month progresses, and it is likely to be lost in the bright dusk twilight sky before the end of May.

MARS, now moving direct in Leo, continues to fade as its distance from Earth increases; its magnitude decreases from 0.0 to +0.5 during the month. The planet is visible in the southern sky as soon as night falls and sets in the early morning hours. Figure 12 shows the path of Mars against the background stars during the year to November.

JUPITER is in conjunction with the Sun on 13 May. By the very end of the month, the planet may just be glimpsed low in the east-north-

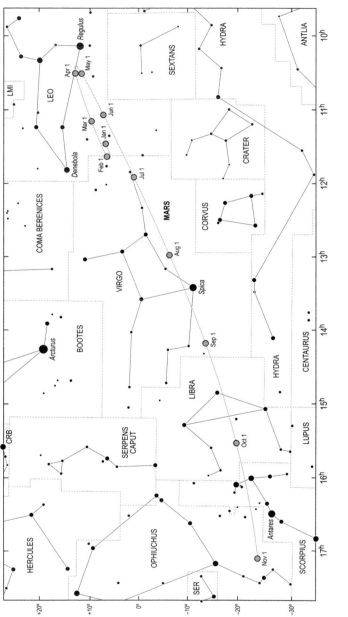

Figure 12. The path of Mars as it moves through the zodiacal constellations from Leo to Scorpius and into Ophiuchus between January and early November 2012.

eastern twilight sky before dawn by observers in the tropics. Its magnitude is −2.0. The planet will not be visible from northern temperate latitudes this month.

SATURN is visible in the south-south-east as darkness falls, and is observable for most of the night, moving retrograde in Virgo. It fades very slightly from magnitude +0.3 to +0.5 during the month.

Hydra. Now that the vast constellation of Argo Navis has been broken up, Hydra, the Watersnake, is the largest constellation in the entire sky. In mythology it represents the multi-headed monster which was killed by Hercules as the second of his twelve labours. Despite its huge size there is only one bright star, Alphard, which is of the second magnitude. It is known as the 'Solitary One' because there are no other conspicuous stars anywhere near it. It is easy enough to find because the Twins of Gemini, Castor and Pollux, point to it.

Hydra begins not far from Gemini and sprawls across the sky, ending below Virgo (Figure 13). Apart from Alphard, the brightest stars are Gamma Hydrae (magnitude 3.0), and Zeta (magnitude 3.1) in the Watersnake's head. There are, however, quite a number of interesting telescopic objects. One of these is the very red Mira-type variable R Hydrae, located not far from Gamma. R Hydrae has a period of 389 days, and a magnitude range between 3.5 and 11. It is mostly within the range of binoculars. Its position is RA 13h 29.7m, declination −23° 17′. This year, maximum is predicted to occur on 26 November, which is not at all convenient for observers! Another interesting object is the 'Ghost of Jupiter' (NGC 3242), a particularly attractive planetary nebula. Its position is RA 10h 24.8m, declination −18° 38′. The magnitude is only 12, so it is not a very easy object with a small telescope.

Two small constellations lie along the Watersnake's back; one of these is Crater, the Cup, which has been identified as the wine goblet of the god Bacchus. It is not conspicuous because the only star above the fourth magnitude is Delta, which is of spectral type G8 and is slightly orange. Also along the Watersnake's back is the much more conspicuous Corvus, the Crow, which is easy to identify because its four main stars are all about the third magnitude and form a conspicuous quadrilateral. Corvus is one of Ptolemy's original forty-eight constellations. According to legend, the god Apollo became enamoured of Coronis, mother of the great doctor Aesculapius, and sent a crow to watch her

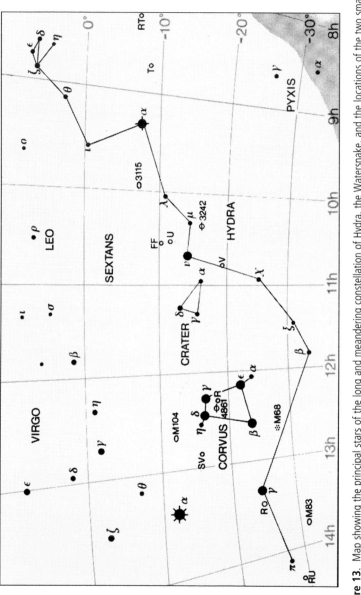

Figure 13. Map showing the principal stars of the long and meandering constellation of Hydra, the Watersnake, and the locations of the two small patterns – Crater, the Cup, and Corvus, the Crow – which lie between Hydra and the stars of Virgo.

and report on her behaviour. In fact, the crow's report was decidedly unfavourable, but Apollo rewarded the bird with a place in the sky!

A Stellar Astrophysicist Remembered. The German astronomer Martin Schwarzschild, son of the noted astrophysicist Karl Schwarzschild, was born in Potsdam on 31 May 1912, one hundred years ago this month. He was educated in Göttingen, and went to the university there. He left Germany in 1936 and taught in Oslo, Norway before travelling to the United States where he researched and taught at Harvard and Columbia universities. After serving in the US Army during World War II, he joined his life-long friend, Lyman Spitzer, at Princeton University in 1947, and it was here that he spent most of his professional life, and where his main work was carried out. Schwarzchild made significant advances in many areas of astrophysics dealing with stellar structure and evolution. His 1958 textbook, *Structure and Evolution of the Stars*, taught a generation of astrophysicists how to use the then emerging computer technology to create models of stars.

In the 1950s and 1960s he took responsibility for the Stratoscope balloon projects, which carried astronomical instruments high into the atmosphere. Stratoscope 1 carried a 30-centimetre telescope to a height of 30 kilometres in order to obtain better images of the Sun, particularly solar granulation and sunspots, confirming the existence of convection in the solar atmosphere. He followed with Stratoscope 2, which sent back infrared spectra of the planets, red giant stars and the nuclei of galaxies. He predicted that the convection cells in red giant stars would be very large and would dominate the stellar surface, and this was eventually confirmed by images of the star Betelgeux. In his later years, Schwarzschild made significant contributions towards understanding the dynamics of elliptical galaxies.

Schwarzchild received many honours in Norway, Denmark and Belgium as well as the United States. He became Vice President of the International Astronomical Union, and continued his work until shortly before his death in Langhorne, Pennsylvania on 10 April 1997. The main belt asteroid 4463 Marschwarzschild is named after him.

June

Full Moon: 4 June *New Moon:* 19 June

Solstice: 20 June

MERCURY is at greatest eastern elongation very early next month, so the planet will be visible in the north-western sky, on evenings in the last three weeks of June, for observers in equatorial and southern latitudes. Unfortunately, the planet is not suitably placed for observation from northern temperate latitudes this month. Figure 14 shows the changes in azimuth and altitude of Mercury at the end of evening civil twilight, about 30 minutes after sunset, from latitude 35°S. The changes in the brightness of the planet are indicated by the relative sizes of the white circles marking Mercury's position at five-day intervals: Mercury

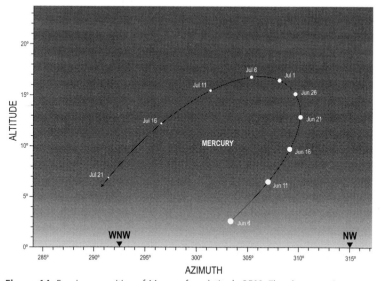

Figure 14. Evening apparition of Mercury from latitude 35°S. The planet reaches greatest eastern elongation on 1 July. It will be at its brightest in early June, before elongation.

is at its brightest before it reaches greatest eastern elongation. The planet's brightness decreases from magnitude −1.1 to +0.5 during June. The diagram gives positions for a time at the end of evening civil twilight on the Greenwich meridian on the stated date. Observers in different longitudes should note that the actual positions of Mercury in azimuth and altitude will differ slightly from those given in the diagram due to the motion of the planet.

VENUS passes rapidly through inferior conjunction on 5–6 June, when it may be seen in a rare transit across the face of the Sun. The last such transit was on 8 June 2004, but if you miss the one this month, you will have to wait over 105 years for the next one! Following inferior conjunction, Venus moves swiftly into the morning sky before dawn. For observers in equatorial and more southerly latitudes, the planet may be glimpsed low in the twilight sky shortly before sunrise from the middle of June onwards. By the end of the month, from these latitudes, the planet will be rising over two hours before the Sun and located quite close to the fainter Jupiter in the dawn twilight sky; Venus, magnitude −4.4 and Jupiter, magnitude −2.0. The phase of Venus will have increased to 16 per cent by the end of the month. Unfortunately, Venus will remain inconveniently low for observers in northern temperate latitudes this month.

MARS is visible in the south-western sky as soon as darkness falls and sets around midnight. The planet's direct motion carries it from Leo across the border into neighbouring Virgo on 21 June. Mars fades from magnitude +0.5 to +0.8 during the month.

JUPITER, magnitude −2.0, is a conspicuous object in the early morning sky for observers in equatorial and more southerly latitudes, but observers further north will probably not be able to pick out the planet until the second half of the month. Jupiter is now moving direct in Taurus, and passes south of the Pleiades star cluster in mid-June. On 17 June, the waning crescent moon will be quite close to Jupiter in the dawn twilight sky. By the end of the month, the much brighter planet Venus will lie in the same part of the sky, but lower down, closer to the horizon.

SATURN starts the month moving retrograde in Virgo, but after reaching its second stationary point on 25 June, it resumes its direct motion once more. The planet fades from magnitude +0.5 to +0.7 during June.

PLUTO reaches opposition on 29 June, in the constellation of Sagittarius, at a distance of 4,674 million kilometres (2,904 million miles). It is visible only with a moderate-sized telescope since its magnitude is +14.

This Month's Transit of Venus. Transits of Venus occur in pairs separated by eight years, after which there are no more for over a century. The last transit, on 8 June 2004, was beautifully seen from my observatory at Selsey under a perfectly clear sky (Figure 15). Venus was a very prominent small disk against the Sun, and the phenomenon of the 'black drop' was well seen. Of course, all of the usual safety precautions have to be carefully followed when observing a transit of Venus just as with observing the Sun.

Whether or not this month's transit of Venus will be seen depends on your location. The entire event lasts over six-and-a-half hours. The transit will not be seen at all from north-eastern, central and southern parts of South America, Portugal and south-western parts of Spain, most of west Africa, and western parts of central and southern Africa. The transit will be visible in its entirety from western parts of the Pacific Ocean, most of Australia (except the west), New Zealand, Japan, most of central, northern and north-eastern Asia, Alaska and north-western Canada. Most of North America, the Caribbean region and Central America will see the beginning of the transit in the late afternoon and evening of Tuesday, 5 June, but the Sun will set before the transit ends. Much of Europe, Africa, the Middle East and Asia will miss the early part of the transit but will see the end on the morning of Wednesday, 6 June.

Of course, one way to be sure that the Sun is above the horizon at the time of the transit is to go where the midnight Sun remains above the horizon, and I am sure that quite a number of people will make their way to the far north of Norway, where they can only hope for a cloud-free sky.

In earlier days, transits of Venus were used to measure the distance between the Sun and the Earth, but now this method has been completely superseded, and transits are regarded as little more than fascinating spectacles. I advise you to make the most of this year's transit. If you fail to see it, you will have to wait until December 2117!

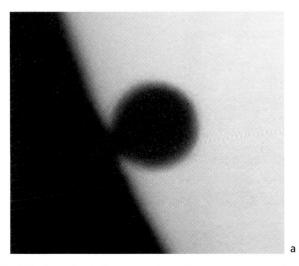

a

Figure 15. a) The 'black drop' effect, captured just after second contact, during the transit of Venus which took place on 8 June 2004. A small black 'teardrop' appears to connect the dark disk of Venus to the limb of the Sun, making it impossible to accurately time the exact moment of second contact. The phenomenon is also visible just before third contact. b) Later during the transit, Venus is visible in silhouette against the brilliant disk of the Sun as an intensely dark, almost black spot. (Both images courtesy of Martin Mobberley.)

b

The Rival of Mars. June is probably the best time to find the fiery red star Antares in the constellation of Scorpius, the Scorpion. The star may be mistaken for the Red Planet, Mars, a fact demonstrated by its name, Antares, or 'Ant-Ares', which means 'rival of Mars' – Ares being the Greek name for the god of war. Antares (magnitude 1.0) is a huge red supergiant star, one of the largest known. It is a cool star, with a surface temperature of only about 3,400°C, so it radiates most of its energy at infrared wavelengths.

The star must be enormous; its diameter is probably around 1,000 million kilometres, so, if placed within our Solar System, it would stretch well out across the Asteroid Belt, two thirds of the way from the Sun to Jupiter's orbit. Antares is 550 light years away, and at least 10,000 times as luminous as the Sun, but the amount of dimming by interstellar dust is not well known so the star may be considerably more luminous. Antares has a small companion, which is really described as being green, though this is mainly due to contrast. At fifth magnitude, the small star is not faint, but it is remarkably difficult to find because it is so overpowered by the glare of the primary.

Scorpius is a magnificent constellation, but it is always low down as seen from Britain, and part of it barely rises. You need exceptional conditions and a completely clear horizon to see the Scorpion's 'sting', even from southernmost parts of the British Isles.

The Co-discoverer of Neptune. One notable astronomer deserves to be remembered this month. Johann Gottfried Galle of Germany was born on 9 June 1812, one hundred years ago. He was born in Pabsthaus and studied at Berlin University from 1830 to 1833. In 1835, following completion of the Berlin Observatory, Galle started work as an assistant to Johann Franz Encke, its director. There, Galle was able to use the high-quality Fraunhofer 9-inch (24-centimetre) refractor, and in June 1838 he discovered Saturn's Crêpe Ring (or C-ring). He was also a successful comet hunter and in the short period from 2 December 1839 to 6 March 1840 he discovered three new comets.

However, Galle is probably best remembered as being the co-discoverer of the planet Neptune. The position of the planet had been worked out by the French mathematician Urbain Le Verrier from its perturbations on the planet Uranus, which had been discovered by William Herschel in 1781. Le Verrier gave a position for the unknown planet, and asked observers in Paris to search for it. Nothing was done quickly, and patience was never Le Verrier's strong point. He therefore wrote to Galle in Berlin, and asked him to use the telescopes there to search in the positions indicated. Galle went to Johann Encke and asked for permission to use the main telescope. Encke was sceptical, but then shrugged his shoulders and said, 'Let us oblige the gentleman from Paris.' The discussion was overheard by a young astronomer, Heinrich d'Arrest, who asked permission to join in the hunt. On the first dark night they went into the observatory and started checking. Galle, at the telescope, called out the stars, and d'Arrest checked them off on the star map. Suddenly, d'Arrest remembered that there was a new, very detailed star map of the area, not yet widely distributed. He found it, used it, and almost at once Galle called out a star and d'Arrest replied, 'That star is not on the map!' Also, the object seemed to show a tiny disk which no star can have.

Encke joined them in the dome, and they followed the object until it set. On the next night they observed it again, and found it had moved by just the amount the unknown planet would be expected to do. Encke promptly contacted Le Verrier, and told him, 'The planet whose position you predicted actually exists!'

In many books it is stated that the discoverer of Neptune was Le Verrier, who had given the position for it. This is not strictly correct, because the two discoverers were Galle and d'Arrest.

This was only a minor part of Galle's work. In 1851, he moved to Breslau (nowadays Wrocław) to become professor of astronomy and director of the local observatory. He enjoyed a worldwide reputation for careful observing and made a number of pioneering investigations. He died at Potsdam in Germany on 10 July 1910. Two craters, one on the Moon and one on Mars, the minor planet 2097 Galle, and a ring of Neptune have been named in his honour.

July

Full Moon: 3 July   *New Moon:* 19 July

EARTH is at aphelion (furthest from the Sun) on 5 July at a distance of 152 million kilometres (94.5 million miles).

MERCURY is at greatest eastern elongation (26°) on 1 July. The planet is still on view at dusk, in the north-western twilight sky, for observers in the tropics and Southern Hemisphere during the first half of July. From more northerly latitudes the planet is unlikely to be seen low down in the bright evening twilight sky. Figure 14, given with the notes for June, shows observers in latitude 35°S the changes in azimuth and altitude of Mercury on successive evenings when the Sun is 6° below the horizon, about 30 minutes after sunset in this latitude. The changes in the brightness of the planet are indicated on the diagram by the relative sizes of the white circles marking Mercury's position at five-day intervals: Mercury is at its brightest before it reaches elongation. During its period of visibility this month, Mercury fades from magnitude +0.6 to +2.1. After mid-July, Mercury's elongation from the Sun rapidly decreases as the planet draws in towards inferior conjunction on 28 July.

VENUS, magnitude −4.4, rises over two hours before the Sun at the beginning of July from equatorial and more southerly latitudes, from where the planet makes a lovely pairing with the fainter Jupiter in the dawn twilight sky. The two bright planets are situated among the stars of Taurus, the Bull, between the Pleiades star cluster and the reddish star Aldebaran (Alpha Tauri), with Venus among the scattered stars of the V-shaped Hyades cluster at the beginning of the month. Unfortunately, at this time, Venus is rather low in the dawn twilight sky for observers further north. The situation improves as the month progresses since Venus draws rapidly out from the Sun and by the end of the month the planet is rising over three hours before the Sun from northern temperate latitudes. The apparent separation of Venus and Jupiter increases as the month proceeds, because the eastwards motion of Venus is faster than that of Jupiter, and by the end of the month they are well separated. The crescent Moon will make an interesting grouping with Jupiter, Venus and Aldebaran on the morning of 15 July, as shown in Figure 16. The phase of Venus increases from 17 per cent to 41 per cent during July.

MARS is visible as soon as darkness falls but sets before midnight. The planet is moving direct through Virgo, some way to the west of Saturn, which is very slightly the brighter of the two planets. Mars fades a little from magnitude +0.9 to +1.1 during the month.

JUPITER is a conspicuous object in the early morning sky, moving direct amongst the stars of Taurus. From northern temperate latitudes it begins the month rather low down just north of east in the dawn twilight, with the much brighter Venus below it. On the morning of 15 July, the waning crescent Moon will lie very close to Jupiter in the sky,

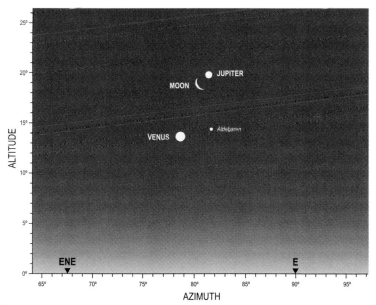

Figure 16. An interesting grouping of Venus, Jupiter and the crescent Moon against the background stars of Taurus in the early morning sky on 15 July 2012, as viewed from latitude 52°N. The angular diameters of the planets and the Moon are not to scale.

with Venus some way below, as shown in Figure 16. By the end of July, Jupiter will be rising before midnight, and positioned just beyond the northern arm of the V-shaped Hyades star cluster. Jupiter's brightness increases very slightly from magnitude −2.0 to −2.1 during July.

SATURN is visible as an early evening object in Virgo throughout July, setting before midnight. Its brightness decreases from magnitude +0.7 to +0.9 during the month.

The Little Lion. Now and then it is entertaining to seek out some of the more obscure constellations. Leo Minor is one of Ptolemy's 'originals', which seems rather strange, because it is entirely unremarkable, with only one star above the fourth magnitude. No mythological legends seem to be attached to it.

For some reason that I cannot pretend to explain, only one star in the constellation has a Greek letter designation; this is Beta, magnitude 4.2. The only star with a proper name is 46, Præcipua, the brightest star in the constellation at magnitude 3.8. It is of spectral type K0, and therefore decidedly orange. Apart from these two, all remaining stars of the Little Lion are magnitude 4.5 or fainter; the brightest of these is 21 (magnitude 4.5).

To locate Leo Minor, find the triangle of stars south of the Pointers in Ursa Major: Psi, Lambda and Mu Ursæ Majoris. Lambda and Mu provide an interesting colour contrast; Lambda (Muscida), magnitude 3.4, is of type G4 and therefore slightly yellowish; Mu (Tania Australis), magnitude 3.1, is a red giant of type M0. The three brightest stars of Leo Minor lie close to Tania Australis, and make up a decidedly dim triangle.

What can be seen in Leo Minor? Well, there are three Mira-type variables, two of which can rise to above the seventh magnitude; one of these, R (RA 09h 45.6m, declination +34° 31′), has a range from magnitude 6.3 to 13.2 in a period of about 372 days. The other, RW (RA 10h 16.1m, declination +30° 34′), has a range from magnitude 6.9 to 10.1, but its period does not seem to be known accurately. It is very red (spectral type N), so it is worth following. There are also various galaxies, of which the brightest is the Sc system NGC 3344, integrated magnitude 9.9 at RA 10h 43.5m, declination +24° 55′.

A favourite quiz question is, 'What is peculiar about Alpha Leonis Minoris?' Answer: 'It doesn't exist. There is no Alpha Leonis Minoris!'

The Milky Way. Given a dark, really clear sky, the Milky Way is at its very best from northern temperate latitudes this month. From Perseus in the north-east, it arches right across the sky, up through Cassiopeia, passing close to the zenith, through Cygnus and Aquila, and on down to Sagittarius and Scorpius in the south. It is a pity that from Britain we never have a really good view of the brightest part of the Milky Way, towards the galactic centre, which lies too far south in the sky. I well remember standing in New Zealand, on a very clear, dark night; the star clouds in Sagittarius were high up – and they cast shadows.

Against this, I also remember a visit to my observatory from a group of boys from a school in London. As darkness fell, one of the boys pointed upward. 'Look, sir, what are all those?' The point I am making here is that *those boys had never seen the Milky Way, and had never really seen the stars.* From where they lived, the glare and glow of artificial lights drowned out almost everything in the sky apart from the Sun, the Moon and the occasional bright planet. If the spread of towns, cities and out-of-town housing estates goes on, there may be a whole generation of children who will never see the true sky at all. I hope that this will not happen, and the Campaign for Dark Skies of the British Astronomical Association has being doing sterling work to make people aware of the growing problem of light pollution in its various forms. But unless action is taken fairly soon, I have an uneasy feeling that we may one day lose our view of the stars from all but a few remote parts of the British Isles.

The Third Astronomer Royal. The English astronomer James Bradley (Figure 17), who became the third Astronomer Royal (following Flamsteed and Halley) was born at Sherborne, near Cheltenham, Gloucestershire, in March 1693. He graduated from Balliol College, Oxford; he was ordained in 1719, and became vicar of Bridstow in Monmouthshire, but his main interest was in astronomy, and in 1721 he resigned his vicarage to become Savilian Professor of Astronomy at Oxford. He is best known for two fundamental discoveries in astronomy, the aberration of light (1725–8), and the nutation of the Earth's axis (1728–48). These two discoveries are widely regarded as some of the most brilliant and useful of the eighteenth century.

Robert Hooke had made an unsuccessful attempt to measure the parallax of the star Gamma Draconis, which passes more or less overhead from London. Together with Samuel Molyneux, a wealthy

Figure 17. Portrait, by Thomas Hudson, of the English astronomer James Bradley, the third Astronomer Royal and discoverer of the aberration of light and the nutation of the Earth's axis. (Image courtesy of National Portrait Gallery, London.)

amateur astronomer who lived at Kew, Bradley decided to try again. A 24-foot 'zenith telescope' was built in 1725 and fixed vertically to a chimney in the Molyneux mansion – it meant cutting holes in the roof and between floors! Amazingly, it worked well. Although Gamma Draconis showed no small annual cyclical motion due to parallax (which is not surprising, since the star is almost 150 light years away), it was found to show an unexplained annual 20 arc second oscillation about its mean position in the sky. In 1728, while sailing on the River Thames, Bradley noted how the wind vane on the boat's mast shifted its orientation with the boat's motion – and realized that the apparent direction from which a star's light reaches the observer is altered by the motion of the Earth. He had discovered the aberration of light, and it gave Bradley a means to derive an improved estimate for speed of light, which was within 2 per cent of the modern value.

After publication of his work on the aberration of light, Bradley continued to observe, to develop and verify his second major discovery, the nutation of the Earth's axis. This is a small cyclical motion superimposed upon the steady 26,000-year precession of the Earth's axis of rotation, caused mainly by the gravitational effect of the 18.6-year rotation period of the Moon's orbit. Bradley tested his discovery of nutation by making observations during an entire 18.6-year revolution of the nodes of the Moon's orbit.

Bradley later compiled an important star catalogue, and also found time to modernize and improve the instruments at the Greenwich Observatory. He died at Chalford, Gloucestershire, on 13 July 1762, 250 years ago this month.

August

Full Moon: 2 and 31 August   *New Moon:* 17 August

MERCURY passed through inferior conjunction towards the end of July. It now moves out to the west of the Sun, reaching greatest western elongation (19°) on 16 August. This is not a particularly favourable elongation for observers at any latitude, but the planet should be glimpsed in the dawn twilight, low in the east-north-east, around the time of greatest elongation, at the start of morning civil twilight, roughly thirty minutes before sunrise. Observers in the tropics and north of the equator should have the best view. Mercury will be at its brightest, and consequently easier to spot, after it reaches greatest western elongation, its magnitude increasing from +2.1 on 7 August to +0.1 at elongation and −1.4 by the end of the month.

VENUS reaches greatest western elongation (46°) on 15 August, and is a brilliant object in the early morning sky, although the planet fades very slightly from magnitude −4.4 to −4.2 during August. By the end of the month, Venus will be rising four hours before the Sun from northern temperate latitudes, although the period of visibility is considerably less from locations farther south on account of the planet's northerly declination. Its rapid eastwards motion carries Venus from Taurus into Gemini during the month. The phase of the planet increases from 42 per cent to 58 per cent in August. The waning crescent Moon will appear close to Venus in the morning sky on 14 August.

MARS is visible in the early evening, low in the west-south-western sky, but sets less than two hours after the Sun. Its rapid eastwards motion carries Mars past Saturn in mid-August; both planets are in Virgo. On the evening of 14 August, Saturn, Mars and the star Spica (Alpha Virginis) will all be in line, and it will be interesting to compare the brightnesses and colours of the three objects; Saturn (magnitude +0.8) should look yellowish, Mars (magnitude +1.1) reddish, and Spica (magnitude +1.0) a bluish-white. From northern temperate latitudes the planet will be rather low in the twilight at dusk by the end of the month, although much better placed for observers situated further south.

JUPITER is a brilliant object in Taurus, brightening from magnitude −2.1 to −2.3 during August, and rising in the late evening by the end of the month. The waning crescent Moon will appear close to Jupiter in the morning sky on 12 August.

SATURN, magnitude +0.8, is visible low in the west-south-western sky as darkness falls, but sets less than two hours after the Sun by the end of the month. The planet is moving direct in Virgo, and will be quite close to the slightly fainter planet Mars around the middle of the month. From northern temperate latitudes, Saturn will be rather low in the twilight at dusk by the end of August, although easier for observers farther south.

NEPTUNE is at opposition on 24 August, in the constellation of Aquarius. It is not visible with the naked eye since its magnitude is +7.8. At opposition Neptune is 4,336 million kilometres (2,694 million miles) from the Earth. Figure 18 shows the path of Neptune against the background stars during the year.

The Brightest Stars. Vega, in Lyra, the Lyre, is almost overhead during August evenings. Nowadays most catalogues rate it as the fifth brightest star in the entire sky. But is it?

There is absolutely no doubt that the three brightest stars are Sirius, Canopus and Alpha Centauri (lumping the two components of Alpha Centauri together). Then we come to four stars which are very nearly equal. Measurements made over time do not agree.

According to the Harvard magnitude values, made around 1880:
Arcturus 0.0, Capella 0.2, Vega 0.2, Rigel 0.3.
According to the Oxford values, made about the same time:
Capella 0.1, Arcturus 0.3, Vega 0.8, Rigel 1.0.
According to Cambridge, 1990:
Arcturus −0.04, Vega 0.03, Capella 0.08, Rigel 0.12.
According to the astrometric satellite Hipparcos, currently:
Arcturus −0.05, Vega 0.03, Capella 0.08, Rigel 0.18.

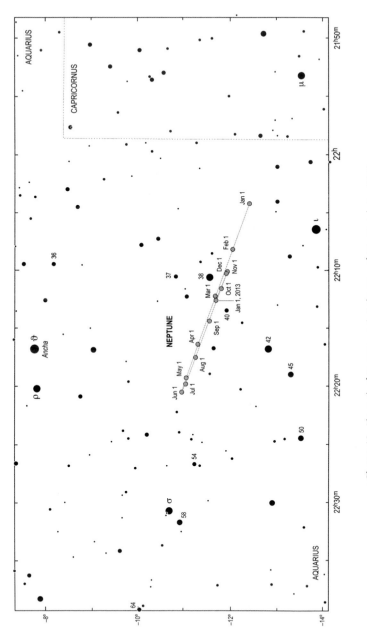

Figure 18. The path of Neptune against the stars of Aquarius during 2012.

But look at the change if we give absolute magnitudes. A star's absolute magnitude is the apparent magnitude it would have if viewed from a standard distance of 10 parsecs (32.6 light years), and so absolute magnitude is a measure of the star's real luminosity. Now Rigel really stands out from the others: Rigel −7.1, Arcturus +0.2, Capella +0.4, Vega +0.6.

Rigel emits most of its light in the ultra-violet and, if this is taken into account, its luminosity is about 85,000 times more than the Sun, whereas Vega could match only 37 Suns.

What about Deneb and Altair, the other two stars making up the 'Summer Triangle'? Altair's apparent magnitude is 0.76, while Deneb shines down modestly at 1.25. But now look at the absolute magnitudes! Altair +2.2, Deneb −8.7! Deneb is indeed a 'cosmic searchlight'. Hipparcos makes it over 250,000 times more luminous than our Sun.

The 'Wonderful' Star. One of the most famous variable stars in the sky reaches maximum brightness this month: Mira, Omicron Ceti. You will have to stay up after midnight to observe it, but it is worth the effort.

The first record of it, as a third-magnitude star, was due to the Dutch observer Fabricius in 1596. It was seen again by Bayer in 1603, and he even gave it its Greek letter. Rather surprisingly, it was not until 1638 that it was recognized as a variable star by Phocylides Holwarda. In 1642, Johannes Hevelius named it Mira, Latin for 'Wonderful'.

Although once like our Sun, Mira is now at the end of its life, and has evolved into a cool red giant star that is highly variable in brightness. Its variations are intrinsic; swelling and shrinking every eleven months or so, Mira sheds vast amounts of material through its powerful 'wind' of gas and dust. The currently accepted value for the star's distance is about 420 light years. Mira's diameter may be 700 times larger than our Sun; if Mira were at the centre of our Solar System, it would extend out more than 480 million kilometres, well beyond Mars's orbit and nearly two thirds of the way to Jupiter. Even at its peak (taking into account the huge amount of infrared radiation produced), it is less than ten thousand times as luminous as the Sun.

The accepted period is 332 days but, as with all long-period variables of this type, neither the period nor the maximum magnitude is identical from one cycle to another. At some maxima, Mira never becomes brighter than the fourth magnitude, but generally it reaches 3, and it is

Figure 19. a) Mira A is a red giant variable star in the constellation Cetus. This ultraviolet-wavelength image mosaic, taken by NASA's Galaxy Evolution Explorer, shows a comet-like 'tail' stretching 13 light years across space. Mira is moving from left to right. The 'tail' consists of hydrogen gas blown off from the star, with the material at the furthest end of the 'tail' having been emitted about 30,000 years ago. The tail-like configuration of the emitted material appears to result from Mira's unusually high speed relative to the Milky Way galaxy's ambient gas – about 130 kilometres per second. Mira itself is seen as a small white dot inside the glowing bulb at far right. **b)** A close-up view of the glowing bulb with Mira visible as the dot inside (far right). Like a boat travelling through water, a bow shock, or build-up of gas in front of it, forms ahead of Mira in the direction of its motion. Gas in the bow shock is heated and then mixes with the cool hydrogen gas in the wind that is blowing off Mira. This heated hydrogen gas then flows around behind the star, forming the wake. When the hydrogen gas is heated, it transitions into a higher-energy state, which then loses energy by emitting ultraviolet light – a process called fluorescence. The Galaxy Evolution Explorer has special instruments that can detect this ultraviolet light.

reported that in 1779 it apparently rivalled Aldebaran. Good comparison stars in Cetus are Alpha (magnitude 2.5), Gamma (3.5), Zeta (3.7), Delta (4.1)and Pi (4.2); also Alpha Piscium (3.9). At minimum it falls to about magnitude 10.

Mira (usually called Mira A) has a binary companion VZ Ceti (Mira B), which varies between magnitudes 9.5 and 12 and has a possible period of about 13 years. The orbital period of the two stars about each other is 400 years, separation 70 astronomical units. Mira's companion

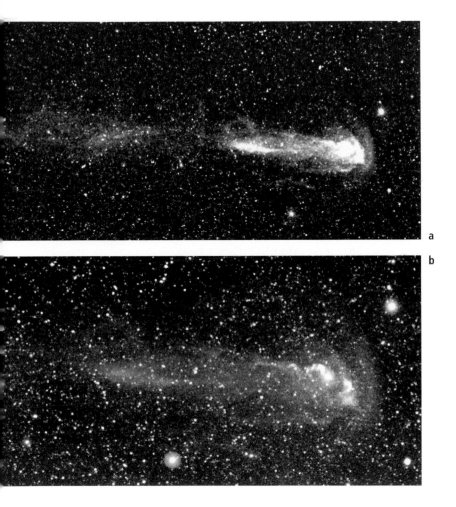

a

b

is most likely a burned-out star called a white dwarf that is surrounded by material captured from Mira's wind.

The Hubble Space Telescope has shown that Mira A itself is not perfectly spherical and is shaped more like a rugby ball; there is a hook-shaped appendage extending in the direction of the binary component, which is no doubt responsible for it. Ultraviolet studies with the GALEX satellite have shown that as Mira A moves along, it sheds material from its outer envelope, leaving a 'tail' thirteen light years long

(Figure 19). The fact that the gas is emitting UV radiation indicates that the material is being heated by the collision with the interstellar medium, and is slowly losing that energy. The material blown off will eventually merge with the ISM and form new stars.

Mira-type variable stars are very common. Some, such as Chi Cygni and R Cassiopeiae, U Orionis and R Leonis, can reach naked-eye visibility.

Charting the Stars by Photography. John Franklin-Adams, one of the Victorian 'grand amateur' astronomers, and a pioneer in photographic star-charting, was born in Peckham on 5 August 1843. Following a very successful business career, he took up astronomy as a serious hobby in 1890, buying a 4-inch telescope on a tripod stand. However, this instrument was soon transferred from Wimbledon, where he then lived, to Machrihanish in Argyllshire, Scotland, where he went for golf and recreation, and in 1897 he bought a house there to which he added an observatory. The 4-inch telescope was mounted on a stone pier, and a 6-inch equatorial, a transit telescope and clocks were added. Franklin-Adams visited Machrihanish as often as he could, and he began to develop his growing interest in photography in connection with astronomy. It was while at Machrihanish that the idea of mapping the distribution of stars in the Milky Way using photography was conceived, along with the idea of charting the entire heavens, both north and south. New lenses of 4-inch, 6-inch and 10-inch diameter were supplied by Cooke & Sons for the work.

In 1902, Franklin-Adams became ill and he was advised to spend some time in the drier climate of South Africa to aid his recovery. This he did, taking the opportunity to pursue his idea of photographing the southern sky. The 10-inch equatorial was set up in the grounds of the Cape Observatory where it was used in 1903 and 1904 to take photographs, each of two hours' exposure, to map the stars of the southern sky. He returned to England in 1904, his health restored, and moved into a new house at Hambledon near Godalming, again building a sizeable observatory. Besides the 10-inch equatorial, there were many new instruments, and he set about taking a set of photographs similar to those obtained from the Cape, but with exposure times of 2 hours 20 minutes, because of the difference in sky transparency. He thus compiled a photographic atlas of the entire sky.

Extensive preparations were made for measuring and counting the stars on the photographic plates and many practical difficulties had to be overcome as the work proceeded. There was steady progress until 1910 when Franklin-Adams's illness worsened and he had to give up both his observatory and his astronomy. Sadly, Frankin-Adams did not live to see the full outcome of his work. He died in Enfield, London, on 13 August 1912, one hundred years ago this month. But work continued on measuring and counting the stars on the plates, including a new set of plates of the Southern Hemisphere stars that had been completed by Mr Innes at the Cape. Subsequently, Franklin-Adams's *Chart of the Heavens*, the first all-sky photographic atlas, was published posthumously in 1913–14. There are 206 charts covering the entire sky, each 15° square, and showing stars as faint as seventeenth magnitude.

Franklin-Adams also took part in a number of successful total eclipse expeditions. Asteroid 982, discovered by H. E. Wood on 21 May 1922 from Johannesburg, was named Franklina in his honour.

September

New Moon: 16 September   *Full Moon:* 30 September

Equinox: 22 September

MERCURY passes through superior conjunction on 10 September and remains unobservable until the end of the month when it becomes visible in the western sky after sunset. For observers in the latitudes of the British Isles, the planet is not suitably placed for observation at all this month, but for those in equatorial and southern latitudes, the planet will become visible just south of due west, at the end of evening civil twilight, in the last week of September. Figure 22, given with the notes for October, shows, for observers in latitude 35°S, the changes in azimuth and altitude of Mercury on successive evenings when the Sun is 6° below the horizon, about 25 minutes after sunset in this latitude. The changes in the brightness of the planet are indicated on the diagram by the relative sizes of the white circles marking Mercury's position at five-day intervals. Mercury will be at its brightest early in the apparition; it fades from magnitude −0.8 to −0.4 during the last week of September.

VENUS remains a spectacular object in the early morning sky, although the planet fades very slightly from magnitude −4.2 to −4.1 in September. The planet continues to rise four hours before the Sun from northern temperate latitudes, although the period of visibility is considerably less from locations farther south. Venus's rapid eastwards motion carries it from Gemini, through Cancer and into Leo in September. The phase of the planet increases from 59 per cent to 70 per cent during the month. The waning crescent Moon will appear close to Venus in the morning sky on 12 September.

MARS, magnitude +1.2, continues to be visible in the early evening in the south-western sky. From northern temperate latitudes the planet is low in the twilight at dusk, and setting less than two hours after the

Sun, but it is rather better placed for observers situated further south. Its eastwards motion carries Mars from Virgo and into Libra during September.

JUPITER remains a brilliant object in Taurus, brightening from magnitude −2.3 to −2.5 during September, and rising in the mid-evening by the end of the month. The last quarter Moon will appear close to Jupiter early on 8 September.

SATURN, magnitude +0.8, is in Virgo. From northern temperate latitudes the planet may be glimpsed low down in the west-south-western sky as darkness falls early in September, but will be lost in the twilight before the end of the month. However, Saturn may still be seen rather low down in the western sky after sunset by those living in the tropics and the Southern Hemisphere.

URANUS is at opposition on 29 September, in the constellation of Pisces. Uranus is barely visible to the naked eye as its magnitude is +5.7, but it is easily located in binoculars. Figure 20 shows the path of Uranus against the background stars during the year. At opposition Uranus is 2,852 million kilometres (1,772 million miles) from the Earth.

Alpheratz. The main constellation of the autumn sky is Pegasus. In mythology, Pegasus was a flying horse, but in the sky his four main stars make up a square – an asterism known as the 'Great Square of Pegasus'. It is high up in the southern sky on autumn evenings in the Northern Hemisphere, passes more or less overhead from locations much further south, and is very easy to find, though star maps tend to make it look smaller and more prominent than it really is. The four main stars of the square are between magnitudes 2 and 3.

The star in the upper left-hand (north-eastern) corner of the square is named Alpheratz, and in itself it is a very ordinary star, of magnitude 2.1, lying 97 light years distant. It is a subgiant star, with a diameter twice that of the Sun, and the spectral type is B8, which means that it has a surface temperature of around 13,000 °C and there is no obvious colour; most people will call it white. Over one hundred years ago, between 1902 and 1904, the American astronomer Vesto Slipher made a series of measurements and found that Alpheratz is actually a

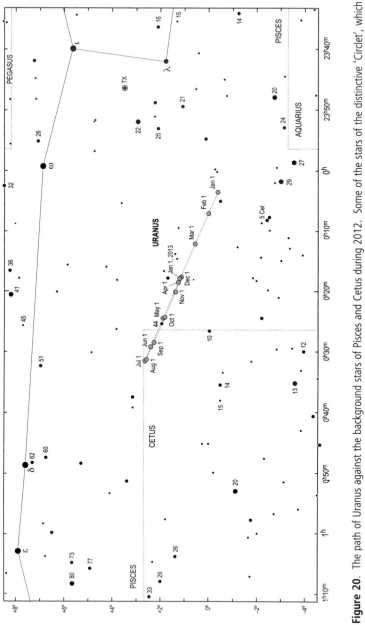

Figure 20. The path of Uranus against the background stars of Pisces and Cetus during 2012. Some of the stars of the distinctive 'Circlet', which marks the head of the western fish in Pisces, and may be found south of the Great Square of Pegasus, are shown towards the far right of the chart.

spectroscopic binary – that is, a very close double star that can only be investigated through the study of its spectrum. The fainter star in the binary system was first resolved by Xiaopei Pan and his team in 1988 and 1989, using a stellar interferometer at the Mount Wilson Observatory in California. It is now known that the two stars in the system orbit each other in a period of 96.7 days. The primary (the larger, brighter star) has a mass of about 3.6 times that of the Sun and a luminosity around 200 times greater than the Sun. The smaller companion, the secondary, has a mass of 1.8 times that of the Sun and luminosity about 13 times that of the Sun.

In 1906, Norman Lockyer found that the star has a number of unusual lines in its spectrum. Subsequently, many of these were found to be due to the elements mercury and manganese. It is the primary star of the binary that is chemically peculiar; it is the brightest in a class of unusual mercury-manganese stars. There is a companion to the binary system, of magnitude 11. However, it has no connection to the binary pair. It was discovered by William Herschel in 1781. It simply appears close to the binary pair in the sky.

Another strange thing about Alpheratz has nothing directly to do with the star itself. Quite clearly, it belongs to the square of Pegasus and in the older catalogues it was listed as Delta Pegasi. For some reason which eludes me, and I think defies common sense, Alpheratz has been removed from Pegasus and been given a free transfer to the neighbouring constellation of Andromeda, so that it is now listed as Alpha Andromedae. Andromeda is mainly a line of stars leading away from the square, but there is no defiant pattern, whereas Alpheratz is so obviously a member of the square! Indeed, the name Alpheratz is commonly taken to mean the 'the navel of the horse', recognizing the star's historical position in Pegasus. However, this is a decision of the International Astronomical Union, and therefore must be accepted.

An Italian in Paris. Giovanni Domenico Cassini was born in Perinaldo, near Sanremo, Italy, on 8 June 1625, and became professor of astronomy at the University of Bologna until 1669, when he went to Paris and, in 1671, became first director of the new observatory there. He enthusiastically took to his new home in France, even adopting a French version of his name, Jean-Dominique Cassini. He discovered four of Saturn's satellites (Iapetus, Rhea, Dione and Tethys), and recognized that the variation in brightness of Iapetus (it is always brighter at western elongation than at eastern) must be due to dark material covering one hemisphere of the satellite. In 1675, he discovered the main division in Saturn's rings, which is still known as the Cassini Division (Figure 21).

Cassini drew up new tables of Jupiter's satellites, made an independent discovery of the planet's Great Red Spot (in about 1665) and, with a colleague, made pioneering observations of Mars in 1672 which

Figure 21. Nine days before it entered orbit, the Cassini spacecraft captured this exquisite view of Saturn's rings. The images that make up this composition were obtained from Cassini's vantage point beneath the ring plane with the narrow angle camera on 21 June 2004, at a distance of 6.4 million kilometres from Saturn. The brightest part of the rings, curving from the upper right to the lower left in the image, is the B ring. Separating this from the outer, less bright, A ring, is the 4,000-kilometre-wide Cassini Division. (Image courtesy of NASA/JPL/Space Science.

enabled him to establish its distance from Earth. From that Cassini was able to calculate the distance from Earth to the Sun and estimate the dimensions of the Solar System. Cassini determined the distance from Earth to the Sun as 140 million kilometres, which is somewhat smaller, but still rather close to the modern value (about 150 million kilometres).

Cassini remained in Paris for the rest of his life, and died there on 14 September 1712, exactly two hundred years ago this month.

October

New Moon: 15 October   *Full Moon:* 29 October

Summer Time in the United Kingdom ends on 28 October

MERCURY is at greatest eastern elongation (24°) on 26 October, and is visible to the south of due west in the evening twilight sky for observers in equatorial and southern latitudes throughout October. Unfortunately, the planet is not suitably placed for observation this month from the latitudes of the British Isles, but for Southern Hemisphere observers this is the most favourable evening apparition of the year. Figure 22 shows the changes in azimuth and altitude of Mercury at the end of evening civil twilight, about 25 minutes after sunset, for observers in latitude 35°S. The changes in the brightness of the planet are indicated by the relative sizes of the white circles marking Mercury's position at five-day intervals: Mercury is at its brightest before it reaches greatest eastern elongation, with the planet fading from magnitude −0.4 to 0.0 during October. The diagram gives positions for a time at the end of evening civil twilight on the Greenwich meridian on the stated date. Observers in different longitudes should note that the actual positions of Mercury in azimuth and altitude will differ slightly from those given in the diagram due to the motion of the planet.

VENUS is a lovely object in the early morning sky, although the planet fades very slightly from magnitude −4.1 to −4.0 in October. The planet moves from Leo into neighbouring Virgo during the month. For observers in the latitudes of the British Isles, the planet still rises about three-and-a-half hours before the Sun, but from the tropics and more southerly locations, the period of visibility is considerably less. The phase of the planet increases from 71 per cent to 81 per cent during October. The waning crescent Moon will appear close to Venus in the morning sky on 12 October.

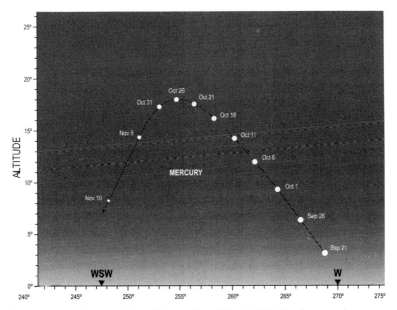

Figure 22. Evening apparition of Mercury from latitude 35°S. The planet reaches greatest eastern elongation on 26 October. It will be at its brightest in late September, before elongation.

MARS, magnitude +1.2, is low in the south-western sky at dusk, and setting just under two hours after the Sun from northern temperate latitudes, although the period of visibility is rather longer from locations further south. The waxing crescent Moon appears close to Mars on 18 October. The planet's motion carries it from Libra, through Scorpius and into Ophiuchus during the month. It will pass north of the reddish star Antares (Alpha Scorpii), the so-called 'Rival of Mars' (see notes for June) on 22 October, and this will present an interesting opportunity for observers in the tropics and the Southern Hemisphere to compare the colours and brightnesses of the two objects.

JUPITER is a lovely object, rising in the early evening from the latitudes of the British Isles – a little later from locations farther south – and visible all night long. It brightens from magnitude −2.5 to − 2.7 during the month. The planet is in Taurus, between the 'horns' of the Bull, reaching its first stationary point on 4 October; thereafter its motion is retrograde.

SATURN passes through superior conjunction on the far side of the Sun on 25 October. It may be glimpsed low down in the western sky at dusk by those living in the tropics and the Southern Hemisphere at the very beginning of October, but thereafter it will be unobservable.

Cepheus. Adjoining the well-known constellation of Cassiopeia is the much less conspicuous pattern of Cepheus, which from northern temperate latitudes lies more or less overhead on October evenings. In mythology, Cepheus was the king of Ethiopia and husband of Cassiopeia. A line from the two stars Shedir and Chaph on one side of the familiar 'W' of Cassiopeia towards the north brings us to the brightest star in Cepheus, known as Alderamin (Alpha Cephei). Apart from Alderamin (magnitude +2.4), Cepheus contains no other star above third magnitude, and no really obvious pattern – some say that it resembles a house with a high-pointed roof – but it does contain two variable stars of special interest: Delta and Mu Cephei (Figure 23).

Delta Cephei is located at the apex of a small triangle of stars at the south-eastern corner of the constellation; the other two stars in that triangle are Zeta and Epsilon Cephei, which, at magnitudes 3.6 and 4.2 respectively, are roughly at either end of Delta's range. Although Delta Cephei does not itself have a proper name, it has given its constellation name to an entire class of variable stars known as 'Cepheids'. Located about 900 light years distant, Delta Cephei's variability has been known ever since 1784, when it was discovered by John Goodricke in York. The variations are intrinsic. The star is pulsing, in a highly regular period of 5.36634 days (5d 8h 47.5m), and the magnitude ranges from 3.5 to 4.4; the rise to maximum takes about one-and-a-half days and the fall to minimum about four days.

All Cepheids are dying high-mass stars that are no longer in equilibrium, such that they expand and contract in an incredibly regular way. Their periods – that is to say the interval between successive maxima – run from about a day to over fifty days. Cepheid variable stars are of immense importance to astronomers, because the period of a Cepheid depends upon the star's true luminosity – the period-luminosity relationship. This means that once we have observed a Cepheid and have derived both its period and its average apparent magnitude, we can use the period to determine the true luminosity and, by comparing this with the apparent luminosity (the apparent magnitude), the star's distance may be derived. Cepheid variables are so luminous that they can

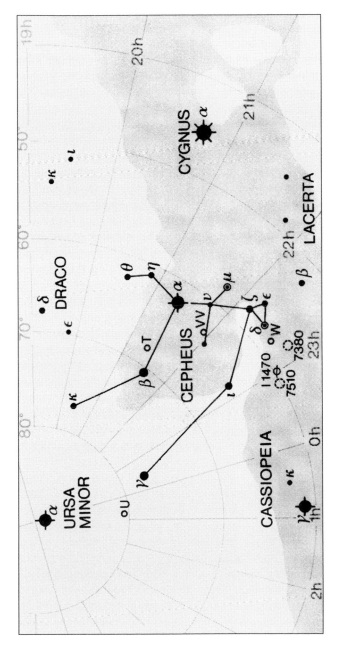

Figure 23. Map showing the principal stars of Cepheus, including the two interesting variable stars Delta (δ) Cephei and Mu (μ) Cephei, both indicated by a small circle with central dot on the map.

be seen across vast distances; they are easily seen in other galaxies, which means that the distances to those galaxies may be determined. The observation of Cepheids in relatively nearby galaxies (out to about 100 million light years) was a key project of the Hubble Space Telescope, one that enabled an accurate cosmological distance scale to be established.

The existence of a Cepheid period-luminosity relationship was discovered one hundred years ago this very year by the American astronomer Henrietta Swan Leavitt. She began work in 1893 at Harvard College Observatory, but in those days women were not allowed to operate telescopes and she spent her time cataloguing variable stars. In 1912, having discovered a number of Cepheid-type variables in the Small Magellanic Cloud (which were all at roughly the same distance), she noticed, when sorting them according to their periods, that the brighter the Cepheid, the longer it took to vary. In addition to recognizing this period-luminosity relationship, Leavitt discovered about 2,400 variable stars in the course of her research. It is fair to say that during her lifetime she never received proper recognition for her work. At least the asteroid 5383 Leavitt and the crater Leavitt on the Moon are named after her.

Some years later, the astronomer Harlow Shapley extended Leavitt's work to include Cepheids located in galaxies other than the Small Magellanic Cloud. He found that the period-luminosity relationship could be applied to all Cepheids, although it was later discovered that there are several different classes of Cepheid variables (with different period-luminosity relationships), and Delta Cephei is now known as a Type I (Classical) Cepheid.

The other notable variable star in the constellation is Mu Cephei. It does have a special name, Erakis, which is rarely used, and which is rather a pity because the star is so remarkable. It is not conspicuous as seen with the naked eye; the star varies in a slow semi-regular fashion from about magnitude 3.4 to about 5.1. Use binoculars to see it and you will instantly see it is fiery red. It is, in fact, one of the reddest stars known and William Herschel named it the 'Garnet Star'. Its distance is highly uncertain because the star is associated with a large region of nebulosity known as IC 1396 or the Garnet Star Nebula (Figure 24). Early distance estimates were in the range 800–1,200 light years, but more recent estimates are considerably further, up to 5,000 light years.

Figure 24. A visible light view of the Elephant's Trunk Nebula, an elongated dark globule within the emission nebula IC 1396 in the constellation of Cepheus. The globule is a condensation of dense gas that is barely surviving the strong ionizing radiation from a nearby massive star. The globule is being compressed by the surrounding ionized gas. The dark globule is seen in silhouette at visible-light wavelengths, backlit by the illumination of a bright star located to the left of the field of view. (Image courtesy of Canada-France-Hawaii Telescope/J.-C. Cuillandre/Coelum, and the Digitized Sky Survey.)

A value somewhere in the range 2,400–2,800 light years has also been suggested.

Mu Cephei is one of the largest and the most luminous stars known in the Milky Way – a type of star known as a red hypergiant. Its diameter is thought to be over 1,650 times that of the Sun, so that if it was placed in the Sun's position its radius would reach out to between the orbits of Jupiter and Saturn. It is around forty thousand times more luminous than the Sun.

Mu Cephei is near the end of its career. It is fusing helium into carbon, whereas a main sequence star like the Sun fuses hydrogen into helium. There is no doubt that in the future Mu Cephei will explode as a supernova. Interestingly, Mu Cephei is one of several red hypergiant variable stars in the constellation of Cepheus; others are W Cephei A, V354 Cephei and RW Cephei.

November

New Moon: 13 November   *Full Moon:* 28 November

MERCURY passes through inferior conjunction on 17 November and rapidly moves out to the west of the Sun, reaching greatest western elongation in early December. The planet is consequently visible as an early morning object from the last week of November until the third week of December. From northern temperate latitudes, in late November, Mercury will be seen in the east-south-eastern sky about the time of the beginning of morning civil twilight. Figure 27, given with the notes for December, shows, for observers in latitude 52°N, the changes in azimuth and altitude of Mercury on successive mornings when the Sun is 6° below the horizon, about 40 minutes before sunrise in this latitude. Mercury is at its brightest after it reaches greatest western elongation and during the last week of November the planet brightens from magnitude +1.2 to −0.1.

VENUS, magnitude −4.0, remains a lovely object in the early morning sky. At the beginning of November, from the latitudes of the British Isles, the planet still rises about three-and-a-half hours before the Sun, but this has decreased to less than three hours by the end of the month. From the tropics and more southerly locations the period of visibility is considerably less. The phase of Venus increases from 81 per cent to 88 per cent during November. The waning crescent Moon will appear close to Venus in the morning sky on 11 November, and Venus passes just south of the much fainter Saturn on the morning of 27 November. The planet moves from Virgo into Libra at the end of the month.

MARS, magnitude +1.2, remains inconveniently low in the south-western sky at dusk, and setting just under two hours after the Sun from northern temperate latitudes, although the period of visibility is somewhat longer from locations further south. The planet's motion carries it from Ophiuchus into Sagittarius during the month. The waxing crescent Moon will pass north of Mars on 15 December.

JUPITER is approaching opposition and by the end of the month it is visible all night long. The planet is moving retrograde in Taurus, its brightness increasing very slightly from magnitude −2.7 to −2.8 during November.

SATURN passed through superior conjunction in late October and becomes visible low in the south-eastern sky before dawn towards the end of November. The planet is in Virgo, magnitude +0.6. Saturn will appear very close to the much brighter Venus on the morning of 27 November.

Star Clusters in Cassiopeia. Cassiopeia is one of the most notable constellations of the northern sky, and it is well placed at this time of the year, passing more or less overhead on November evenings. Its five main stars, which form the familiar, slightly squashed 'W' shape, are not very bright, but they are so far north in the sky over North America and northern Europe that they never set.

Looking at the principal stars in Cassiopeia so that it appears as a 'W' shape, the middle star of the 'W' is Gamma Cassiopeiae or Cih, and there are two stars of comparable brightness to Cih on one side (Alpha or Shedir and Beta or Chaph), and two rather fainter stars on the other side (Gamma or Ruchbah and Epsilon or Segin). Epsilon (magnitude 3.4) is the faintest of the five stars in the 'W' and Ruchbah (magnitude 2.7) is the second faintest.

The Milky Way passes through Cassiopeia, and the region abounds with rich starfields which are a fine sight in binoculars or a low-power telescope. Quite near the star Ruchbah, one finds a number of open star clusters all visible within the same binocular field (Figure 25). First we have the wonderful seventh-magnitude open cluster, NGC 663. This is a prominent binocular cluster of well over 100 stars (mainly of magnitude 9 and fainter), covering an area about half the size of the full Moon – equivalent to a diameter of about 30 light years at the cluster's estimated distance of just under 7,000 light years. Overall it has a roundish, slightly elliptical shape with lines of ninth- and tenth-magnitude stars, although some say that in a telescope (under slightly higher magnification) it is shaped rather like a horseshoe.

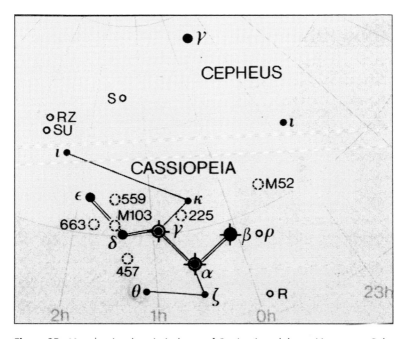

Figure 25. Map showing the principal stars of Cassiopeia and the positions – near Delta (δ) Cassiopeiae – of the three interesting open star clusters NGC 663, M103 and NGC 457 described in the text, all indicated by a small dashed circle on the map. Three other open star clusters NGC 559, NGC 225 and M52 (not described in the text) are also shown here.

Close to NGC 663, and slightly nearer Ruchbah, is the irregular open cluster NGC 581 or M103. It is one of the 'extra' objects added to Messier's list from Pierre Méchain's observations, although apparently not observed by Charles Messier himself. At magnitude 7.4, M103 is not quite as bright as NGC 663, with less than half its apparent diameter; at the cluster's estimated distance of over 8,000 light years, this is equivalent to a diameter of around 14 light years. In binoculars, M103 is easy to find and identify; it appears as a hazy, somewhat fan-shaped patch. M103 has around 170 stars, mainly of magnitudes 11 to 14, packed into a small area of sky only 6 arc minutes across. With a low power telescope, the boundary of the main part of the cluster is marked by a triangle of tenth-magnitude stars and there is also a distinctive orange-red star just off centre.

On the other side of Ruchbah to M103 and NGC 663 is the fine open cluster NGC 457, sometimes also known as the Owl cluster or the ET cluster because of its visual appearance (Figure 26). This is a fairly loose cluster of over 100 stars, mainly of magnitudes 11 to 13, covering an area just less than half the diameter of the full Moon – so its apparent size is intermediate between that of NGC 663 and M103. This is equivalent to a diameter of about thirty light years at the cluster's estimated distance of a little over nine thousand light years, although this is rather uncertain. At magnitude 6.4, the cluster itself is just below naked-eye visibility, but the double star Phi Cassiopeiae, which is just visible to the naked eye, lies on its southern outskirts, although most probably this is only a foreground star and not a true member of the cluster.

Figure 26. The lovely open star cluster NGC 457 in Cassiopeia, which is sometimes called the Owl cluster or the ET cluster because of its visual appearance. (Image courtesy of Ken and Emilie Siarkiewicz/Adam Block/NOAO/AURA/NSF.)

Vega and Capella. The two bright stars Vega and Capella are almost exactly equal in magnitude, and lie on opposite sides of the Pole Star, so that they may pass virtually overhead as seen from Britain and the northern United States. Vega, the brightest star in Lyra, the Lyre, occupies the place of honour during summer evenings; Capella, in Auriga, the Charioteer, occupies it on winter evenings.

During evenings in November the two stars are roughly equal in altitude, Vega sinking, Capella rising, so that it is interesting to compare them. Their colours are very different. Vega, an A-type star, is bluish; Capella is actually a very close binary star and both components are of spectral type G, so it is yellow.

This also means, of course, that Vega is the hotter of the two. Its surface temperature is about 9,500 °C, as against 4,900° and 5,700° for the two components of Capella. On the other hand, the components of Capella are approximately ninety times and sixty-five times the luminosity of our Sun, while Vega can match no more than about forty Suns. The distances are forty-two light years and twenty-five light years respectively, so Vega is much the closer of the two.

Though Capella and the Sun are very similar in spectral type, there is a great difference between them. The two components of Capella are both about three times the mass of our Sun and much larger, with diameters of 19 million and 12 million kilometres, while our Sun's diameter is just 1.4 million kilometres. The most modern apparent magnitude measures give Vega +0.03 and Capella +0.08, so Vega is just the brighter of the two.

The only star in the Northern Hemisphere of the sky to surpass Capella and Vega in brilliance is Arcturus in Boötes (the Herdsman), with an apparent magnitude of –0.04, which is best seen on evenings during the northern spring.

December

New Moon: 13 December   *Full Moon:* 28 December

Solstice: 21 December

MERCURY reaches greatest western elongation (21°) on 4 December and from the tropics and northern temperate latitudes, it is visible as a morning object until the third week of the month. The planet is not so well placed from the Southern Hemisphere. For observers in the Northern Hemisphere this will be a most interesting morning apparition of Mercury on account of its proximity to Venus during the first two weeks of December. The brilliant Venus will be a useful guide to locating the much fainter Mercury. Figure 27 shows, for observers in latitude 52°N, the changes in azimuth and altitude of Mercury and Venus on successive mornings when the Sun is 6° below the horizon. This condition is known as the start of morning civil twilight and, in this latitude and at this time of year, it occurs about forty minutes before sunrise. During the first three weeks of December, Mercury's brightness increases very slightly from magnitude −0.2 to −0.5. The diagram gives positions for a time at the start of morning civil twilight on the Greenwich meridian, on the stated date. Observers in different longitudes should note that the observed positions of Mercury and Venus in azimuth and altitude may differ slightly from those shown in the diagram.

VENUS, magnitude −4.0, is visible in the eastern sky before dawn. At the beginning of the month, Venus rises over two-and-a-half hours before the Sun from northern temperate latitudes and rather less from locations further south, but by the end of the year it will be rising only about one-and-a-half hours before the Sun from all latitudes as its elongation from the Sun decreases. The planet's phase increases from 88 per cent to 94 per cent during the month. The proximity of the brilliant Venus to the much fainter Mercury during the first two weeks of December has already been noted above.

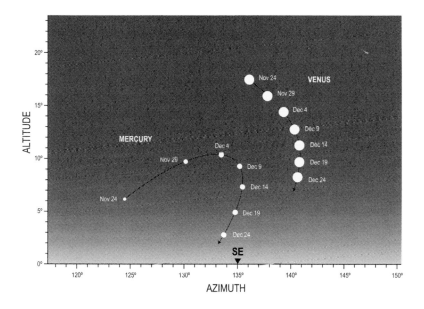

Figure 27. Morning apparition of Mercury from latitude 52°N. The planet reaches greatest western elongation on 4 December. It will be at its brightest in late December, after elongation. The chart also shows the positions of the brilliant Venus, in relation to Mercury, between 24 November and 24 December. Venus is a useful guide to locating the much fainter Mercury during the first two weeks of December. The angular diameters of Mercury and Venus are not drawn to scale.

MARS, magnitude +1.2, moves from Sagittarius into Capricornus during December. The planet sets just under two hours after the Sun, and continues to be rather low down in the south-western sky at dusk.

JUPITER, magnitude −2.8, is at its brightest this month since it is at opposition on 3 December and therefore available for observation throughout the night. However, the planet's northerly declination means that observers in the Northern Hemisphere will have the best view. The planet is moving retrograde in Taurus, with Jupiter in line with the northern arm of the V-shaped Hyades cluster at opposition, when the planet's distance from Earth is 609 million kilometres (378 million miles). The four Galilean satellites, which Galileo first saw in

145

January 1610, are readily observable with a small telescope or even a good pair of binoculars provided that they are held rigidly.

SATURN, magnitude +0.6, moves eastwards from Virgo into neighbouring Libra this month and it continues to be visible in the south-eastern sky before dawn. The apparent tilt of the rings, as viewed from Earth, has increased to almost 19° by year end.

The Geminid Meteor Shower. Active from 7–16 December, but with a slow rise to maximum, the Geminids are currently the most active of the regular annual showers, with rates outstripping even those of the August Perseids for a 24-hour interval centred on their 13–14 December maximum – a real treat for observers prepared to brave the winter cold and damp. In recent years, the Geminid Zenithal Hourly Rate has peaked at 120–140 meteors per hour, although observed rates will generally be rather lower than this. This year, conditions are ideal as far as the Moon is concerned; new Moon occurs on 13 December, so there will be absolutely no interference from moonlight this year.

In 2012, Geminid meteor activity is expected to peak during the early evening (around 1900 hours) on Thursday, 13 December. The maximum is broad, however, and respectable Geminid rates will be noted on the previous night, particularly before dawn on the morning of 13 December, but the evening of 13 December and early morning hours of 14 December are likely to yield the highest observed rates. There is also the added bonus of an increased proportional abundance of bright events after maximum; past observations show that bright Geminids become more numerous some hours after the rates have peaked, a consequence of particle-sorting in the meteor stream. This would be during the late evening and early morning hours of December 13/14.

The Geminid radiant (at RA 07h 32m Dec +33°, just north of the star Castor) rises early in the evening and reaches a respectable altitude well before midnight, so observers who are unable to stay up late can still enjoy the show. However, the early morning hours of both Thursday and Friday (13 and 14 December), are likely to see the greatest Geminid activity, when the shower radiant is high in the sky.

Geminid meteors enter the atmosphere at a relatively slow 35 kilometres per second, and, thanks to their robust (presumably rocky/asteroidal as opposed to dusty/cometary) nature, tend to last longer than most in luminous flight. Unlike swift Perseid or Orionid meteors, which last only a couple of tenths of a second, Geminids may be visible for a second or longer, sometimes appearing to fragment into a train of 'blobs'. Their low speed and abundance of bright events makes the Geminids a prime target for digital astrophotography.

Unusual in being associated with an asteroid – (3200) Phæthon – rather than a comet, the Geminid meteor shower has grown in intensity since the 1980s as a result of the meteor stream orbit being dragged gradually outwards across that of the Earth. A consequence is that we currently encounter the most densely populated parts of the meteor stream. This happy situation is unfortunately only temporary, and in a few more decades, Geminid displays can be expected to diminish in intensity. Here, observers have an excellent opportunity to follow, year on year, the evolution of a meteor stream. In recent years, from the British Isles, the Geminids have shown typical peak observed rates of sixty to seventy meteors per hour in good skies. Wrap up warmly and enjoy what should be a great display.

The Principal Stars of the Hunter. By December evenings, the brilliant winter constellations have once again come into view. Of these, Orion, the Hunter, is pre-eminent; with its characteristic outline, its prominent stars and its surrounding retinue, it dominates the southern aspect of the sky.

As with all constellations, we see the stars of Orion in just two dimensions, and it is not possible just by looking up at them to know which are the nearer ones to us, and which are the furthest away. But thanks to the efforts of astronomers over a great many years and, in more recent times, the analysis of data obtained by the European Space Agency's astrometry satellite, Hipparcos, we now have reliable distances for almost 120,000 stars.

For the nearer stars, most distances are known to better than 10 per cent, but the uncertainties in the quoted values increase for the more distant stars. This is why one sees such a wide range of figures quoted in different books and journals. Let us look at the distances of the seven principal stars in Orion: Betelgeuse and Bellatrix which mark Orion's shoulders; Mintaka, Alnilam and Alnitak, the three stars in Orion's Belt; and Saiph and Rigel, the stars at Orion's feet. For completeness the visual (V) magnitudes are also given for each star:

Star	Name	V Mag.	Distance (light years)
Beta Orionis	Rigel	0.18	860
Gamma Orionis	Bellatrix	1.64	245
Delta Orionis	Mintaka	2.25 var	915
Epsilon Orionis	Alnilam	1.69	1340
Zeta Orionis	Alnitak	1.74	815
Kappa Orionis	Saiph	2.07	720
Alpha Orionis	Betelgeuse	0.45 var	570

So, it turns out that, of these seven stars, Bellatrix, the star at the top right of Orion as it appears from northern Europe and the US, is really the nearest, and Alnilam, the star in the centre of Orion's Belt, is the most remote.

Incidentally, of all the stars in the sky between first and third magnitudes, Mintaka (the faintest of the three Belt stars) is closest to the celestial equator. It lies only one quarter of a degree to the south, so the star rises and sets almost exactly due east and due west.

The Last Men on the Moon. It seems quite incredible to think that it was forty years ago this month that the last men walked on the Moon. The final lunar landing of the Apollo programme was *Apollo 17*, which lifted off from Cape Canaveral on 7 December 1972. The three crew members were Eugene Cernan, Dr Harrison 'Jack' Schmitt and Ron Evans; Schmitt was the first professional geologist to go to the Moon. The landing site was in the Taurus-Littrow region of the Mare Serenitatis, where Cernan and Schmitt touched down in their lunar module on 11 December, having left Evans circling the Moon in the command module. Cernan and Schmitt spent a total of 3 days and 3 hours on the lunar surface, during which time they completed three EVAs (Extra-Vehicular Activities) of 7.2, 7.6 and 7.3 hours, and covered a total distance of 29 kilometres in their lunar roving vehicle (Figure 28). They also collected 110.5 kilogrammes of lunar samples. It was Schmitt who found the famous 'orange soil' near the crater, which was known unofficially as Shorty. The colour proved to be due to small glassy beads. At the end of the third EVA, Eugene Cernan became the last man on the Moon (so far!) when he re-entered the lunar module,

Figure 28. Near the Apollo 17 landing site, Family Mountain (centre background) and the rim of Shorty crater (far right) frame the lunarscape in this image of astronaut-geologist Dr Harrison 'Jack' Schmitt working alongside the lunar roving vehicle. Schmitt and fellow astronaut Eugene Cernan were the last to walk on the Moon. (Image courtesy of Apollo 17 and NASA.)

following Dr Schmitt. The crew splashed down safely in the Pacific Ocean on 19 December 1972. The first stage in lunar-manned exploration was over.

Eclipses in 2012

MARTIN MOBBERLEY

During 2012 there will be four eclipses: an annular eclipse of the Sun, a partial eclipse of the Moon, a total eclipse of the Sun and a penumbral eclipse of the Moon.

1. *An annular eclipse of the Sun* on 20 May will be visible as a partial solar eclipse across eastern Asia, the central and northern Pacific Ocean and North America, with the annular track crossing China, Japan and the north Pacific Ocean, and ending in the south-western USA at sunset. The point of greatest annular eclipse occurs at a position of 49.09° north and 176.28° east in the Bering Sea where the duration of annularity is five minutes forty-six seconds. The final part of the annular track reaches the Californian coastline in the early evening of 20 May (21 May 01:24 UT) just south of the border with Oregon. It then tracks across Nevada, Utah, Arizona and New Mexico, ending at sunset in

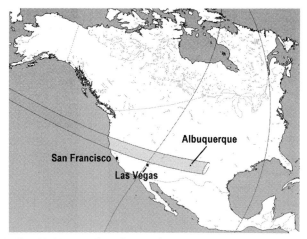

Figure 1. The 20 May annular eclipse ends at sunset in the USA and passes over Albuquerque. (Diagram by the author using WinEclipse by Heinz Scsibrany.)

Texas, just before the track would reach the town of Snyder, at around 01:39 UT. The annular phase will pass directly over Albuquerque at an altitude of five degrees where the annular phase lasts four minutes twenty-six seconds.

2. *A partial eclipse of the Moon* on 4 June will be visible from the Pacific hemisphere of the Earth with the umbra reaching 37 per cent into the

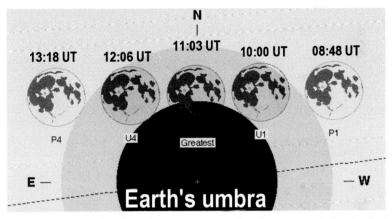

Figure 2. The Moon drifts through the northern part of the Earth's penumbral and umbral shadow on 4 June. (Diagram based on the eclipse prediction by Fred Espenak, NASA GSFC.)

lunar disk at 11:03 UT and obscuring much of the southern lunar hemisphere. However, as the event is only a partial one the orange-red glow of the lunar surface from refracted light, seen at total lunar eclipses, will be lost in the glare from the Moon's northern hemisphere.

3. *A total eclipse of the Sun* on 13 November will be visible as a partial solar eclipse across the southern Pacific Ocean. The track of totality starts at sunrise in the Northern Territory region of Australia, crosses northern Queensland (leaving the Australian coast at Oak Beach, just north of Cairns) and then passes north of New Zealand. The umbral track then crosses the southern Pacific Ocean and ends at sunset before it would reach the South American coastline. The point of greatest eclipse occurs in the south Pacific Ocean 2,000 km east of New Zealand at latitude 39° 56.9′ S and longitude 161° 19.8′ W where the duration of totality is four minutes two seconds and the umbral track will be

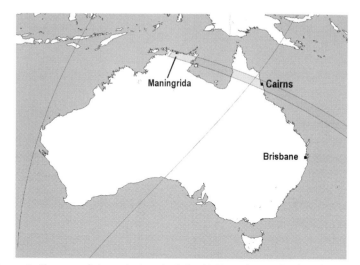

Figure 3. The start of the umbral track of the 13 November total solar eclipse across Australia. (Diagram by the author using WinEclipse by Heinz Scsibrany.)

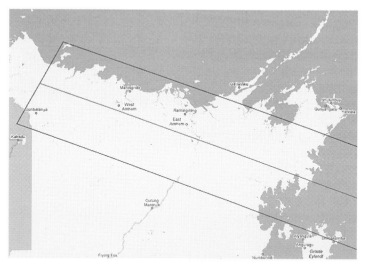

Figure 4: The sunrise umbral track of the 13 November total solar eclipse across the Northern Territory region of Australia. (Diagram based on the eclipse prediction by Fred Espenak, NASA GSFC.)

Eclipses in 2012

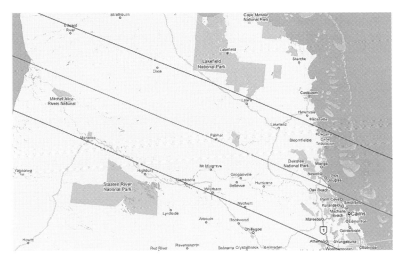

Figure 5: The umbral track of the 13 November total solar eclipse across northern Queensland, Australia. (Diagram based on the eclipse prediction by Fred Espenak, NASA GSFC.)

178.9 km wide. At Oak Beach in northern Queensland the duration of totality will be two minutes five seconds (mid-totality 20:39:14 UT) with the Sun 13.6° above the sea horizon at mid-totality. From that Oak Beach location first contact will occur at 19:44:33 UT, just ten minutes after sunrise and with the Sun one degree above the horizon.

4. *A penumbral eclipse of the Moon* on 28 November will be visible from much of the Earth's surface, excluding South America, the Eastern USA, the Atlantic Ocean and western Africa. By its nature the penumbral dimming will be very subtle and only the northern edge of the Moon will show any hint of being darker to the naked eye. In theory, the end of the event will be visible from the UK at moonrise, but in practice, the dimming will be insignificant.

Occultations in 2012

NICK JAMES

The Moon makes one circuit around the Earth in just over twenty-seven days and as it moves across the sky it can temporarily hide, or occult, objects that are further away, such as planets or stars. The Moon's orbit is inclined to the ecliptic by around 5.1° and its path with respect to the background stars is defined by the longitude at which it crosses the ecliptic passing from south to north. This is known as the longitude of the ascending node. After passing the node the Moon moves eastward relative to the stars reaching 5.1° north of the ecliptic after a week. Two weeks after passing the ascending node it crosses the ecliptic moving south and then it reaches 5.1° south of the ecliptic after three weeks. Finally, it arrives back at the ascending node a week later and the cycle begins again.

The apparent diameter of the Moon depends on its distance from the Earth but at its closest it appears almost 0.6° across. In addition the apparent position of the Moon on the sky at any given time shifts depending on where you are on the surface of the Earth. This effect, called parallax, can move the apparent position of the Moon by just over 1°. The combined effect of parallax and the apparent diameter of the Moon mean that if an object passes within 1.3° of the apparent centre of the Moon as seen from the centre of the Earth, it will be occulted from somewhere on the surface of our planet. For the occultation to be visible the Moon would have to be some distance from the Sun in the sky and, depending on the object being occulted, it would have to be twilight or dark.

For various reasons, mainly the Earth's equatorial bulge, the nodes of the Moon's orbit move westwards at a rate of around 19° per year, taking 18.6 years to do a full circuit. This means that, while the Moon follows approximately the same path from month to month, this path gradually shifts with time. Over the full 18.6-year period all of the stars that lie within 6.4° of the ecliptic will be occulted.

Only four first-magnitude stars lie within 6.4° of the ecliptic. These are Aldebaran (5.4°), Regulus (0.5°), Spica (2.1°) and Antares (4.6°). As the nodes precess through the 18.6-year cycle there will be a monthly series of occultations of each star followed by a period when the star is not occulted. In 2012 there will be six occultations of first-magnitude stars, all of them of the star Spica. Three of these occur at a solar elongation greater than 30°.

In 2012 there will be fourteen occultations of bright planets, one each of Venus and Mars, four of Mercury and eight of Jupiter. Nine of these events take place at a solar elongation of greater than 30°.

The following table shows events potentially visible from somewhere on the Earth when the solar elongation exceeds 30°. More detailed predictions for your location can often be found in magazines or in the *Handbook of the British Astronomical Association*.

Object	Time of Minimum Distance (UT)		Minimum Distance	Elongation	Best Visibility
Jupiter	15 July 2012	03:06	0.5°	−46°	Europe
Spica	25 July 2012	16:18	1.2°	+81°	Antarctic
Jupiter	11 Aug 2012	20:37	0.1°	−67°	Philippines
Venus	13 Aug 2012	19:54	0.6°	−46°	NE Russia
Spica	21 Aug 2012	21:49	0.9°	+55°	Antarctic
Jupiter	8 Sep 2012	11:01	0.6°	−91°	South America
Mars	19 Sep 2012	20:37	0.1°	+51°	South Atlantic
Jupiter	5 Oct 2012	20:51	0.9°	−117°	Southern Indian Ocean
Jupiter	2 Nov 2012	01:10	0.9°	−145°	South Atlantic
Jupiter	29 Nov 2012	00:56	0.6°	−175°	Mid Atlantic
Spica	9 Dec 2012	11:58	0.7°	−54°	South Pacific
Jupiter	26 Dec 2012	00:13	0.4°	+154°	Mid Atlantic

Comets in 2012

MARTIN MOBBERLEY

Thirty-four short-period comets should reach perihelion in 2012. All these returning comets orbit the Sun with periods of between five and eighteen years and many are too faint for amateur visual observation, even with a large telescope. Bright, or spectacular, comets have much longer orbital periods and, apart from a few notable exceptions such as 1P/Halley, 109P/Swift-Tuttle and 153P/Ikeya-Zhang, the best performers usually have orbital periods of many thousands of years and are often discovered less than a year before they come within amateur range. For this reason it is important to regularly check the best comet websites for news of bright comets that may be discovered well after this *Yearbook* is published. Some recommended sites are:

British Astronomical Association Comet Section: www.ast.cam.ac.uk/~jds/
Seiichi Yoshida's bright comet page: www.ast.cam.ac.uk/~jds/
CBAT/MPC comets site: www.minorplanetcenter.org/iau/Ephemerides/Comets/
Yahoo Comet Images group: http://tech.groups.yahoo.com/group/Comet-Images/
Yahoo Comet Mailing list: http://tech.groups.yahoo.com/group/comets-ml/

The CBAT/MPC web page above also gives accurate ephemerides of the comets' positions in right ascension (RA) and declination (Dec.).

Six periodic comets are expected to reach a magnitude of twelve or brighter during 2012 and so should be observable with a visual telescope, or with amateur CCD (change-coupled device) imaging equipment, in a reasonably dark sky. In addition the long-period comets C/2009 P1 (Garradd) and C/2006 S3 (LONEOS) will reach perihelion in December 2011 and April 2012 respectively and the former object, already bright at the end of 2011, should start the year as a seventh- or eighth-magnitude object. The infamous comet 29P/Schwassmann-Wachmann 1, now (in 2011) near aphelion, is usually too faint for visual observation even in large amateur telescopes but is renowned for going into outburst several times per year when it can reach magnitude eleven. The distances of comets from the Sun and the Earth are often quoted in Astronomical Units (AU) where 1 AU is the average Earth–Sun distance of 149.6 million km or 93 million miles.

The comets in question are listed below in the order they reach perihelion.

Comet	Period (years)	Perihelion	Peak Magnitude
29P/Schwassmann-Wachmann 1	14.7	2004 July 3	11 when in outburst
C/2009 P1 (Garradd)	long	2011 Dec 23	7 in Jan 2012
P/2006 T1 (Levy)	5.3	2012 Jan12	7 in Jan 2012
78P/Gehrels 2	7.2	2012 Jan 13	12 in Jan 2012
21P/Giacobini-Zinner	6.6	2012 Feb 13	10 in Feb 2012
C/2006 S3 (LONEOS)	long	2012 Apr 16	12 in Apr 2012
96P/Machholz	5.3	2012 Jul 14	2 in July 2012
185P/Petriew	5.5	2012 Aug 13	11 in Aug 2012
P/1994 X1 (McNaught-Russell)	18.3	2012 Dec 4	12 in Dec 2012

WHAT TO EXPECT

As 2012 starts, things are looking quite promising with the brightest comets being objects that were already visible at the end of 2011. These are the comets C/2009 P1 (Garradd) and P/2006 T1 (Levy). The former comet is predicted to reach perihelion (q = 1.55 AU) on 23 December 2011 and so, all being well, will have been enthusiastically observed by amateurs at the tail end of that year. Hopefully C/2009 P1 will still be a seventh-magnitude object at the start of 2012 and it should be a nice sight in the pre-dawn morning sky half a degree south-west and north-west of the globular cluster M92, in Hercules, on 3 February and 4 February respectively. The comet will be circumpolar from UK latitudes from the start of February right through until the end of April, peaking in declination on the evening of 11 March, just shy of +71 degrees Dec., on the Ursa Minor/Draco border. From mid-March to mid-April C/2009 P1 is in western Ursa Major and, hopefully, should still be around eighth magnitude as it heads for the border with Lynx.

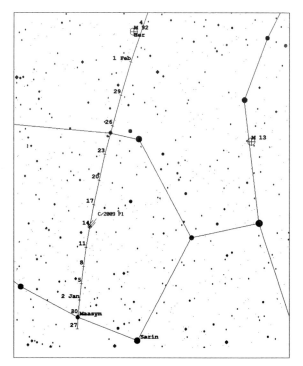

Figure 6. The track of C/2009 P1 (Garradd) through Hercules in January.

The other comet predicted to have a binocular magnitude as 2012 starts, P/2006 T1 (Levy), reaches perihelion on 12 January, at a solar distance q of 1.01 AU. This rare visual discovery by the tireless comet-sweeping Arizona amateur David Levy was made in October 2006 using a 0.41-metre reflector. Many suspect that this comet experienced an abnormal outburst at discovery as otherwise, being periodic, it should have been found at previous returns by the professional patrols. On this, its first return to the inner Solar System after its discovery, the comet is predicted to stay within 0.2 AU of the Earth for the final two weeks of January, which may allow it to reach seventh magnitude at this time, although it will be sinking rapidly into the Northern Hemisphere twilight from UK latitudes in late January as it leaves Pisces on the 16th, crosses Cetus, and enters Eridanus on the 22nd. So, all things considered, even without any bright comets being discovered long after this book goes to print, January 2012 is looking quite promising for the comet observer. Sadly, the same cannot be said for the rest of 2012, although for those with decent telescopes and CCDs there are some interesting targets and the new discoveries that are being made every month to be observed.

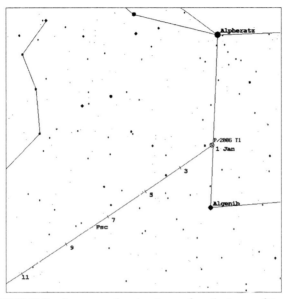

Figure 7. P/2006 T1 (Levy) moves south and east, away from the Square of Pegasus, low in the January evening sky.

Two much fainter targets in the January to March period will attract the attention of experienced observers, namely 78P/Gehrels 2 and 21P/Giacobini-Zinner. The first of these, 78P, also reaches perihelion on 12 January, although at twice the solar distance of P/2006 T1 (Levy). On that date they will only be 8° apart in south-eastern Pisces, low in the early evening sky, although at twelfth magnitude 78P will be much the fainter comet. In late February, 78P drifts into Aries but by then it will have faded to twelfth magnitude at best. 21P/Giacobini-Zinner is an old favourite of many experienced observers and its predicted magnitude of 10 in February looks hopeful. However, its elongation from the Sun will be dropping rapidly as the year begins and only those with an excellent western evening horizon will stand any chance of observing this object.

Comet 29P/Schwassmann-Wachmann 1 is the most regularly out-bursting comet in the sky. While it is not in the same explosive league as 17P/Holmes (famous for its half-millionfold outburst in October 2007), it has been observed flaring very regularly in recent years, possibly simply because it has been monitored more closely than in the photographic era. When in its dormant state, 29P sits at around sixteenth to seventeenth magnitude: a challenging object even for CCD imagers. However, in recent years, it has reached eleventh magnitude in outburst, putting it within easy visual range of large amateur telescopes or more modest telescopes equipped with CCD cameras. From January to late April, comet 29P is in northern Corvus, crossing the border into southern Virgo on 19 April. The region is roughly seventy minutes of right ascension due west of the brilliant star Spica and so the comet is at opposition in late March, but at a declination of −12° it will be low in UK skies and the comet is now at aphelion too. Nevertheless regular monitoring is recommended.

On 16 April the large comet C/2006 S3 (LONEOS) will reach perihelion in Serpens, just north of the border with Sagittarius. However, despite being an impressive comet, C/2006 S3 will not be bright, only reaching magnitude 12 or so. In addition, at a declination of −13 degrees, it will be an object fairly low down in the pre-dawn sky from the far Northern Hemisphere. Nevertheless, it is still worth attempting to image this comet as it definitely warrants the term 'impressive'. Even on the date of perihelion, C/2006 S3 will be a huge 5.13 AU from the Sun and 4.83 AU from the Earth. To have a magnitude of 11 or 12 at the distance of Jupiter from the Sun implies it may have an absolute magnitude as high as 1.0 or 2.0; not as active as Hale-Bopp, but still very energetic. This is a big comet. If only it were coming closer!

Moving into the summer months the predicted second magnitude figure for 96P/Machholz looks exciting until you take into account its tiny perihelion distance of just 0.12 AU. This guarantees that the comet will get nice and hot, but at a solar elongation of less than two degrees it will be invisible to Earth-based astronomers. Even Southern Hemisphere observers will be hard pressed to see this comet in the morning sky, before perihelion, and in the evening sky, after perihelion.

Two faint comets which should be within the visual range of large amateur telescopes in late 2012 are also worthy of mention. These are the periodic comets 185P/Petriew and P/1994 X1 (McNaught-Russell).

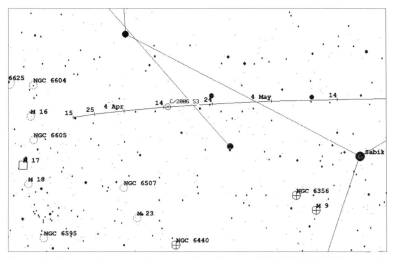

Figure 8. The path of the large (but very distant) comet C/2006 S3 (LONEOS) from mid-March to mid-May. In mid-March it is not far from the Eagle nebula M16.

185P comes to perihelion on 13 August at a healthy 0.93 AU from the Sun. It will be passing rapidly (three arc minutes per hour) from Taurus to Gemini in the pre-dawn sky, not far from Venus. Despite being only eleventh magnitude, 185P is a rare twenty-first-century amateur visual discovery made by Vance Petriew at the August 2001 Saskatchewan star party, and so for that reason alone it is worth tracking down.

P/1994 X1 was discovered jointly by the world's greatest ever comet hunter, Rob McNaught, and by Ken Russell, both of the Anglo–Australian Observatory at that time. With an orbital period of eighteen years this will be the comet's first observed return to the inner Solar System and the perihelion date is predicted to be 4 December, at which time it will be 1.28 AU from the Sun and 0.86 AU from the Earth while in northern Aquarius. A peak magnitude of 12 is possible although advanced amateurs may be able to image it in the autumn months. There were no visual observations of it when it was discovered in 1994 so it could have a few surprises in store.

The tables overleaf list the right ascension and declination of, arguably, the five most promising comets at their peak as well as the distances, in AU, from the Earth and the Sun. The elongation, in degrees from the Sun, is also tabulated along with the estimated visual magnitude, which can only ever be a rough guess as comets are a law unto themselves!

C/2009 P1 (Garradd)

Date 2012	RA (2000)			Dec.			Distance from Earth	Distance from Sun	Elong- ation from Sun	Mag.
	h	m	s	°	′	″	AU	AU	°	
Jan 1	17	30	21.7	+26	48	55	1.937	1.555	52.9	7.4
Jan 11	17	29	06.7	+30	15	40	1.825	1.571	59.3	7.3
Jan 21	17	25	51.4	+34	45	30	1.697	1.599	67.2	7.2
Jan 31	17	18	45.0	+40	35	40	1.562	1.637	76.3	7.1
Feb 10	17	04	03.7	+48	00	45	1.434	1.686	86.2	7.1
Feb 20	16	33	12.8	+56	55	13	1.330	1.743	96.4	7.0
Mar 1	15	24	33.4	+65	58	13	1.272	1.808	105.4	7.1
Mar 11	13	07	35.4	+70	43	35	1.279	1.879	111.0	7.3
Mar 21	10	48	45.9	+66	56	33	1.357	1.956	111.6	7.6
Mar 31	09	38	54.9	+58	55	54	1.499	2.038	107.6	8.0

P/2006 T1 (Levy)

Date 2012	RA (2000)			Dec.			Distance from Earth	Distance from Sun	Elong- ation from Sun	Mag.
	h	m	s	°	′	″	AU	AU	°	
Jan 1	00	09	10.4	+20	53	54	0.256	1.020	90.9	7.6
Jan 6	00	37	36.7	+16	24	07	0.232	1.011	90.3	7.4
Jan 11	01	10	50.2	+10	44	13	0.212	1.008	90.4	7.2
Jan 16	01	48	46.8	+03	53	34	0.198	1.009	91.6	7.0
Jan 21	02	30	37.8	−03	43	03	0.192	1.015	93.7	7.0
Jan 26	03	14	39.9	−11	14	16	0.195	1.025	96.6	7.1
Jan 31	03	58	31.1	−17	44	46	0.207	1.041	99.8	7.3

78P/Gehrels 2

Date 2012	RA (2000)			Dec.			Distance from Earth	Distance from Sun	Elong- ation from Sun	Mag.
	h	m	s	°	′	″	AU	AU	°	
Jan 1	00	34	44.0	+02	41	50	1.767	2.011	89.2	12.8
Jan 6	00	42	53.0	+03	19	26	1.817	2.009	86.3	12.9
Jan 11	00	51	21.7	+03	59	32	1.868	2.009	83.4	12.9
Jan 16	01	00	08.3	+04	41	47	1.920	2.009	80.6	13.0
Jan 21	01	09	11.7	+05	25	50	1.971	2.010	77.9	13.0
Jan 26	01	18	30.5	+06	11	19	2.023	2.011	75.2	13.1
Jan 31	01	28	03.5	+06	57	53	2.075	2.014	72.6	13.2
Feb 5	01	37	49.2	+07	45	09	2.128	2.018	70.1	13.2
Feb 10	01	47	46.5	+08	32	47	2.180	2.022	67.6	13.3
Feb 15	01	57	54.6	+09	20	28	2.232	2.027	65.2	13.4
Feb 20	02	08	12.7	+10	07	54	2.285	2.033	62.8	13.5
Feb 25	02	18	40.2	+10	54	48	2.337	2.040	60.5	13.5

C/2006 S3 (LONEOS)

Date 2012	RA (2000)			Dec.			Distance from Earth	Distance from Sun	Elong- ation from Sun	Mag.
	h	m	s	°	′	″	AU	AU	°	
Apr 1	18	01	35.3	−13	38	57	4.848	5.132	100.9	12.5
Apr 6	17	58	04.8	−13	34	31	4.755	5.132	106.6	12.5
Apr 11	17	54	05.0	−13	29	57	4.664	5.131	112.4	12.4
Apr 16	17	49	35.3	−13	25	14	4.578	5.131	118.3	12.4
Apr 21	17	44	35.5	−13	20	23	4.497	5.131	124.3	12.4
Apr 26	17	39	06.2	−13	15	23	4.422	5.132	130.4	12.3

P/1994 X1 (McNaught-Russell)

Date 2012	RA (2000)			Dec.			Distance from Earth	Distance from Sun	Elong- ation from Sun	Mag.
	h	m	s	°	′	″	AU	AU	°	
Nov 1	20	32	33.2	+10	40	46	0.857	1.359	94.3	12.5
Nov 11	21	04	35.4	+06	36	32	0.842	1.320	91.8	12.3
Nov 21	21	41	18.5	+02	39	57	0.841	1.293	89.6	12.2
Dec 1	22	21	31.7	−00	54	29	0.855	1.281	87.9	12.2
Dec 11	23	03	40.7	−03	52	19	0.887	1.283	86.4	12.3
Dec 21	23	46	07.0	−06	03	00	0.938	1.300	85.1	12.5
Dec 31	00	27	22.9	−07	23	52	1.007	1.330	83.9	12.8

Minor Planets in 2012

MARTIN MOBBERLEY

Some 500,000 minor planets (or asteroids) are known. They range in size from small planetoids hundreds of kilometres in diameter to boulders tens of metres across. More than 200,000 of these now have such good orbits that they possess a numbered designation and 15,000 have been named after mythological gods, famous people, scientists, astronomers and institutions. Most of these objects live between Mars and Jupiter, but some 6,000 have been discovered between the Sun and Mars and more than 1,000 of these are classed as potentially hazardous asteroids (PHAs) because of their ability to pass within 8,000 000 kilometres of the Earth while also having a diameter greater than 200 metres. The first four asteroids to be discovered were (1) Ceres, now regarded as a dwarf planet, (2) Pallas, (3) Juno and (4) Vesta, which are all easy binocular objects when at their peak, as they all have diameters of hundreds of kilometres. The discovery of (5) Astræa, (6) Hebe and (7) Iris followed.

In 2012 all of the first seven numbered minor planets reach opposition at some point during the year. The dwarf planet (1) Ceres reaches opposition late in the year, on 16 December in north-eastern Taurus, just a few degrees north of the Crab Nebula, where it will be an easy binocular object of magnitude 6.7.

Minor planet (2) Pallas reaches magnitude 8.3 three months earlier, in late September, while in north-western Cetus. The third numbered minor planet Juno is at its best on 20 May, when it will reach magnitude 10.2 in Serpens just above the border with Ophiuchus and Libra. (4) Vesta is a much brighter target in mid-December when it vies for attention with (1) Ceres, also in Taurus, and, at magnitude 6.5, it wins the brightness contest!

The fifth numbered minor planet Astræa peaks much earlier in the year, in mid-March, when it just scrapes to magnitude 9.0 while at opposition in the north western corner of Virgo. A few weeks earlier (6) Hebe also reaches opposition, half a magnitude fainter, in central

Leo, a few degrees north of the well-known Messier galaxies M95, M96 and M105. Finally, (7) Iris reaches opposition in early May, although in this case the magnitude 9.5 asteroid will be far better placed for Southern Hemisphere observers as it will be at −21 degrees declination in southern Libra.

Ephemerides for the best-placed or brightest minor planets at opposition in 2012

Dwarf planet (1) Ceres

Date 2012	RA (2000)			Dec.			Distance from Earth	Distance from Sun	Elong-ation from Sun	Mag.
	h	m	s	°	′	″	AU	AU	°	
Nov 1	06	15	18.1	+22	19	38	2.007	2.704	125.2	8.0
Nov 11	06	13	47.3	+22	51	48	1.898	2.696	135.6	7.7
Nov 21	06	09	18.8	+23	28	33	1.807	2.689	146.8	7.5
Dec 1	06	02	07.3	+24	08	12	1.739	2.681	158.5	7.3
Dec 11	05	52	48.6	+24	48	08	1.697	2.674	170.7	7.0
Dec 21	05	42	23.3	+25	25	23	1.684	2.667	176.1	6.8
Dec 31	05	32	09.4	+25	57	52	1.700	2.660	164.0	7.1

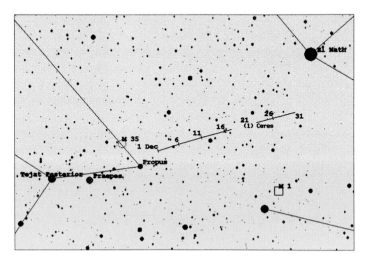

Figure 9. The dwarf planet Ceres is due north of the Crab Nebula, M1, in Taurus in December.

(2) Pallas

Date 2012	RA (2000) h	m	s	Dec. °	′	″	Distance from Earth AU	Distance from Sun AU	Elong- ation from Sun °	Mag.
Sept 1	00	38	47.4	−01	51	04	2.087	3.004	150.1	8.8
Sept 11	00	33	23.8	−04	14	46	2.013	2.983	161.1	8.5
Sept 21	00	26	34.2	−06	47	24	1.969	2.962	169.8	8.3
Oct 1	00	19	00.1	−09	18	48	1.956	2.941	167.2	8.3
Oct 11	00	11	30.0	−11	38	35	1.974	2.919	156.9	8.5
Oct 21	00	04	53.8	−13	38	03	2.021	2.897	145.5	8.7
Oct 31	23	59	54.6	−15	11	53	2.091	2.874	134.3	8.9

(3) Juno

Date 2012	RA (2000) h	m	s	Dec. °	′	″	Distance from Earth AU	Distance from Sun AU	Elong- ation from Sun °	Mag.
May 1	16	17	25.3	−04	28	36	2.427	3.353	152.4	10.3
May 11	16	10	05.0	−03	39	17	2.386	3.353	160.0	10.2
May 21	16	01	56.2	−02	58	38	2.373	3.353	162.6	10.2
May 31	15	53	41.6	−02	29	46	2.388	3.351	158.4	10.2
Jun 10	15	46	04.0	−02	14	33	2.430	3.350	150.3	10.4
Jun 20	15	39	38.5	−02	13	32	2.496	3.347	140.9	10.5

(4) Vesta

Date 2012	RA (2000) h	m	s	Dec. °	′	″	Distance from Earth AU	Distance from Sun AU	Elong- ation from Sun °	Mag.
Nov 1	05	39	00.5	+17	24	26	1.784	2.570	133.5	7.2
Nov 11	05	34	11.5	+17	25	09	1.699	2.571	144.6	7.0
Nov 21	05	26	31.1	+17	28	37	1.636	2.571	156.3	6.8
Dec 1	05	16	39.9	+17	35	02	1.598	2.571	168.1	6.6
Dec 11	05	05	40.5	+17	44	29	1.589	2.571	174.5	6.5
Dec 21	04	54	51.0	+17	57	15	1.609	2.570	164.4	6.7
Dec 31	04	45	29.1	+18	13	54	1.656	2.569	152.5	6.9

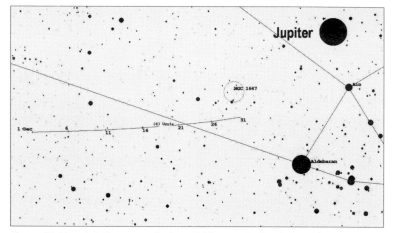

Figure 10. Vesta is in Taurus, not far from Aldebaran or Jupiter, in December.

(5) Astræa

Date 2012	RA (2000)			Dec.			Distance from Earth	Distance from Sun	Elong- ation from Sun	Mag.
	h	m	s	°	′	″	AU	AU	°	
Feb 1	11	55	17.0	+03	10	02	1.295	2.099	133.6	10.0
Feb 11	11	54	39.5	+03	58	47	1.225	2.106	144.2	9.8
Feb 21	11	51	01.4	+05	07	48	1.175	2.114	155.4	9.5
Mar 2	11	44	54.9	+06	30	18	1.146	2.123	166.9	9.3
Mar 12	11	37	22.8	+07	55	54	1.142	2.134	175.0	9.0
Mar 22	11	29	42.1	+09	13	30	1.163	2.145	166.6	9.3
Apr 1	11	23	12.5	+10	13	23	1.208	2.157	155.4	9.6
Apr 11	11	18	54.6	+10	50	07	1.275	2.171	144.5	9.9

(6) Hebe

Date 2012	RA (2000)			Dec.			Distance from Earth	Distance from Sun	Elong- ation from Sun	Mag.
	h	m	s	°	′	″	AU	AU	°	
Feb 1	11	12	02.9	+11	09	07	1.843	2.718	146.3	9.8
Feb 11	11	05	53.9	+12	48	48	1.794	2.734	158.0	9.7

Feb 21	10	57	58.5	+14	34	13	1.772	2.749	168.8	9.5
Mar 2	10	49	08.5	+16	16	00	1.780	2.763	170.9	9.5
Mar 12	10	40	27.1	+17	45	18	1.818	2.777	161.2	9.7
Mar 22	10	32	52.7	+18	55	57	1.883	2.791	149.9	9.9

(7) Iris

Date 2012	RA (2000)			Dec.			Distance from Earth	Distance from Sun	Elong- ation from Sun	Mag.
	h	m	s	°	′	″	AU	AU	°	
Apr 1	15	08	42.9	−22	56	27	2.093	2.931	140.2	10.2
Apr 11	15	02	16.8	−22	32	10	2.014	2.933	151.4	10.0
Apr 21	14	53	55.0	−21	54	38	1.960	2.935	162.9	9.8
May 1	14	44	24.7	−21	05	31	1.933	2.936	173.5	9.5
May 11	14	34	45.0	−20	08	37	1.935	2.937	170.8	9.6
May 21	14	25	53.3	−19	09	0/	1.966	2.936	159.7	9.8
May 31	14	18	38.9	−18	12	45	2.023	2.935	148.4	10.0

As well as observing bright binocular asteroids some advanced amateur astronomers with large telescopes and CCDs are often interested

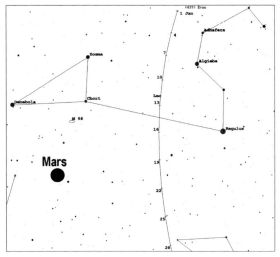

Figure 11. Eros moves rapidly through Leo in January.

in imaging the potentially hazardous asteroids (PHAs), typically tens or hundreds of metres in diameter, that sail close to the Earth and those bigger objects that, while they are not a conceivable hazard, are coming unusually close. The brightest object in this latter category, 433 Eros, passes within 27 million kilometres of the Earth on 31 January. At the start of January, Eros can be found as a magnitude 9.4 object just behind (to the west of) Leo's sickle and moving steadily south. At the time of closest approach, on 31 January, it will be a magnitude 8.6 binocular object in Sextans.

This 34 x 11 x 11 km potato-shaped minor planet was visited by the NEAR Shoemaker spacecraft in 2000 and the probe landed on the surface of Eros on 12 February 2001.

Figure 12. The asteroid 433 Eros, imaged by the NEAR Shoemaker probe in 2000.

Only two other asteroids are predicted to be brighter than magnitude 12 at their closest approach to the Earth in 2012, namely 2007 PA8, briefly, in late October/early November and 4179 Toutatis in December. 2007 PA8 should reach magnitude 11.5 while travelling through the southern constellation of Columba and 4179 Toutatis will reach magnitude 10.5 while moving rapidly east through Pisces, Aries and Taurus.

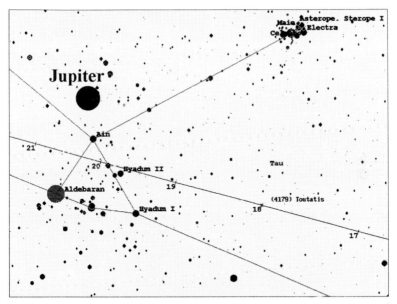

Figure 13. 4179 Toutatis passes rapidly through the Hyades, midway between Jupiter and Aldebaran, in mid-December.

A list of the most interesting close asteroid approaches in 2012 is presented in the following table. It should be borne in mind that the visibility of close-approach asteroids is highly dependent on whether they are close to the solar glare and on which hemisphere they are being observed from. The table gives the minimum separation between the Earth and the asteroid in AU (1 AU = 149.6 million km) and the date of that closest approach and also the constellation in which the object can be found. The brightest magnitude achieved and the corresponding date and constellation are also given. Some of the objects listed are as faint as magnitude 17 or 18 at their best and so present a real challenge, even to advanced amateur astronomers.

Asteroid desig./ name	Min sep (AU)	Date Closest in 2012	Const. Closest	Peak Mag.	Brightest Date	Const. Brightest
(7341) 1991 VK	0.06504	Jan 25.99	Sgr	14.8	16 Feb	Lup
(433) Eros	0.1787	Jan 31.46	Sex	8.5	3 Feb	Sex
1998 HE3	0.03192	May 10.75	Leo	16.1	7 May	Vir
(4183) Cuno	0.1218	May 20.80	Phe	12.3	1 Jun	Tel
2005 GO21	0.04396	Jun 21.76	Ori	14.7	14 Jun	Eri
(37655) Illapa	0.09513	Aug 12.33	Tau	15.7	9 Aug	Cet
(4581) Asclepius	0.1079	Aug 16.37	Per	18.5	8 Aug	And
(4769) Castalia	0.1135	Aug 28.71	Oph	14.1	17 Aug	Sgr
(99907) 1989 VA	0.1644	Nov 4.40	Ari	14.8	6 Nov	Ari
2007 PA8	0.04329	Nov 5.70	Col	11.5	2 Nov	Eri
(4179) Toutatis	0.04633	Dec 12.28	Cet	10.5	16 Dec	Ari
(88213) 2001 AF2	0.1812	Dec 15.71	Ret	17.9	22 Dec	Eri
(33342) 1998 WT24	0.1779	Dec 23.32	Vir	17.5	11 Dec	Vir

Accurate minor planet ephemerides can be computed for your location on Earth using the MPC ephemeris service at: www.minorplanetcenter.org/iau/MPEph/MPEph.html

Meteors in 2012

JOHN MASON

Meteors (popularly known as 'shooting stars') may be seen on any clear moonless night, but on certain nights of the year their number increases noticeably. This occurs when the Earth chances to intersect a concentration of meteoric dust moving in an orbit around the Sun. If the dust is well spread out in space, the resulting shower of meteors may last for several days. The word 'shower' must not be misinterpreted – only on very rare occasions have the meteors been so numerous as to resemble snowflakes falling.

If the meteor tracks are marked on a star map and traced backwards, a number of them will be found to intersect in a point (or a small area of the sky) which marks the radiant of the shower. This gives the direction from which the meteors have come.

Bright moonlight has an adverse effect on visual meteor observing, and within about five days to either side of Full Moon, lunar glare swamps all but the brighter meteors, reducing, quite considerably, the total number of meteors seen. The good news for 2012 is that very few of the major annual meteor showers will be hampered by moonlight: only the Quadrantids, the Eta Aquarids and the rise to maximum of the Perseids will be adversely affected. Observations of the Quadrantids will be hindered by a waxing gibbous Moon, which sets in the early morning hours. For Perseid maximum, the Moon is a waning crescent rising in the early morning hours. However, it is often possible for visual observers to minimize the effects of moonlight by positioning themselves so that the Moon is behind them and hidden behind a wall or other suitable obstruction.

There are many excellent observational opportunities in 2012. The peak of the Quadrantids occurs at about 06h, so the final rise to maximum may be observed without interference from moonlight for several hours before dawn. The maxima of the April Lyrids, Orionids, Taurids (in early November), Leonids and Geminids are all observable with little or no interference from moonlight. Indeed, the autumn months of 2012 look like being a bumper time for meteor observers!

The April Lyrids have been rather neglected in recent years, but with no interference from moonlight in 2012, and a peak occurring in the early morning hours as the radiant climbs, it is hoped that observers will make a point of covering the shower this time around. Both the Orionids and the Taurids have produced unexpected activity in recent years – unusually high rates for the Orionids in 2006, and an atypical number of bright Taurids in 2005 – so observations of these two showers in October and November, respectively, are always worthwhile. Enhanced Leonid rates are now unlikely, but observations are important in the 'normal' years between the great displays. The Geminids are now the richest of the annual meteor showers and this year's maximum coincides with New Moon, so high observed rates are likely in the pre-dawn hours of 13 December and that evening as the radiant climbs. The main Geminid peak is expected at around 19h, but past observations show that bright Geminids become more numerous some hours after the rates have peaked, a consequence of particle-sorting in the meteoroid stream.

The following table gives some of the more easily observed showers with their radiants; interference by moonlight is shown by the letter M.

Limiting Dates	Shower	Maximum	Radiant RA	Dec.	
			h m	°	
1–6 Jan	Quadrantids	4 Jan, 00h	15 28	+50	M
19–25 April	Lyrids	22–3 April	18 08	+32	
24 Apr–20 May	Eta Aquarids	5–6 May	22 20	−01	M
17–26 June	Ophiuchids	19 June	17 20	−20	
July–August	Capricornids	8,15, 26 July	20 44	−15	
			21 00	−15	
15 July–20 Aug	Delta Aquarids	29 July, 6 Aug	22 36	−17	M
			23 04	+02	
15 July–20 Aug	Piscis Australids	31 July	22 40	−30	M
15 July–20 Aug	Alpha Capricornids	2–3 Aug	20 36	−10	M
July–August	Iota Aquarids	6–7 Aug	22 10	−15	M
			22 04	−06	
23 July–20 Aug	Perseids	13 Aug, 04h	3 04	+58	M

16–31 Oct	Orionids	20–22 Oct	6	24	+15
20 Oct–30 Nov	Taurids	2–7 Nov	3	44	+22
			3	44	+14
15–20 Nov	Leonids	18 Nov, 08h	10	08	+22
Nov–Jan	Puppid-Velids	early Dec	9	00	−48
7–16 Dec	Geminids	14 Dec, 14h	7	32	+33
17–25 Dec	Ursids	22 Dec	14	28	+78

Some Events in 2013

ECLIPSES

There will be five eclipses, two of the Sun and three of the Moon.

25 April: Partial eclipse of the Moon – Europe, Africa, Middle East, Asia and Australasia

9–10 May: Annular eclipse of the Sun – northern Australia, Solomon Islands, central Pacific Ocean

25 May: Penumbral eclipse of the Moon – Canada (except north-west), United States, Central and South America, western Europe, western and southern Africa

18 October: Penumbral eclipse of the Moon – North America (except Alaska), Central and South America, Europe, Africa, Middle East, Asia (except Japan)

3 November: Annular/Total eclipse of the Sun – central Atlantic Ocean, Gabon, Congo, DR Congo, northern Uganda, north-western Kenya, Ethiopia, Somalia

THE PLANETS

Mercury may be seen more easily from northern latitudes in the evenings about the time of greatest eastern elongation (16 February) and in the mornings about the time of greatest western elongation (18 November). In the Southern Hemisphere the corresponding most favourable dates are 9 October (evenings) and 31 March (mornings). The planet may also be spotted from both hemispheres in the evenings around the time of greatest eastern elongation on 12 June and in the mornings around the time of greatest western elongation on 30 July.

Venus is visible in the mornings at the start of the year. It passes through superior conjunction on 28 March and will be visible in the evenings from June until the end of December. It reaches greatest eastern elongation (47°) on 1 November.

Mars does not come to opposition in 2013. The planet is in conjunction with the Sun on 18 April, and becomes visible in the morning sky later in the year. It will next be at opposition on 8 April 2014, when the planet will be in Virgo.

Jupiter does not come to opposition in 2013. It will next be at opposition on 5 January 2014, when the planet will be in Gemini.

Saturn is at opposition on 28 April in Libra.

Uranus is at opposition on 3 October in Pisces.

Neptune is at opposition on 27 August in Aquarius.

Pluto is at opposition on 2 July in Sagittarius.

Part Two

Article Section

The Barwell Meteorite

HOWARD G. MILES

First published in the *1968 Yearbook of Astronomy*. Revised 2011.

On Christmas Eve 1965, just as it was getting dark and when most people were making their final preparations for the following day, there suddenly appeared in the sky above the city of Coventry an orange-red ball of light with a long yellow tail moving rapidly in a northerly direction. This was followed a short time later by a very loud bang, just like that made by an aircraft breaking the sound barrier. Some people phoned the police, some the local airport, but no one could give a satisfactory explanation. The general feeling was that it would all be explained in the local press after the holidays.

FLASHES AND BANGS FROM THE SKY

Meanwhile, in the village of Barwell, just north of the hosiery town of Hinckley, strange things were happening. One of its residents, Mr Crow, was walking home from work along a road called The Common when he saw a flash in the sky and then heard a bang. A few moments later, he heard something swish down out of the sky and land nearby with a thud. He then heard four or five more objects come down in quick succession, the last one landing in the road close to him, a piece of it breaking a window in the house of his neighbour, Mr Grewcock. Mrs Grewcock heard the glass breaking and called to her husband. He went to the front of his house and found pieces of rock and white powder covering the pavement. He went to pick up a piece of the rock but it was too hot to handle. Apart from clearing up the mess, neither he nor Mr Crow did anything about it until after Christmas. Further down the road, another resident, Mr England, noticed on Christmas morning a dent in the bonnet of his car which he had left in the drive the previous afternoon. He suspected that children were responsible and threw the lump of rock, which he assumed was responsible for the damage, on to the waste ground on the opposite side of the road.

With the festivities over and the return to normal life, Mr Grewcock reported the incident to the police. On the morning of 27 December, he explained to Police Constable Scott what he and Mr Crow had seen and gave him the lump of rock which he thought had broken the window. Later on, Constable Scott took a statement from Mr Crow and collected more 'samples'. He submitted his report and the samples to his seniors and the samples were forwarded to the museum at Leicester and then on to the British Museum in London.

Stories of the strange object in the sky over Coventry and the surrounding areas appeared in the local press and on 28 December a

Figure 1. Artist's impression by the late Paul Doherty of the brilliant fireball which preceded the meteorite fall at Barwell on Christmas Eve 1965.

spokesman at Elmdon Airport, Birmingham, suggested that the 'fiery object' might have been a meteor. The author, who lives in Coventry, saw the flash of the fireball as it swept across the sky (Figure 1) and heard the sonic boom. In fact, it shattered an electric light bulb in his house. The idea that it might have been a meteor suggested the possibility of the re-entry of an artificial satellite which would produce identical phenomena. Consequently, he immediately started to collect details of the event. Gradually the news from Barwell leaked out and it was soon realized that the events at Barwell and the objects seen in the sky over Warwickshire were very closely connected. On 6 January, twenty of the country's leading geologists descended on the village in search of further fragments of the stones, which had now been identified as meteorites. Appeals were made to the local residents to hand in any specimens they had collected. Very soon just over 50 lb (22.7 kg) had been recovered, but later, when very favourable rewards were offered, the total weight went up to just over 97 lb (44.1 kg).

THE HEAVIEST RECORDED FALL IN ENGLAND

The total weight of the recovered fragments makes the Barwell fall the heaviest recorded meteoritic fall in England and the second heaviest in the British Isles. The following table lists all of the recorded falls since the turn of the twentieth century.

1902	Crumlin, Co. Antrim, N. Ireland	9 lb 5½ oz (4.2 kg)
1914	Appley Bridge, Lancashire	33 lb (15 kg)
1917	Strathmore, Perthshire, Scotland	29 lb 4 oz (13.3 kg)
1923	Ashdon, Essex	3 lb (1.4 kg)
1931	Pontlyfni, Gwynedd, Wales	5 oz (0.14 kg)
1949	Beddgelert, Gwynedd, Wales	1 lb 12oz (0.794 kg)
1965	Barwell, Leicestershire	97 lb approx (44 kg)

Note: In the original article, the total weight of each meteorite fragment fall was given in imperial units (lb, oz), but here the metric equivalents are given alongside. Since publication of the original article, the following British and Irish meteoritic falls have been recorded:

1969	Bovedy, Co. Londonderry, N. Ireland	5.46 kg
1991	Glatton, Cambridgeshire	0.767 kg
1999	Leighlinbridge, Co. Carlow, Ireland	0.271 kg

The heaviest recorded fall in Britain and Ireland occurred at Limerick, Ireland, in 1813, where the total weight of all the meteorite fragments recovered was 106 lb (48.1 kg).

On 8 January, I visited Barwell and was shown by Constable Scott the various places where the fragments had been found, and in the afternoon searched the fields to the west of The Common and found eight small fragments. By this time I had collected sufficient information to give a reasonable value for the track of the fireball, but even at this early stage there was contradictory evidence which at that time could not be explained. It was obvious that many more reports were needed and an intensive campaign was started using the Press, radio and TV. Dr Meadows of Leicester University collected further reports. I interviewed most of the people who lived in the Coventry area who submitted reports, because it was considered that this was the only way in which the maximum information could be obtained. Much of South Warwickshire was combed for further observations, and many useful reports were collected. In addition, reports were received from a wide area stretching from Wiltshire in the south to Shropshire in the north, and from Port Talbot in the west to Colchester and Grantham in the east. Within this area, there were well-defined regions where no or very few reports were received. This was due mainly to the large amount of cloud that existed. This cloud cover was very patchy, resulting in one observer reporting only very small breaks in the cloud layer and yet another only a short distance away reporting a relatively clear sky.

Most of the observations made from Coventry described an object through breaks in the cloud, and it soon became obvious that these people had not all seen the same object. Some reported virtually an overhead passage, whilst others described an object passing low to the east. A second fireball was suspected, and later this was confirmed when an observer from Rugby saw an object more or less overhead and, a few moments later, saw a second object passing over Coventry. Up till this time it had been assumed that the 'Coventry' fireball had been responsible for the fall at Barwell. It was therefore necessary to check this, because from the information available at that time it could feasibly have been this second object. The correctness of the original theory was confirmed by two methods.

An examination of the positions from which meteoritic material had been recovered showed that if they conformed to the usual pattern, in which they were restricted to an ellipse in which the direction of the major axis coincided with the direction of flight of the fireball and that the smaller fragments were to be found towards the end nearest to the direction in which the fireball approached, then there was a well-defined part of this ellipse in which no meteorites had been found. A thorough search of this part was made by the author and others, including one of the editors of this *Yearbook*, Patrick Moore, and three large fragments were found (Figure 2).

This restored the balance, and showed that the major axis pointed in the direction of Coventry and not Rugby. Second, the reduction of the reports of the many observations from the areas of Coventry and Rugby received later showed, without doubt, that the fireball which passed over Coventry had been responsible for the fall.

Even with the identification of the second object, there were many other observations, some from Coventry, some from Loughborough

Figure 2. Fragments of the Barwell meteorite recovered by the author and Patrick Moore in early January 1966.

and other regions of Leicestershire, and others from Staffordshire, which could not be fitted into this picture. The matter came to a head when observers from Coventry reported seeing a fireball from their west-facing windows. The identification of this third fireball cleared all the principal difficulties raised from reports from Warwickshire and Leicestershire. There were, however, still three well-defined groups of observations which could not be fitted into the above picture.

Although some of the Staffordshire reports could be linked with the third fireball, others from this area and from Cheshire and Shropshire could not be so associated with any confidence. These reports, together with observations from Port Talbot, Nottingham and Bristol, indicated that a fourth object had crossed the sky on a track roughly parallel to the others but passing well to the west of Coventry.

The second group of observations came from an area which included the counties of Buckinghamshire and Berkshire. These reported a single object falling directly downwards in the north, some saying that it was to the east of the north line. These, together with some reports from Northamptonshire and Leicester, suggest the

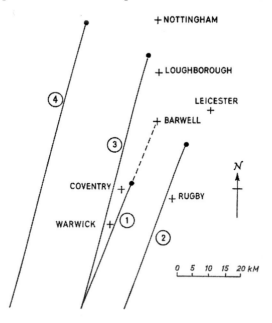

Figure 3. Ground tracks of the four fireballs described in the text.

possibility of the existence of a fifth object, although it is by no means conclusive. It is strange, however, that none of these southern reports mentioned more than one object, although this could be explained by the presence of extensive cloud.

The last group all reported an object travelling in a south-westerly direction in a line extending from Grantham in Lincolnshire to Stroud in Gloucestershire. Unfortunately, these reports have not yet been explained, although it is fairly certain that they had no connection with the Barwell incident.

IDENTIFYING THE VARIOUS FIREBALLS

To help in the identification of the various fireballs (Figure 3), a code letter has been assigned to the individual members, the actual code being similar to that used in the official reports.

Fireball 1

This travelled in a NNE direction, passing over the eastern boundary of Coventry, and resulted in the fall at Barwell. The track of this object has been determined to quite a high degree of accuracy. Several observers reported a virtually overhead passage, and on interview were able to indicate whether the track was slightly to the west or to the east. An observer from Barwell saw the fireball descending directly downwards in the south-west, although she did not see the extinction point. Reports from the Coventry area indicated that the fireball went extinct at a point just beyond the north-east boundary of Coventry. The area between the city and Barwell was searched very carefully for observers. Although over a hundred people were interviewed, not one was found who had seen anything, although some had heard thunder. This is in accordance with the usual behaviour of a medium-sized object. As the meteor enters the atmosphere it is slowed down by air resistance. This resistance increases considerably when it penetrates the lower and denser regions. A certain point is reached, usually near the tropopause, i.e. the lower boundary of the stratosphere, when the speed of the object is no longer sufficient to generate enough frictional heat to keep the meteor glowing. At this point, most of the cosmic speed has been lost and the meteor moves solely under the influence of the Earth's gravitational force. In this case, the extinction occurred at a height of about ten kilometres.

As with all the other objects, the description of the fireball varied considerably, due, no doubt, to some observers seeing the object through breaks in the cloud and others seeing it through thin cloud. It is fairly certain that just before extinction was reached, it developed a long yellow tail. Observers from Warwick and Kenilworth generally reported only a short one. The colour of the fireball depended on the direction in which it was seen. Those who saw it after it had passed over likened it to the after-burn on a jet engine. Most agreed that it was about half the size of the Sun and about as bright as the setting Sun. As it passed over Kenilworth, one observer described how the flood waters appeared red in colour.

Fireball 2

This travelled on approximately a parallel track to the previous fireball, but to the east, passing nearly over Rugby and becoming extinct just south of Leicester. This object was seen over a much wider area, although there were some regions where very few observations were reported. It descended at roughly the same angle (20° to the horizontal) and had a similar extinction height, although its behaviour at this point was entirely different. Two observers reported the object blowing up, forming four major fragments, one piece shooting upwards and the other three downwards.

The appearance of this fireball was similar to that of fireball 1, although there were some important differences. It was reported generally as orange-red in colour, but earlier along the track it was reported as much whiter. The dust trail left behind was definitely corkscrew in shape, and persisted for several minutes. The dust trail for No. 1 was described as straight, and it did not persist for very long. Just north-west of Rugby the characteristics of the bolide[1] changed when it appeared to explode. The descriptions after this do not agree very well, but this may have been due to interference by cloud.

The sonic effects from this fireball were particularly severe, and were recorded over a wide area. It was the sonic boom from this object that caused the difficulties in the early analysis of the Coventry reports. A large percentage of the reports from Coventry mentioned either a delay

1. Although the International Astronomical Union (IAU) has no official definition for the term 'bolide', astronomers generally use the term to identify an exceptionally bright fireball, particularly one that explodes (sometimes called a detonating fireball).

of about thirty seconds between the sighting and the boom or a delay of one and a half minutes. It is now believed that the thirty-second delay reports are linked with fireball 1 and the one-and-a-half-minute delay reports are associated with No. 2, the extra time being required for the sound to travel the distance from track 2 to Coventry.

Fireball 3

This travelled on a slightly diverging track from No. 1, but passing over the western boundary of Coventry. It had an extinction point just south of Nottingham. The fireball had a well-developed yellow tail and left a smoke trail, except towards the end of its flight. The quality of the observations of this object was not as high as those of the previous ones, but the few good reports were sufficient to locate the track with a reasonable degree of accuracy. It is assumed that this object was responsible for the reports of flashes in the north-west from some observers at Barwell. In contrast with the other objects, there were no definite reports of a sonic boom attributed to this fireball.

If the calculated tracks of nos. 1 and 3 are produced backwards, they converge on a point near Stow-on-the-Wold, Gloucestershire. The implication that these two objects were the result of the explosion of a single object was confirmed by an observer from Birmingham who witnessed the separation.

Fireball 4

The existence of this object had been suspected for a long time, but the information collected was of such a nature that it was impossible to dismiss other possible explanations. A high-quality report from Bristol indicating that an object had been seen travelling in a NNE direction but passing to the west of Bristol, together with one or two other key observations, put the question beyond doubt. This bolide was reported by several people as being a bluish-green in colour and having a short yellow tail. As with the other objects, some reports were received describing the fireball as shedding sparks and others mentioned a dust trail. The extinction was probably in a position about ten kilometres up but to the west of Nottingham.

With the exception of the divergence of tracks 1 and 2, the tracks of the fireballs were approximately parallel. It is thought likely that they were all products of a single body which broke up initially in the upper regions of the atmosphere. The break-up near Stow, the break-up just prior to extinction of No. 2, and the many reports of sparks falling from the fireballs suggest that fragmentation took place readily. It is possible that the objects described above were only the major fragments of a very complex break-up. This may explain some of the odd reports which had to be rejected because of lack of confirmation. There have been no reports of meteoritic material resulting from any of the fireballs other than fireball 1. Therefore it is likely that the other objects were relatively small pieces which were completely or nearly completely destroyed during their passage through the atmosphere. If they did reach the surface of the Earth, the chances of their being discovered are now virtually non-existent.

THE METEORITIC FALL AT BARWELL

The angle of descent of the fragments just prior to landing must have been quite steep. One piece, which penetrated the roof of a factory, descended at about 80° from the south-east. The positions of others were so near to walls and houses that they must have fallen nearly vertically. The actual direction of fall, however, bears no relationship to the direction of flight of the original fireball, because aerodynamic forces can produce quite complex motions on irregularly shaped fragments. They also produce wide ranges in the rate at which the individual pieces fall to the ground. For example, one piece was found lying on the surface and another piece of similar size and falling on to a similar surface was buried to a depth of about twenty inches (half a metre).

There is no obvious regularity in the size distribution within the fall area, although the very small pieces have only been found within a relatively well-defined area in the south-west. This lack of gradation suggests a final break-up late in flight, an idea which is supported by the lack of secondary burning on the surface of the fragments. A study of the fusion crust shows that the bolide did not rotate during its flight through the atmosphere. Further, the character of the fusion markings indicates that the weight prior to break-up was about 200 lb (over 90 kg), indicating that only about half of the meteorite has been found.

The meteorite belongs to the commonest type, known as olivine-hypersthene (L6) chondrite, a stony variety consisting mainly of magnesium iron silicate with about 8 per cent free nickel iron. The freshly fallen samples looked very similar to newly broken concrete but after the passage of time they took on a brownish colour due to the rusting of the free iron. The meteorite is very brittle and crumbles very easily. Although about 100 lb (45 kg) is suspected to be still lying where it fell, the likelihood of it ever being found is fairly remote. Even after a few months, there was evidence of erosion, with the sharp edges tending to become rounded. The following table lists all of the main meteoritic fragments recovered:

1. 4 pieces totalling 3.9 oz (0.11 kg)
2. 6 lb 4.5 oz (2.85 kg) – in a hole 9 inches deep in clay
3. 3 lb 11 oz (1.67 kg) – in a hole 1 foot deep in well-dug soil
4. 2 pieces – 7 lb 3 oz (3.26 kg) and 3 lb 14.5 oz (1.77 kg)
5. Estimated total of 2 lb (0.9 kg) – fragments on road
6. 1 lb (0.45 kg) – on car bonnet
7. 5 lb 7 oz (2.47 kg) – on surface on grassland
8. 3 pieces – 14 oz, 4 oz and 1 oz (total 0.54 kg) – passed through roof of works
9. 2 pieces – 2 lb 9 oz (1.16 kg) and 7 lb 12 oz (3.52 kg) – on drive to house

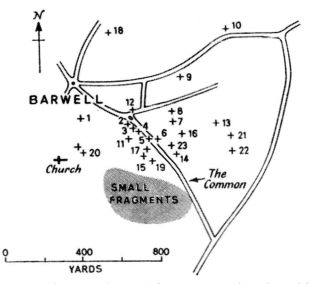

Figure 4. The location of the principal meteoritic fragments recovered in and around the village of Barwell. Note that the identification numbers shown on this map relate to the table listing the main fragments recovered. The area in which small pieces weighing less than an ounce (< 0.03 kg) were found is shown shaded.

10. 7 lb 6 oz plus fragments totalling 1lb 12oz (total 4.14 kg) – on tarmac drive
11. 1lb 8oz (0.68 kg) – on surface by hedge
12. 3 oz (0.085 kg)
13. 1 lb 9 oz (0.71 kg) – in hole 3 inches deep in grassland
14. 4 lb 15 oz (2.24 kg)
15. 17 lb 6 oz (7.89 kg) – in hole approx 2 feet 6inches deep in grassland
16. 7 lb 14 oz (3.58 kg)
17. 9 oz (0.26 kg)
18. 2 pieces totalling 5 lb 14 oz (2.67 kg)
19. 2 lb 13 oz (1.28 kg)
20. 2 pieces totalling 1 lb (0.45 kg) on grassland
21. 1 lb 4oz (0.57 kg) – in shallow hole in grassland
22. 15 oz (0.43 kg) – in shallow hole in grassland
23. 15 oz (0.43 kg) – in shallow hole in grassland

Note: The total weight of all the fragments listed in this table amounts to 44.1 kg. The numbers alongside each entry in the table relate to the map of Barwell showing the location of the principal fragments (Figure 4).

Although much will be learned from this fall, it is not expected to produce any startling discoveries. Nevertheless, all meteoritic falls must be investigated fully, because at the moment meteorites are the only material from outer space that can be studied in our laboratories. The Barwell meteoritic fall has certainly aroused considerable interest in scientific circles.

Brighter than a Million Suns

DAVID A. ALLEN

First published in the *1979 Yearbook of Astronomy*. Revised 2011.

CHILE, JANUARY 1972

I was using a 40-inch telescope at Cerro Las Campanas. The equipment at the Cassegrain focus was an infrared photometer, a device which simply measures the brightness of stars at wavelengths several times that of normal light. In 1972, the southern skies, accessible from Chile, were little-studied at infrared wavelengths, so the thrill of exploration pervaded the observing run.

I set the telescope to a sixth-magnitude star in the constellation Carina and put my eye to the eyepiece. I still vividly recall that moment, for the field of view is perhaps the most beautiful you can see through a telescope of that size. The whole field was filled with a misty, green luminescence – a great cloud of excited gas, woven into veil-like curving streams. In the lower right, a keyhole-shaped dark patch clearly sat in front of the misty nebula. Inky black it was, as though something opaque was stuck to the back of the eyepiece. But as I pressed the slow-motion buttons, it moved with the star field: the keyhole was, in fact, a massive cloud of dust and gas at a distance of several thousand light years, obscuring the nebula behind it. At the top of the field of view lay a loose cluster of stars, mostly blue. And there, in the centre of the field, was the star I wanted to observe. But it was quite unlike a star – more a celestial poached egg, a round orange blob a second or two across and set in a frothy white nebula.

I manipulated the poached egg into the appropriate position on the graticule; I pulled a lever which tilted a mirror so that its light no longer reached the eyepiece but landed on the infrared detector. Immediately the dome was filled with a loud, electronic buzz. I knew the reason even before I walked over to the mountain of electronics that inevitably accompanies the infrared observer at his work. The buzz was emanating from the chart recorder: the pen, which normally drew a wiggling red line down the middle of the chart, was pinned against the right-hand edge of the paper, plaintively complaining that too strong a signal was coming down the snaking black cables from the detector.

So the poached egg is a very bright infrared source. Sadly, I wasn't the first to discover this fact: Gerry Neugebauer and Jim Westphal, of the California Institute of Technology, had done so four years previously. In fact, they had shown that at the infrared wavelengths of 10 and 20 μm this is the brightest source in the sky aside from Solar System objects. What is this distinguished object? It is known as η (eta) Carinae.

Carina is in itself a large constellation, but it is only a part of the disbanded Argo Navis. When Argo was broken up, the Bayer letters were retained with the individual stars, so that α and β went to Carina, γ and δ to Vela, and so on. We are concerned with Carinae, formerly η Argûs, and since eta is the seventh letter of the Greek alphabet, it is a fair bet that η was about the seventh brightest star in Argo. But so large a constellation as Argo, straddling a rich section of the Milky Way, must have had more than seven bright, naked-eye stars; indeed, from the magnitudes of δ, ε, ζ, θ, ι, and κ Argûs we can deduce that η Argûs was thought by Bayer to be third-magnitude. Why, then, is η Carinae now barely visible to the naked eye, even from so clear a place as the Andean foothills of Chile? To answer this we must delve into a little history.

A BRIEF HISTORY OF ETA CARINAE

I believe that Bayer's *Uranometria* offers the first estimate of the magnitude of η Carinae; this is usually quoted as fourth, despite the argument above. In any case, in the late seventeenth century, the star was rated fourth magnitude by Halley. It was soon shown to be variable, meandering irregularly between second and fourth magnitudes. Nobody was particularly disturbed by this behaviour, since variable stars were pretty well known by then. But people began to take notice when it exceeded its normal brightness range in the early nineteenth century. In December 1837, John Herschel found it to have crossed the zero magnitude point, and five years later it became the second brightest star in the sky at about magnitude −1.0.

Having reached its peak, seemingly satisfied with second brightest, it began its protracted descent to one-thousandth of its peak brilliance, and it has stayed between about sixth and seventh magnitude ever since.

The nebulosity that reminded me of an egg white, but which is usually called the homunculus because of its shape, was not noticed until about seventy years ago when Innes catalogued what appeared to him to be companion stars embedded in a faint nebula. It soon became apparent that these were knots of nebulosity moving radially outwards from the central orange object. The homunculus is now about ten arcseconds across, large enough for a fairly accurate determination of its expansion and hence of the date of its genesis. This date does not appear to be the same for all portions of the homunculus, and this suggests that there were several ejections in different directions spread over a few years. The average date, determined by Bob Gehrz and Ed Ney from recent photographs, is 1835 – just the time that the greatest brightening occurred. Gehrz and Ney took some rather attractive colour photographs of η Carinae: these were reproduced on the July 1972 cover of *Sky and Telescope*.

The homunculus is not the only nebulosity associated with η Carinae. A faint outer ellipse was discovered by David Thackeray on photographs taken with the Radcliffe 74-inch telescope in South Africa. This, too, is expanding. And η Carinae lies in the midst of the Carina Nebula, NGC 3372, a giant cloud of ionized hydrogen, the second brightest nebula in the sky (after Messier 42 in Orion) and easily visible to the naked eye despite lying at about four times the distance of the Orion Nebula. NGC 3372 is excited by a large number of hot, blue stars, some of which formed the spangled cluster which I saw in the forty-inch view, lying to the north of η Carinae.

When a star undergoes such remarkable antics as had been shown by η Carinae, the services of that most valuable of all astronomical equipment, the spectrograph, are brought to bear on it. The first spectrum of η Carinae was taken in 1892: this showed the star to be a normal supergiant of spectral type F5, quite similar to the star Canopus which, half a century earlier, it had outshone. Three years later another spectrogram was secured, and a most dramatic change was found to have occurred. Instead of the dark Fraunhofer absorption lines of a normal stellar spectrum, η Carinae showed a series of bright emission lines. The strongest of these were due to hydrogen but, in addition, there was a great number of lines of once-ionized iron, plus a few more lines of chromium, nickel, titanium, copper, nitrogen and sulphur. So drastic a change within a period of three years is in itself remarkable, but is made more so by the fact that since 1895 this spectrum has remained virtually unaltered. Another surprising aspect of this spectrum is the absence of lines due to oxygen, one of the most abundant of elements and one which usually features prominently in the spectra of nebulae.

THE DISTANCE OF ETA CARINAE

In documenting the remarkable properties of this star it is now necessary to consider its distance. Here we get onto uncertain ground. Distance determination in astronomy is always difficult and unreliable, and this is particularly so in the constellation of Carina where we happen to be looking directly along the inner spiral arm of our Galaxy, so that there are stars and nebulae at a whole range of distances.

The most reliable estimate of the distance is that for the cluster of blue stars, Trümpler 16, which illuminates NGC 3372. Even here, however, different researchers have come up with values ranging between 3,500 and 12,000 light years. The most recent (and hence, hopefully, most reliable) values tend to be the lower, so I will adopt 6,000 light years as the best guess. In addition to this uncertainty of almost a factor of two, there is still debate about whether η Carinae does lie in Trümpler 16 and NGC 3372 rather than in front of or behind them. Gehrz and Ney estimated the distance by assuming that the outward expansion of the homunculus was at the velocity inferred from its spectrum by Thackeray. They derived 6,800 light years, with a fair amount of uncertainty. This seems to confirm that we are using the right sort of distance.

Now consider that in 1843 η Carinae was the second brightest star in the sky. Sirius, the brightest, is a mere nine light years away. Even Deneb, the most distant first magnitude star, is only 1,500 light years distant. To compete with these, η Carinae must have been an extremely luminous object. We can estimate its total luminosity and find a value about six million times that of the Sun. Six million times . . . not only is this very much brighter than conventional novae, but it exceeds by almost a factor of ten the most luminous star known, S Doradûs in the Large Magellanic Cloud.

Today, of course, η Carinae appears much fainter. However, it now pours out most of its energy at infrared wavelengths. The mechanism is well understood. The homunculus is a shell of gas and dust that surrounds the star. Because we view the central object through this dust, it appears dimmed. The dust absorbs the visible and ultraviolet radiation of the star, but in so doing it is heated. Hot dust radiates in the infrared, but it radiates only as much energy as it absorbs from the star. Thus in our determination of the luminosity of the central object we must include the contribution from the infrared radiation. When we add this we find that η Carinae is as luminous now as it was in the 1840s: it still radiates several million times as much energy as the Sun. Since 1843, η Carinae has output about as much energy as is released by the most energetic supernova explosion, or is emitted by most stars in their entire lifetimes.

THE CENTRAL STAR

It is particularly unfortunate that the major outburst of η Carinae occurred only a few decades before the spectrograph became readily available. A better knowledge of the outburst itself, and particularly of the star before its outburst, would greatly have assisted our understanding of the star. As it is, we can only examine the object in its present state, largely veiled behind a curtain of dust. What, then, can we say about it?

First, we can determine its luminosity, as I have just described. We also can estimate its temperature in a number of ways. From the emission lines we have one value, since stars of different temperatures will ionize atoms to different extents. From the colour of the star we get another estimate. The deep orange hue is due largely to the hydrogen emission, but we can make allowance for this and for the other emission lines. We can also correct for the reddening effect of the dust in the homunculus. A third value comes from X-ray satellites. Very hot stars are X-ray sources, but η Carinae has not been detected, so this gives an upper limit to its temperature. These three methods indicate that the star has a temperature of around 25,000 degrees Kelvin (about the same in Celsius). This is not exceptionally high: many of the very luminous stars in our and other galaxies are much hotter.

The temperature uniquely defines how much energy is radiated from each area of the star's surface. Knowing the total luminosity we can thus deduce its surface area and, from this, its diameter. We find that η Carinae is something like ten times as big as the sun. Again this is not unduly large – the red supergiants, such as Betelgeuse, are several hundred times the diameter of the Sun.

In all probability it is the mass of η Carinae which is extreme. We don't know, for there is no direct method of determining this parameter. If its density were the same as that of the Sun, its mass would be one thousand times as great. But this is certainly an overestimate: it can be shown theoretically that so large a mass of gas trying to form a star inevitably breaks up into smaller portions. In fact, it is difficult for a star to form with a mass much larger than a hundred times that of the Sun. η Carinae may be somewhere near this limit. Theory also predicts that so massive a star is unstable and will intermittently or continuously eject some of its mass. In particular, there is a rule known as the Eddington limit (after the former Cambridge astrophysicist), which states the maximum allowable luminosity for a stable star of given mass. At one hundred solar masses, the Eddington limit is four million times the luminosity of the Sun. If a star begins to generate nuclear energy at a higher rate than this, it will blow off its outer layers. η Carinae is probably very close to its Eddington limit.

But what of the F supergiant spectrum of 1892? At that date, η Carinae appeared to be a normal star with a surface temperature of about 8,000 degrees K. How do we reconcile this with its present state? The answer is that the 1892 spectrum was not of η Carinae. I don't mean that someone observed the wrong star. What I mean is that, at the time, the shell of material which had been thrown off fifty years earlier, and which eventually formed the homunculus, was opaque. We couldn't see through it to the central star. We saw merely the surface of a cocoon of gas which happened to have cooled to 8,000 degrees K. Within the next few years, by 1895, the shell thinned out sufficiently, due to its expansion, for us to see through to the underlying star with its emission-line spectrum. The transition from opaqueness to transparency can occur quite quickly, and once the central object is visible, the shell need not affect the spectrum at all. Thus, no change in the spectrum would necessarily have been expected between 1895 and the present time.

THE NATURE OF ETA CARINAE?

We cannot yet say with certainty what η Carinae really is. We need additional observations, either through the adoption of novel techniques or by some clever use of existing instrumentation which has not yet been devised. Or we need the star to show some type of change comparable to that of the last century.

But I cannot end a *Yearbook* article on so negative a note; here is a credible guess – for which I claim no originality – which appears to fit the observations. At the centre of the largest clouds of gas in our Galaxy an unusually massive knot tried to collapse to form a star. It succeeded quite recently: perhaps only a few tens of millennia ago. As the star proceeded to produce energy at its natural rate, it became unstable and began a series of minor outbursts. Each outburst involved only a slight increase in its total luminosity – maybe a factor of two or less – but it was enough to cause a fraction of its outer layers to be thrown off as a shell. One such shell was ejected over a period of a few years around 1840. The previous shell has almost disappeared; all that remains of it is Thackeray's elliptical nebula.

As the gaseous shell expanded, it cooled until eventually dust grains could condense within it. We know from the infrared observations that some of the dust is composed of silicates (like sand and stony meteorites), as is the dust around cool red stars such as Betelgeuse. Silicates use up a lot of oxygen in their formation, and this might amount to an explanation for the striking deficiency of oxygen lines in the emission spectrum of η Carinae. In any case, it was the formation of a translucent screen of dust, more than the change in luminosity following its outburst, that caused the star to fade during the second half of the nineteenth century.

This is one possibility. However, we cannot reject an alternative model in which η Carinae is a binary star. It is now well established, primarily due to the work of Bob Kraft of the Lick Observatory, that novae are binary stars. The nova outburst is triggered because gas becomes sucked out of one star by the gravitational attraction of the other. The luminosity, rise in magnitude from minimum, and the time scale of the outburst and subsequent fading are parameters which differ from star to star for reasons which are not entirely understood. Typically, a nova rises in brightness very rapidly, develops the spectrum of an A or F star (that of the cooling shell), and proceeds shortly after maximum to an emission spectrum. This then proceeds through a series of spectral changes as the star fades back to minimum. The first stage of the emission spectrum usually resembles the spectrum of η Carinae.

Some novae progress much more slowly through this sequence than do others. Nova Pictoris (1925) and Nova Delphini (1967) are examples. Some, the so-called slow novae, take decades. RR Telescopii erupted in 1944 and has faded only a few magnitudes since. RT Serpentis (eruption 1909) and, more recently, V1016 Cygni and HM Sagittae, have behaved similarly. There are objects called symbiotic stars which seem to freeze the fading spectra of slow novae: these may be undergoing the same processes as the slow novae, but even more gradually. No examples of any of these types of object are anywhere near as luminous as η Carinae, but that alone does not preclude η Carinae from being a uniquely slow and bright specimen of mass transfer in a binary system.

For η Carinae is a unique object.

EPILOGUE

(written by Professor Fred Watson in June 2011)

This marvellous article from the *1979 Yearbook* is vintage David Allen (Figure 1). He sucks you in with first-hand observing anecdotes, introduces his subject with a bang as a chart-recorder noisily overloads in the telescope dome, and then deftly weaves a tapestry of mystery as he recounts the history of humankind's understanding of Eta Carinae. As the article progresses, and the strands of mystery are neatly tied up one by one, you begin to feel that we know all there is to know about this enigmatic object. Then, right at the end, he throws in a completely orthogonal idea – one so speculative and outlandish as to beggar belief.

Phew. What a story. No wonder that virtually every edition of the *Yearbook of Astronomy* from 1969 to 1995 kicked off its 'Articles' section with one of David's fine pieces. But, in fact, the excellence of his writing is only half the story. Like so much of David's work, his Eta Carinae article was extraordinarily prescient. The barmy idea that ended David's 1979 article – to wit, that Eta Carinae might actually be a multiple star rather than a single object – was confirmed in 2005 by US scientist Rosina Iping and her colleagues, using observations made with NASA's FUSE spacecraft. Once again, David was well ahead of his time.

Figure 1. David Allen in 1993, when he was awarded the Australian Museum Eureka Prize for the Promotion of Science. (Image courtesy of David Malin.)

It was in the eighth edition of the *Yearbook of Astronomy* that the youthful David Allen made his debut with the already well-established readership of this august publication. He had only recently begun his PhD studies at the University Observatories in Cambridge, but that first article in the 1969 edition provided a remarkably accurate cameo preview of his subsequent career. Entitled simply 'Infrared Astronomy', it gave an insightful and provocative overview of what was then a brand new branch of science.

Rereading that article before writing this Epilogue, I was struck by two things in particular that foreshadowed his life's work. One was David's detailed diagram of the complete electromagnetic spectrum, from gamma rays to radio waves – an uncommon sight in popular astronomy books of the 1960s. The other was that he dealt with celestial objects of every kind, ranging from the Moon (which was actually his PhD topic) to the newly discovered quasars. Although David always specialised in infrared astronomy, his later career encompassed observations made in every region of the electromagnetic spectrum.

And his research touched every class of object. When I was invited to write a celebration of his life for the *1996 Yearbook of Astronomy*, I was astonished to discover that a bibliography of David's research publications stretched to well over twenty single-line-spaced pages, and covered every aspect of astronomy – ranging from the Solar System to cosmology.

From 1975 until his tragic death in 1994 at the age of forty-seven, David worked at the Anglo–Australian Observatory (now the Australian Astronomical Observatory), and it was here that he achieved stardom in two other areas of astronomical endeavour. The first was a succession of world-leading infrared instruments for the Anglo–Australian Telescope (AAT), beginning with IRPS (the infrared photometer spectrometer) and culminating in the award-winning IRIS (an infrared imaging spectrometer). That legacy lives on today with IRIS2, a wide-angle infrared camera that is the AAT's workhorse imager.

The final strand in David's many-stringed bow was the one you already know about – his ability to bring esoteric scientific concepts to a broad general audience. Whether writing for a popular-level readership or appearing with Patrick Moore on *The Sky at Night*, David expressed himself with a clarity that veteran Australian broadcaster Robyn Williams once described as 'almost poetry'. And that extended beyond mere words to the extraordinary prescience I mentioned before. As early as his 1969 article, David was talking about 25-metre telescopes – machines that are even now only just on the horizon.

Which brings me back, at last, to Eta Carinae. What do we know now that we didn't know in 1979? Quite a lot, as it turns out – but probably little that would have surprised David. The famous Hubble Space Telescope image of Eta Carinae shown here (Figure 2) was made rather more than a year after his death – how he would have relished that. Today, the confirmed binary (or, possibly multiple) nature of the central star makes Eta Carinae just a little easier to understand. At least we are no longer dealing with an object that conventional wisdom says shouldn't exist because it's too massive. The presence of a faint companion in a 5.52-year orbit around the main star gives some hope that we might one day have better estimates of their respective masses.

At around one hundred times the mass of the Sun, however, Eta Carinae itself is still a monster. And one possibility that David does not deal with in his article is that we are seeing the progenitor of a super-

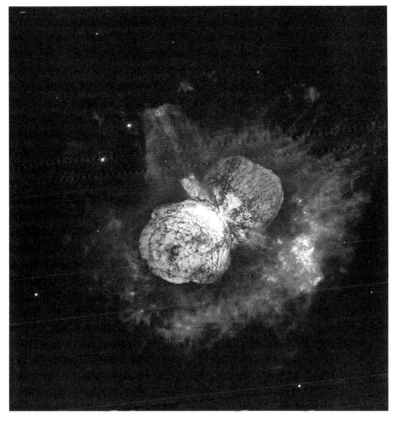

Figure 2. The Hubble Space Telescope's dramatic portrait of Eta Carinae shows the expanding lobes of the Homunculus Nebula and the equatorial disk between them, which is also expanding. Both are the result of an outburst event observed in the 1840s. (Image courtesy of Nathan Smith (University of California, Berkeley), and NASA.)

nova explosion that could burst into Earth's daytime skies at any point within the next million years. Such a ringside view of a supernova in action would thrill the world's astronomers, providing hard data on events that are currently only theoretically understood. And it could be soon. Eta Carinae has been brightening steadily for the last seventy years or so – although that could simply be the result of veiling dust being gradually dispersed by intense stellar winds, rather than an intrinsic brightening.

X-ray observations with NASA's Chandra spacecraft have shown that over the past thousand years or so, Eta Carinae has probably experienced several energetic outbursts, or so-called 'supernova imposter' events. The one that peaked in the 1840s, described in David's 1979 article, remains the best-studied, however. And there is one further remarkable strand that links this event to the redoubtable David Allen. As someone who loved his adopted country of Australia – its people, flora and fauna – he would no doubt have been delighted by the recent discovery that this outburst was observed by the indigenous Boorong people of Victoria, and incorporated into their Dreamtime sky stories. It is fitting that this most enigmatic of celestial objects should have been adopted by one of the most ancient cultures on Earth.

The Night Sky – AD 50,000

STEVEN A. BELL

First published in the *1984 Yearbook of Astronomy*. Revised 2011.

This article explores the future of the constellation patterns and the stars that make up those patterns. Tables of the brightest stars for 1,000,000 BC and AD 1,000,000 are also compiled.

One of the most useful items for the beginner in astronomy is a good star chart. This will help him or her to recognize the major constellation patterns and the stars which make up those patterns.

The chart which they see before them will look very much the same as Ptolemy saw it when he drew up his original forty-eight patterns nearly two thousand years ago. Since then these patterns have been regarded as 'fixed' because the motions of the stars in most cases are so small that they can only be detected over a period of about a century. However, over longer periods of time, these motions will change the patterns of the night sky to such an extent that certain constellations will fade below naked-eye visibility whilst others will be scattered over vast areas of the sky.

How will the night sky of the future differ from that of the present? For instance, what will the night sky look like in fifty thousand years hence and how will time affect the patterns we are familiar with today? To take this question to an extreme, which stars were the first magnitude objects in the skies of a million years ago and which ones will become the bright stars of a million years hence?

PRODUCING SIMPLE STAR MAPS

About a year ago, my university Astronomical Society required a simple star map to help members get acquainted with the night sky. I decided that the quickest and simplest method to do this was to write a computer programme to calculate the positions of a selected number of stars. This selection was made up of those stars involved in the constellation patterns visible from St Andrews at different times during the year. The map had to show not only the position of the stars in azimuth and altitude but also their magnitude as well as their place in a particular constellation pattern.

Approximately four hundred stars were involved in these patterns and plotting such a number by hand would have been extremely tedious. Consequently, a completely computerized method of creating and plotting the star map was the most desirable answer.

Simple formulae are readily available to calculate the azimuth and altitude of a star. These formulae require the position of a star in right ascension and declination, the latitude of the site from which the observations are being made and the local sidereal time.

An algorithm to provide the local sidereal time was found in the book *Practical Astronomy with your Calculator*,[1] which proved to be very simple to use. Knowing the time and date, Greenwich sidereal time could be calculated, which gives the local sidereal time by subtracting the longitude of the site measured positively westwards. It is worth pointing out that this programme uses readily available formulae and data with the emphasis on the simplicity with which it can be used.

From the results obtained using this initial attempt, it became clear that the method could be extended to all eighty-eight constellations, and star maps could be created for almost anywhere in the world. I say 'almost' because the formulae used to do this break down for the north and south poles. To display all 88 constellations 806 stars are needed.

The positions used for this programme were those of the epoch 1950.0 and it was obvious that to make the map a little more accurate positions would be needed that were corrected for precession. It was at this point that the idea of seeing how the constellations altered over long periods of time occurred to me. This would require knowledge of the movements of the stars which eventually cause the distortion and break-up of the patterns we see today.

---

1. *Practical Astronomy with your Calculator*, 2nd Edition, Peter Duffett-Smith, Cambridge University Press, 1981.

STAR MAPS FOR THE FAR FUTURE

The equatorial system of co-ordinates, namely right ascension and declination, is not fixed in space and moves relative to a true inertial frame of reference owing to the effects of precession. This change in the co-ordinate system is caused by the combined gravitational effects of the Sun and the Moon on the non-spherical figure of the Earth. Hence the celestial pole describes a circle of radius 23 degrees 27 minutes around the pole of the ecliptic in a period of just under 26,000 years. The resulting slow westward movement of the vernal equinox brings about a change in the right ascension and declination for every object in the sky.

Simple formulae to correct for precession tend to break down round the poles, and it was for this reason that rigorous precession formulae devised by Newcomb at the end of the last century were used in the updated programme. It must be said that these formulae are only really applicable for a few centuries and were never intended for protracted use.

On a longer time-scale, the cumulative effects of the individual stellar motions become more noticeable. Some stars are approaching the Sun and will become brighter, while others are receding and will fade. Each star in the Galaxy has its own motion reflected by its proper motion and radial velocity. Nearby stars, such as Alpha Centauri and Arcturus, have large proper motions simply because of their proximity to the Sun. Some stars, however, belong to moving groups which started life at a certain time and place in the Galaxy. A good example of this are five members of Ursa Major, namely Beta, Gamma, Delta, Epsilon and Zeta, which maintain the well-known shape of this constellation for some considerable time. To obtain the position of a star in the future, one must have knowledge of its proper motions in right ascension and declination, its distance and also its radial velocity. It is necessary to assume that the star moves in a straight line through space. Using a change of co-ordinate system this calculation can be done quite easily. This method can be used for up to about a million years into the future as well as the past. Since the Galaxy rotates once in about 250 million years, the approximation that a star moves in a straight line in the period suggested is a reasonable one. The limit is set by the effects of galactic rotation but this complication does not significantly change the star's position during the 2 million-year period in question. It is therefore ignored in this programme.

The source of the data used for this version of the programme was the *Sky Catalogue 2000.0*[2] which contains the necessary data for the epoch 2000.0. Where gaps existed in the information required, e.g., radial velocities and distances, reference was made to more specialized catalogues. Combined magnitudes have been quoted for double stars that occur in the constellation patterns, and where variable stars occur these magnitudes have been quoted for maximum light. It has also been assumed that stars remain constant in output over the period considered and, in the case of the variable stars, their maxima have also been assumed constant.

Having incorporated these routines into the programme, precession could be used for up to ten thousand years into the future or the past and what could be called 'space motions' used for the rest of the time. Over substantial periods of time, precession is neglected in favour of the more noticeable effects of the intrinsic stellar motions. The programme could now give reasonably accurate positions for the stars at any time between AD 1,000,000 and 1,000,000 BC.

The map produced is a 180° by 90° rectangular plot which gives a reasonable representation of the night sky up to an altitude of about 60°. Above this limit constellations are liable to be stretched depending on their shape. More sophisticated map projections could, no doubt, be used but limitations in the plotting system currently available for this programme preclude their use.

The patterns used to delineate the constellations follow those used in *Sky Atlas 2000.0*.[3] There are many different ways to join up the stars of the night sky and those used by this atlas seem to depict the figures they are supposed to represent quite well. The magnitudes of the stars are simply plotted as dots; the larger the dot the brighter the star.

In spite of the inherent difficulties in the map projection being used here, every reasonable attempt has been made to make the star map appear as realistic as possible. One such refinement is the correction for atmospheric extinction. This effect is caused by the scattering of star light by the Earth's atmosphere which is most noticeable when a bright star is near the horizon. It appears considerably dimmer than it would

2. *Sky Catalogue 2000.0*, edited by Alan Hirchfeld and Roger Sinnott, Cambridge University Press & Sky Publishing Corporation, 1st edition, 1984.
3. *Sky Atlas 2000.0*, Wil Tirion, Cambridge University Press & Sky Publishing Corporation, 1981.

if it was high in the sky. This extinction can be represented quite well by trigonometrical functions of the altitude of the object being observed, although the relation begins to break down at altitudes of less than 10°. For this range of altitudes extinction tables are available for most observatories in the world and a representative average has been used for a sea level site in this programme.

The four star maps in this article have been produced to show the sky as it looks at the moment, and as it will appear in AD 50,000. The maps shown have been produced for Central England (latitude 52° north and longitude 0°) and Eastern Australia (latitude 35° south and longitude 150° east). Two of them use the precession routine, namely one for Christmas Day 1984 in England and the other for 1st July 1984 in Australia. The other two use the 'space motions' routine and are set up for 1st July AD 50,000 in Australia, and 1st January AD 50,000 in England. All the maps are chosen to appear as the sky does at 22:30 local time after having been corrected for atmospheric extinction. The direction in which they point is given at the bottom left-hand corner.

THE BRIGHTEST STARS – PRESENT AND FUTURE

Having outlined the way in which the star maps are created, it is time to go back to the questions asked at the beginning of the article. I thought it might be of interest to find out what happens to the first magnitude stars of today. Some of these stars are a substantial distance away from the Sun and little change can really be expected. It is among the nearby stars that changes will become most noticeable. The reference to a star being in a given constellation is based on the boundaries that exist at the present.

The brightest star in the night sky today is Sirius, but this has not always been the case. A million years ago it was only a third magnitude star in Lynx, and it will reach its maximum brightness of −1.7 in just over 60,000 years' time. In the years after that it begins to fade, and will eventually become a third-magnitude star once again in the southern constellation of Indus in a million years' time. Its nearest rival, Canopus, remains a first-magnitude star over the 2 million-year period in consideration. Alpha Centauri has a very different story. It became visible to the naked eye 900,000 years ago, and will fade beyond that limit again in another 900,000 years' time. It will be seen to its best advantage in nearly 30,000 years' time when it reaches −1.0. It also

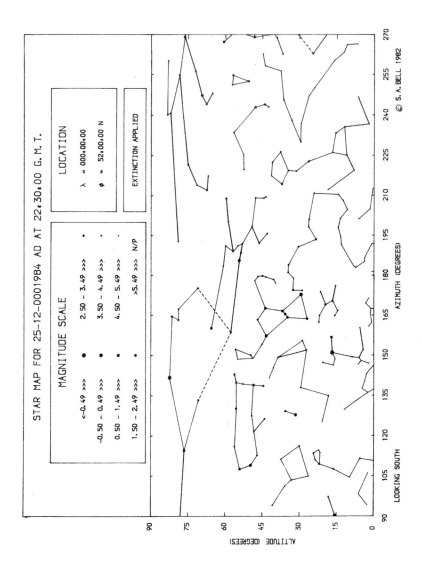

Figure 1. Map created for 25/12/1984 at 22:30 GMT looking south. Location 52° north on the Greenwich Meridian. Extinction applied and patterns drawn.

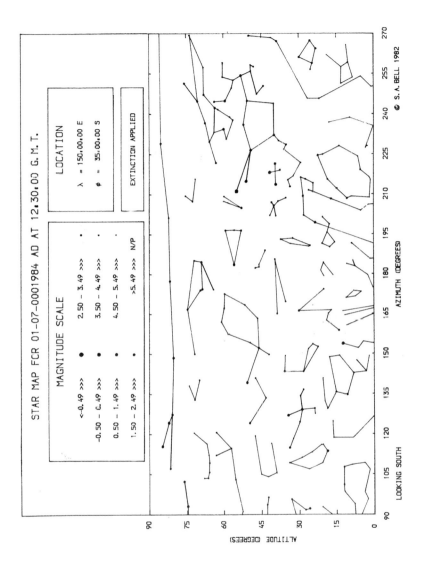

Figure 2. Map created for 01/07/1984 at 22:30 LMT looking south. Location 35° south at longitude 150° east. Extinction applied and patterns drawn.

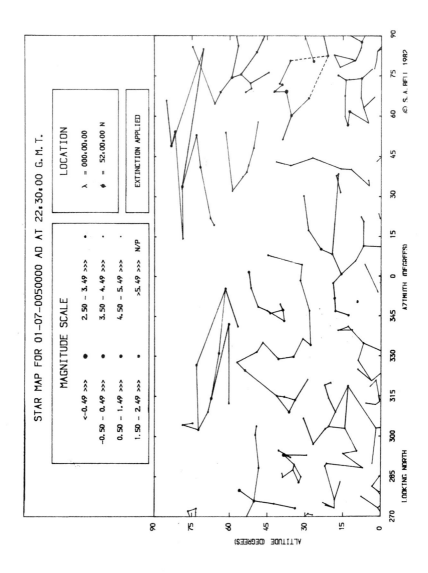

Figure 3. Map created for 01/07/50,000 at 22:30 GMT looking north. Location 52° north on the Greenwich Meridian. Extinction applied and patterns drawn.

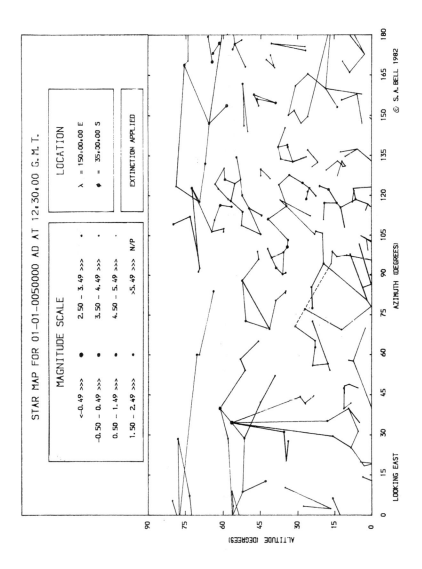

Figure 4. Map created for 01/01/50,000 at 22:30 LMT looking east. Location 35° south at longitude 150° east. Extinction applied and patterns drawn.

moves through a large portion of the sky during that time, starting in the southern constellation of Telescopium and ending up in Auriga.

Vega, one of the 'summer triangle' stars for northern observers, is gradually brightening and reaches its best in 300,000 years' time while in the constellation of Cassiopeia. On the other hand, Capella is now past its best, and will become a very ordinary star in Orion in a million years' time. Both first-magnitude members of Orion fade very slightly, but still retain their first-magnitude status.

Procyon in Canis Minor reaches its best in a mere 10,000 years' time, but even then does not really brighten a great deal. Like Capella, it fades to around the third magnitude in a million years' time. Achernar, the only really significant star in Eridanus, fades to nearly second magnitude by the end of the period examined here, making this constellation a rather dull, meandering line of stars.

Some stars escape the ravages of time to a large extent and remain more or less where they are today. Both Acrux and Mimosa in Crux Australis and Beta Centauri brighten over the 2-million-year period by a small amount, and this trend is also shared by Deneb and Antares. Spica, on the other hand, shows no real change in brightness.

Altair reaches −0.6 in another 150,000 years' time, when it will lie in Andromeda. As for Regulus and Pollux, they are past their best and are heading towards life as second- and third-magnitude stars in Monoceros and Cetus respectively. Fomalhaut in Pisces Austrinus was at its best nearly a quarter of a million years ago, when it reached a magnitude of 0.85 in Capricornus.

The last of the recognized first-magnitude stars of the present is Aldebaran, which was probably the brightest star in the skies above our Stone Age ancestors when it was in the constellation of Cepheus.

Lists can be found of the brightest stars in the night sky in many books. The list for AD 50,000 contains the same stars that are found in the compilations for the present. It should be pointed out that some stars in the middle and at the end of that list do change position although the brightest three retain their current positions. If these listings are compiled for both 1,000,000 BC and AD 1,000,000 then new entries can be expected. Tables I and II show this quite clearly. A rough idea of where the star is in the sky is given by the name in the constellation column.

A point that must be made here when dealing with these compilations for great distances into the past or the future is that there are just over 8,000 naked-eye stars in the night sky at present. The sample used for this programme is 10 per cent of that number, so there is a possibility that a star, or indeed stars, in the remaining 90 per cent may become brighter than those listed in the tables. It is less likely that a star below naked-eye visibility will make a long-lasting impact on the night sky since intrinsically bright nearby stars are something of a rarity.

The stars which appear in both the tables are most likely to be the more distant bright stars. Good examples of this are Canopus, Rigel, and Betelgeuse, which are all very slowly receding from us.

Table 1. The Twenty Brightest Stars for AD 1,000,000

	Star name	Magnitude	Constellation
1	Gamma Draconis	−2.18	Ophiuchus
2	Delta Scuti	−1.32	Scutum
3	Beta Aurigae	−0.68	Pisces
4	Gamma Scuti	−0.67	Telescopium
5	Alpha Carinae	−0.56	Puppis
6	Delta Sagittarii	0.08	Grus
7	Beta Orionis	0.29	Orion
8	Beta Centauri	0.45	Musca
9	Gamma Geminorum	0.69	Virgo
10=	1 Centauri	0.79	Centaurus
10=	Alpha Crucis	0.79	Carina
12	Alpha Scorpii	0.90	Scorpius
13	Alpha Ceti	0.94	Fornax/Phoenix
14	Alpha Orionis	0.97	Orion
15	Alpha Virginis	1.08	Corvus
16.	Beta Ursae Majoris	1.09	Hercules
17.	Epsilon Ursae Majoris	1.11	Hercules
18	Beta Librae	1.15	Hydra
19	Epsilon Herculis	1.21	Leo/Leo Minor
20	Alpha Cygni	1.23	Cygnus

Table 2. The Twenty Brightest Stars for 1,000,000 BC

	Star name	Magnitude	Constellation
1	Kappa Orionis	−4.27	Draco/Ursa Minor
2	Zeta Sagittarii	−2.71	Aquarius
3	Alpha Columbae	−1.51	Auriga
4	Alpha Carinae	−0.83	Dorado/Pictor
5	Alpha Eridani	−0.69	Aquarius/Capricornus
6	Gamma Velorum	−0.25	Vela
7	Beta Orionis	−0.06	Orion
8	Alpha Orionis	−0.03	Orion
9	Zeta Leporis	0.08	Hydra
10	Beta Ursae Majoris	0.29	Ursa Major
11	Zeta Volantis	0.45	Sculptor
12	Beta Centauri	0.80	Circinus/Lupus
13	Alpha Tucanae	0.95	Perseus
14	Alpha Virginis	1.02	Virgo
15	Alpha Crucis	1.03	Centaurus
16=	Epsilon Canis Majoris	1.04	Canis Majoris
16=	Alpha Scorpii	1.04	Ophiuchus/Scorpius
18	Theta Eridani	1.06	Columba/Puppis
19	Theta Aurigae	1.07	Vulpecula
20	Gamma Orionis	1.24	Orion

THE CHANGING CONSTELLATIONS

Finally, let us answer the question of how the constellation patterns change over the next 50,000 years. The types of change can be split into three categories. The first category covers those patterns which are only minimally affected by time, the second covers those which have one or two stars which distort the pattern or change noticeably in magnitude, and the third category deals with a less frequent occurrence where a constellation is so distorted that it becomes unrecognizable. Obviously the third category becomes the most popular over really long periods of time, but over the quoted period the second category is likely to be the most popular. Constellations that have not been mentioned here fit the first category.

Moving broadly alphabetically through the constellations, the first to suffer the effects of time is Aquila. Due to its motion northwards, Altair loses its two fainter companions and becomes half a magnitude brighter than it is at present. On the other hand, Beta moves southwards which gives rise to a 20° separation between the two leading stars.

Auriga, the Charioteer, in general moves bodily south in the sky by a small amount, and the kite-like shape gets a little crushed by the more rapid movement of its northerly members, notably Capella. The Haedi, or 'The Kids', slowly start to split up, mainly caused by the movement of Epsilon. On the other hand, Boötes undergoes a north-south stretch caused mainly by Arcturus moving south. At this time Arcturus will be less than 5° away from Spica, which will provide a striking contrast between these two first-magnitude stars because of the orange colour of Arcturus and the blue-white of Spica.

The faint but discernible pattern of Capricornus, which resembles an inverted triangle, will be disrupted by the movement of Delta, its brightest star. It moves south by about 12°. Like Capricornus, Cepheus is not a very prominent constellation at the moment, but will improve over the next 50,000 years. Eta brightens by nearly a magnitude, while Gamma brightens by half a magnitude.

Alpha Centauri has already shown itself to be a rapidly moving object, and at this time it resides in the head of Hydra at magnitude −0.6. By AD 30,000 this star will become visible in Britain. Cetus, the Whale, now resembles a kite and its tail, since the body becomes a line of stars while the head remains more or less intact. Northern observers, having gained Alpha Centauri, have to lose a first-magnitude star in return. Sirius will no longer be visible in Scotland, and barely struggles above the horizon in southern England. This is unfortunate, since it will then nearly be at its brightest.

The smaller companion to Canis Major, Canis Minor, becomes unrecognizable since the two stars are more than 20° apart. Corona Australis, the Southern Crown, loses its shape and resembles a figure eight more than anything else. The most famous small constellation in the sky, Crux Australis, no longer resembles a cross or a kite depending on your interpretation. It should be renamed Corona Australis, because it resembles a bright curved line of stars.

The small but conspicuous constellation of Delphinus takes the shape of a triangle caused mainly by the movement of Delta. Hercules, however, becomes an unrecognizable jumble of stars. Whatever shape there is today completely disappears. It is best described as an area of moderately bright stars. Lepus the Hare looks very much as though Orion has put his foot on it and crushed it. Moving northwards is Lyra, led by its brightest star, Vega, which is straying away from the faint but discernible rectangle of stars nearby. Returning to the Southern Hemisphere, Pavo, one of the most conspicuous of the southern birds, loses Delta to the nearby constellation of Tucana, making Pavo appear more elongated.

Both the head and the tail of Serpens get distorted, and the head loses one of its brighter members to Libra. Moving southwards, Triangulum Australe becomes a poor geometrical description, since it is better described as a square.

The last constellation in this round-up is Ursa Major, one of the most useful sign-posts in the night sky to anyone trying to find his way around the Northern-Hemisphere constellations. Unfortunately, it will no longer be able to perform this function, partly because other constellations have moved and partly because Alpha and Eta are slowly moving away from the main group. I say 'group' because the other five stars making up the familiar shape are moving through space as a group.

That concludes this glimpse into the skies of AD 50,000. No doubt the skies of the far future will look very different to this and at some time a new identification system will have to be brought in. Maybe someone will create new constellations, although I very much doubt they will have the colourful imagination of our ancestors who first created and named the patterns we have today.

EPILOGUE
(written in June 2011)

When I wrote this article, I was in my final year as an undergraduate at St Andrews University in Fife, Scotland. I did not know it at the time, but my future career was to be in positional astronomy with HM Nautical Almanac Office (HMNAO), joint producers of *The Astronomical Almanac* and *The Nautical Almanac* with the US Naval

Observatory and several other specialist publications. After doing a doctoral thesis in St Andrews on close binary stars and postdoctoral studies on binaries in external galaxies and late-type contact binaries, I joined HMNAO as a Scientific Editor in 1993, based at the Royal Greenwich Observatory (RGO) in Cambridge. In the ensuing eighteen years, I have moved with HMNAO to the Rutherford Appleton Laboratory in Oxfordshire during late 1998 when the RGO closed and to the UK Hydrographic Office in Somerset during 2006, and am now Head of the Office.

The material in this article started my interest in predicting phenomena and computing in general. During my thesis work and ensuing research I wrote observation planning software and data reduction and analysis software, which prepared me well for the job I do now. My work with HMNAO made my interest in eclipses 'official' – I wrote a booklet called *The RGO Guide to the 1999 Total Eclipse of the Sun* and created a web site called *Eclipses Online* (http://astro.ukho.gov.uk/eclipse/). I regularly try and sight the new crescent moon and run a web site called Crescent Moon Watch (http://astro.ukho.gov.uk/moonwatch/), which gathers observations of this phenomenon in an attempt to improve HMNAO's predictions of the sighting of the new crescent moon which, in turn, is used by the Muslim community in determining the start of the month in the Islamic calendar. I still try to observe noctilucent clouds from the somewhat unforgiving skies of Somerset and the aurora borealis when further north.

The article reprinted here stands pretty much as it is, including the references as those were the ones I used at the time. However, improvements in star catalogues, particularly proper motions, radial velocities and parallaxes over the last twenty-five years have had a significant impact on astrometry and positional astronomy. The predictions of magnitude changes for various objects and their positions in the sky of the far future may now be somewhat different to those given in the article. Over the time I have been part of HMNAO, we have also seen improvements in precession and nutation theory and the implementation of sidereal time. In the next few years we may well see civil time separated from the rotation of the Earth with the loss of the leap second which may have a significant impact on the work of almanac producers.

Computer hardware has also moved forward. The graphics used in this article came from a system called GHOST80 which was on a Vax

11/780 in St Andrews. The diagrams look rather crude now as the graphics instruction set and plotting device were rather limited. Unfortunately, the program that generated these diagrams is on an archive tape that is now unreadable as back-up technology has moved rapidly on. I dare say that some of the commercial planetarium-style software now available could improve upon this article.

It only remains for me to say that my interest in astronomy came from my father who was an RAF navigator during the Second World War and helped me learn my way around the night sky, and from the infectious enthusiasm of Patrick Moore in his books and TV programmes. It is an honour to have this article included in the fiftieth anniversary edition of *The Yearbook of Astronomy*.

Robotic Space Exploration: 1962–2012

DAVID M. HARLAND

Introduction

The first *Yearbook of Astronomy* appeared in 1962, the year of the very first successful planetary encounter by a spacecraft. Since that time, the discoveries of robotic probes have ensured that textbooks describing our planetary neighbours have had to be completely rewritten. This article reviews the many highlights of the past half-century of robotic space exploration in a chronological manner.

1962

At the dawn of the space age we knew very little of the planet Venus apart from the fact that it is about the same size as Earth, significantly closer to the Sun, and perpetually enshrouded in cloud. There was speculation that a large proportion of the surface was an ocean and that the land was swampy. When radio telescopes acting as radiometers measured the 'temperature' of the planet in the microwave part of the electromagnetic spectrum, they found it to be rather hot, but it was not clear whether this represented the surface or the ionosphere. When NASA's Mariner 2 flew by the planet at a range of 35,000 km on 14 December, this marked the first successful planetary encounter. Its microwave radiometer found a high temperature that did not vary from the night side across the terminator into daylight. If this represented the surface, then it was of the order of 425 °C, which argued against there being life on the planet.

1963

On 2 April the Soviet Union launched Luna 4 in an attempt to land on the Moon, but a trajectory error caused it to make a flyby at a range of 8,500 km.

1964

The smallest features on the Moon that could be resolved by a telescope of that time were several hundred metres across, and the big question for NASA, charged by John F. Kennedy in 1961 with the task of landing a man on the Moon within the decade, was the nature of the surface at finer scales. On 31 July 1964, Ranger 7 provided the first direct evidence by transmitting television pictures of the Sea of Clouds as it plummeted to destruction. It showed there to be craters, at every scale right down to the resolution of the final picture, of less than a metre. The presence of large boulders implied that the surface would be able to support the weight of a spacecraft.

1965

After observing Mars at its 'great opposition' of 1877, G.V. Schiaparelli in Italy announced the existence of intriguing linearities which he named 'canali'. Their reality and nature was disputed. In 1894 Percival Lowell built an observatory in Arizona equipped with a much larger telescope to research the issue, and went on to publish a number of books advocating a race of intelligent beings on Mars who were making canals to transport water from the polar ice caps to the arid equator. By the start of the space age Lowell's fanciful ideas had been dismissed, but it was thought likely that some form of vegetation, perhaps akin to

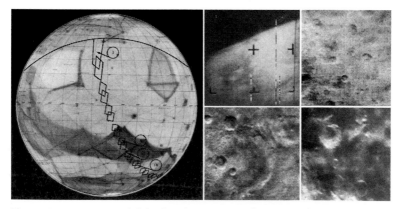

Figure 1. Mariner 4 provided the first close-up views of Mars, revealing it to be heavily cratered. The large crater in frame 11 (lower centre) was named 'Mariner' in celebration of the spacecraft. (Image courtesy of NASA/JPL.)

lichen, was responsible for the seasonal albedo variations. So expectations were rife when Mariner 4 passed by the planet at a range of 9,600 km on 15 July 1965. It produced two shocking results. First, a series of pictures running in a narrow strip concentrated in the southern hemisphere showed it to be heavily cratered, somewhat like the Moon. Second, the manner in which the radio signal was attenuated as the vehicle passed behind the planet indicated that the ground-level pressure of the atmosphere was a mere 4 to 6 millibars, and since infrared measurements by instruments on stratospheric balloons in 1963 had shown this to be the pressure of carbon dioxide at the surface, it was evident that the atmosphere was not only extremely rarefied but also pure carbon dioxide – the nitrogen which had been expected to boost the pressure was absent. The prospects for life were dashed.

1966

On 3 February 1966 an egg-shaped capsule named Luna 9 was landed on the Moon in the western region of the Ocean of Storms. A photometer equipped with a nodding and rotating mirror sent a panorama of the site. To the frustration of the Soviet owners, this transmission was intercepted by Jodrell Bank and the picture was published in the *Daily Express* ahead of the official release. The probe was in a crater whose shallow interior was littered with angular rocks. The Americans were not far behind, and on 2 June landed Surveyor 1 in the Flamsteed Ring, a large crater almost submerged by the Ocean of Storms, somewhat to the east of the Soviet lander. It had a more sophisticated camera and showed a fairly level plain pocked by craters.

1967

The study of Venus greatly advanced on 18 October 1967 when Venera 4 released a capsule that provided the first in-situ report on conditions in the atmosphere by measuring the ambient temperature and pressure as it slowly descended on a parachute over the night hemisphere. The Soviets announced that it had survived to the surface, where the temperature was about 280 °C and the pressure was around 18 bars. Surprisingly, the atmosphere was at least 95 per cent carbon dioxide; nitrogen had been expected to be predominant, as on Earth. The next day, Mariner 5 made a flyby at a range of 4,100 km and its data implied that the probe had *not* survived to the surface. After extensive analysis, the Soviet scientists conceded that Venera 4 had either drained its battery or succumbed to the conditions at an altitude of 27 km. Extrapolating the remote-sensing data down to the ground gave a surface temperature of around 525 °C and a pressure of 75 to 100 bars. Venus had evidently suffered a runaway 'greenhouse effect'.

Figure 2. The Orientale multiple-ring impact basin on the Moon, as revealed by Lunar Orbiter 4. (Image courtesy of NASA/JPL/USGS.)

1968

The year 1968 was dominated by the Moon. On 10 January, Surveyor 7, the last in the series, landed in the ejecta blanket immediately outside the rim of the prominent crater Tycho in the southern highlands. And then on Christmas Day the crew of *Apollo 8* flew ten orbits around the Moon.

1969

This year was momentous for *Apollo 11*, achieving Kennedy's challenge by landing on the Moon on 20 July, but there were also robotic successes at Venus for the Soviets and at Mars for the Americans.

On 16 May the capsule released by Venera 5 arrived at Venus. It was known that the probe would not survive to the surface, but it was given a smaller parachute to enable it to descend faster than its predecessor and thereby provide data from a lower altitude before its battery expired. The transmission ceased at an altitude of 18 km, when the pressure exceeded 27 bars and the ambient temperature was 320°C. The next day Venera 6 arrived and confirmed these results.

Mariner 6 flew by Mars at a range of 3,430 km on 31 July and Mariner 7 did likewise on 5 August, confirming the findings of their predecessor. It was concluded that Mars was a geologically inert world still scarred by impacts suffered early in its history.

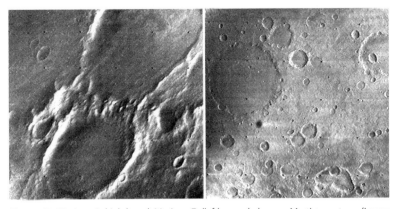

Figure 3. Mariner 6 (right) and Mariner 7 (left) revealed more Martian craters. (Image courtesy of NASA/JPL.)

1970

To augment its manned programme, the Soviet Union developed a robotic spacecraft capable of landing on the Moon, and it was decided to use this to obtain a lunar sample. It had been hoped that Luna 15 would beat *Apollo 11* home, but the automated spacecraft had crashed trying to land shortly after Neil Armstrong and Buzz Aldrin finished their moonwalk. But on 20 September, Luna 16 successfully landed in darkness in the Sea of Fertility. It lowered an arm and drilled a 101-g sample, which it stowed in the return capsule. The ascent vehicle lifted off on a trajectory for Earth. The capsule landed in Kazakhstan on 24 September. This was an impressive technical achievement, but it was eclipsed by the fact that two *Apollo* crews had already returned with a total of 55 kg of lunar material.

The Soviets scored another triumph when Luna 17 landed in the Sea of Rains on 17 November and released the Lunokhod 1 rover. Over the next ten months, hibernating during the lunar night and recharging its batteries at sunrise, it drove a total of 10,500 metres. It transmitted to Earth more than 20,000 individual pictures and 206 panoramas, and performed over 500 penetrometer tests and 25 X-ray analyses of the surface. While it was active, *Apollo 15* visited the Apennine mountains on the eastern shore of the Sea of Rains, using its own rover to undertake geological field trips that ranged up to 6 km from the lander and covered a total of 28 km over a three-day period, during which they collected a variety of samples. This was an interesting contrast between the robotic and human presence in exploration.

The year ended with Venera 7's arrival at Venus on 15 December. The probe had been made to withstand 180 bars and a temperature of 540 °C for 90 minutes. The mass penalty was reduced scientific instrumentation. However, about nineteen minutes into the descent the parachute tore, and as the rip extended over the next few minutes the descent rate increased and the capsule oscillated wildly. Several minutes before reaching the surface, the parachute failed and the probe struck the ground at 16.5 m/s. At this point the transmission appeared to end. Nevertheless, the scientists were delighted to have atmospheric data all the way down. Shortly thereafter, it was realized that the strength of the signal had declined to 3 per cent at impact and the transmission had continued for another 23 minutes. Evidently the probe had come to rest with its antenna tilted well away from Earth. The team were elated to have achieved the first successful landing on another planet. The temperature on the surface was around 465 °C, the pressure was 92 bars, and the wind speeds were less than 2.5 m/s.

1971

The focus for 1971 was Mars. Both spacefaring nations wanted to be the first to insert a spacecraft into orbit around the planet. On 9 May NASA attempted to launch Mariner 8, but the upper stage of the launch vehicle failed and fell into the Atlantic. The following day, a Soviet spacecraft was stranded in parking orbit when the final stage of its launcher failed to make the escape burn. The Soviets successfully dispatched Mars 2 on 19 May, and Mars 3 on 28 May. NASA successfully launched Mariner 9 on 30 May. The Soviets were being more adventurous than the Americans, with spacecraft which would release landers to dive into the atmosphere shortly before the spacecraft entered orbit round the planet. Although it was the last to set off, Mariner 9 was the first to arrive, and achieved orbit on 14 November. Next to arrive was Mars 2. The spacecraft achieved orbit on 27 November, albeit not the one planned. Unfortunately, an inaccurate trajectory caused its lander to enter the atmosphere at a steeper angle than intended, causing it to smash into the surface before its parachute could deploy. Mars 3 entered orbit on 2 December, although again it was not as intended. Its lander completed the entry, descent and landing sequence, and the lander began to transmit a panoramic image but the transmission ceased after a few seconds and no meaningful data was received. No one really knows why it fell silent.

Figure 4. In 1971 *Apollo* 15 explored the rim of Hadley Rille of the Moon. (Image courtesy of NASA.)

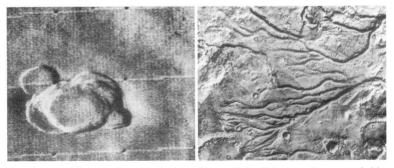

Figure 5. A caldera complex atop Ascraeus Mons and dendritic channels, as revealed by Mariner 9. (Image courtesy of NASA/JPL.)

The orbiters were hindered by the fact that a dust storm was obscuring everything apart from the summits of the tallest mountains. By the time the dust began to settle early in 1972, the two Soviet orbiters were suffering technical difficulties. But Mariner 9 was in excellent health and went on to map the planet. Between them the three flyby missions had imaged a mere 1 per cent of the surface, and it was realized that the impression that they had portrayed of a geologically inert world was misleading. Not only were there mountains, but these also had summit calderas and were actually much larger volcanoes than occur on Earth. The majority of the southern hemisphere, which the flyby missions had observed, was indeed cratered, but amongst the craters were networks of valleys which looked like they had been cut by water drainage. And the presence of water on the surface at some time in the planet's history suggested that the atmosphere had once been thicker and the climate warmer and wetter than it is today. Possibly conditions had been conducive to the development of life, and perhaps that still survived.

1972

While Mariner 9 was mapping Mars, the Soviets dispatched Venera 8. With the previous mission having provided firm knowledge of conditions at the surface, they had lightened the structure of the probe to withstand 105 bars and used the mass this released to reinstate some instruments. It survived its descent to the surface on 22 July 1972, and continued to send data for another 63 minutes. All the previous probes had been targeted at the night-time hemisphere to ensure direct-to-Earth communications, but this one was placed several hundred kilometres over the dawn terminator at a longitude from which it would still be able to transmit to Earth, and it was equipped with a photometer to measure the illumination to assist in the design of imagers for future landers. It identified three main optical regions in the atmosphere – two cloud layers with a thick upper layer of fog from 65 to 49 km, below which was a lighter haze. Below 32 km the illumination was constant, implying that the lower atmosphere was fairly clear. A gas analyser revealed water vapour to be scarce. In 1969 it had been suggested on the basis of measurements of the index of refractivity that the condensates in the atmosphere were acid-laden droplets of water. Venera 8 detected sulphuric acid in the clouds. The fact that such aerosols would reflect sunlight so efficiently explained why the planet had such a high albedo.

1973

With our knowledge of Venus and Mars increasing, NASA began to look farther afield. On 5 December 1973, Pioneer 10 flew by Jupiter at a range of 130,000 km, providing unprecedented views of the banded structure of the planet's atmosphere and making the first in-situ measurements of its magnetosphere. The trajectory was designed to make the spacecraft pass behind the innermost of the main satellites, Io, as viewed from Earth, and the way in which the radio occultation attenuated the signal showed Io to possess an ionosphere.

The Soviets tried again for Mars in 1973. This time they were unable to send combined orbiter/landers because the launch window was not as favourable as in 1971, so they assigned the landers to flyby spacecraft and equipped the orbiters with heavier scientific payloads. A pair of orbiters and two flyby spacecraft were dispatched, but when they arrived in 1974, the results were dismal, largely owing to the use of faulty transistors in their circuitry. Mars 4 failed to perform the insertion burn and sailed past the planet at a range of 2,000 km on 10 February. Two days later Mars 5 achieved the planned orbit, but in doing so developed a pressure leak in its instrument compartment

Figure 6. A close-up view of Jupiter produced by Pioneer 10, showing a moon and its shadow on the planet. (Image courtesy of NASA/Ames.)

which caused it to fall silent on 28 February. This was disastrous, as there were now no orbiters to act as radio relays for the landers. After Mars 6 suffered an early failure in its telemetry system, the commands to the spacecraft were transmitted 'in the blind'. Having made automated manoeuvres, the bus released its payload on 12 March. The lander was equipped to transmit data in real-time during the descent, rather than to store it and send it after landing. Entry and descent by parachute went well, but when the lander was released for impact with the surface, contact was lost. No one knows what went amiss. The Mars 7 bus released its payload on 9 March, but the lander malfunctioned and missed the planet by 1,300 km. In view of the paucity of results from this massive effort, the Soviets decided to give up on Mars for the foreseeable future and instead focus on Venus, where they were having success.

1974

NASA further expanded its investigation of the Solar System in 1974 with a mission to Mercury. In the first use of the 'gravity-assist' technique, a 5,770 km flyby of Venus on 5 February was used to deflect the trajectory of the vehicle in a manner designed to create a flyby of its objective at a range of 700 km on 29 March. It revealed Mercury to have large multi-ring basins and craters similar to those of the Moon, but with significant differences in the types of terrain. In fact, the spacecraft's orbit was such that with small propulsive manoeuvres it was able to revisit the planet on 21 September and 16 March 1975. During the Venus flyby, Mariner 10 obtained the first close-up imagery of the planet, and an ultraviolet filter confirmed telescopic hints of a zonal

Figure 7. Mariner 10 revealed Mercury to be heavily cratered. (Image courtesy of NASA/JPL.)

wind pattern caused by the fact that the upper atmosphere rotates every four days, in stark contrast to the planet itself – which travels around the Sun in less time than it takes to rotate once on its axis.

On 3 December, Pioneer 11 flew 43,000 km over the Jovian cloud tops for a gravity-assist on to a path which was inclined to the ecliptic in such a manner that the spacecraft would cross the inner Solar System, peaking 150 million km above the orbit of Mars in early 1976, and then encounter Saturn in 1979.

1975

The year 1975 saw Helios 1, the first deep-space mission developed as a joint venture between NASA and West Germany, pass within 0.30 AU of the Sun on 15 March to investigate the interplanetary medium within the orbit of Mercury.

Later in the year, the Soviets kicked off a new series of missions to Venus. The entry system was a sphere that contained a new heavy lander. The previous probes had established the atmosphere to be so thick that to reach the surface in a reasonable time this lander was designed to jettison its parachute high in the atmosphere and use a disk-like aerobrake to fall sufficiently slowly for contact with the surface to be survivable. It was also equipped with a larger number of instruments, including cameras. In the past, the spacecraft had simply followed its probe into the atmosphere and burned up, but the new carrier was to make a deflection manoeuvre and enter orbit. The increased bandwidth for the instruments was achieved by having the spacecraft relay the lander's transmission to Earth by its high gain antenna.

Venera 9 arrived on 22 October 1975, and Venera 10 three days later. Both landers provided atmospheric data during the descent and went on to operate on the surface for about an hour. The clouds were a light fog, with a much smaller drop size than normal for Earth. Distinct layers were detected at altitudes of 60 to 57 km, 57 to 52 km and 52 to 49 km. Light scattering data indicated a lesser loading of aerosols from the cloud base at 48 km down to about 25 km, below which the atmosphere was clear. The index of refractivity was consistent with droplets of sulphuric acid. The aerosols scattered light, but there was absorption in the blue and this, along with intense Rayleigh scattering, led to an increasing orange colour with depth. Each lander sent a black-and-white 180° panorama of its landing site, and data on the composition of

ВЕНЕРА−9 ОБРАБОТАННОЕ ИЗОБРАЖЕНИЕ

ВЕНЕРА−10 ОБРАБОТАННОЕ ИЗОБРАЖЕНИЕ

Figure 8. The first images from the surface of Venus, returned by the Venera 9 and 10 landers. (Image courtesy of NASA/Don Mitchell.)

the surface. Venera 9 was on level terrain of slabby, angular rocks that did not show much erosion. Venera 10 was on a smoother plain with larger and more eroded pancake rocks which were interspersed with weathered material. In both cases the surface seemed to have a basaltic composition.

1976

The highlight of 1976 was the arrival of the two Viking missions at Mars. Both spacecraft entered orbit around the planet and released their landers. The Viking 1 lander touched down on 20 July and sent back pictures that showed Chryse Planitia to be strewn with rocks and small dunes of sand. On 3 September, Viking 2 landed in Utopia Planitia on the opposite side of the planet. The task of the Vikings was to use a robotic arm to collect samples of soil to be tested for evidence of the presence of life. The consensus view was that the soil held oxidants that would destroy organic molecules, but the specific oxidants were not identified and the negative result remained contentious. The subsequent discovery of extremophile micro-organisms in terrestrial environments revealed life to be rather more tenacious than the designers of the Viking lander biology tests had presumed. As for the orbiters, they mapped the planet at a higher resolution than Mariner 9, in colour, and the results lent considerable weight to the hypothesis that the planet's climate had been warmer and wetter early in its history.

1976 also saw the Soviets wrap up their lunar programme with Luna 24. It landed in the Sea of Crises on 18 August, drilled a 1.6-metre-long core and returned this to Earth.

Figure 9. The first image from Viking 1 (top) and a subsequent panorama of the landing site on Chryse Planitia. (Image courtesy of NASA/JPL.)

1977

NASA dispatched two Voyager spacecraft in 1977. The first was to use a gravity-assist at Jupiter to reach Saturn. If that mission went well, then its partner would take advantage of a once-in-176-year alignment of the gas giants to attempt an audacious Grand Tour.

1978

In 1978 NASA resumed its study of Venus. This time it sent an orbiter and a bus that deployed four atmospheric probes. The orbiter arrived on 4 December. In addition to long-term monitoring of the atmosphere, it used a radar altimeter to map the topography of almost the entire planet with a surface resolution of 75 km – our knowledge of the planet was so scanty that even such a crude map was insightful. As a result of its exceedingly slow rotation, Venus does not have either polar flattening or an equatorial bulge. Remarkably, apart from a few elevated regions and some shallow depressions, most of the planet lies very close to the planetary mean.

The bus released its probes so that they would arrive on 9 December at various latitudes, two on the day side and two on the night side. One was larger than the others. The large probe deployed a parachute at an altitude of 68 km for a slow descent while sampling the clouds, then discarded it at 47 km to fall freely in the lower atmosphere. The smaller ones did not have parachutes. Because the probes could not sample their environment during the initial deceleration, the bus reported on conditions in the upper atmosphere prior to burning up. Although not designed to survive impact, one of the small probes transmitted from the surface for an hour until its battery expired.

The results were consistent with the data from the Veneras. There was a fine smog in the altitude range 90 to 70 km, the main cloud base was at 48 km, and it was clear down to a thin layer of haze at 31 km, below which it was clear again. The maximum opacity was at 50 km. Because the lower atmosphere was stagnant, the base of the troposphere (the convective part of the atmosphere, which on Earth is the lowest layer) was at an altitude of 48 km, at the base of the cloud deck. The large probe confirmed that at least 75 per cent of the cloud consisted of aerosols of sulphuric acid. The acid is a result of photochemical oxidation by solar ultraviolet at altitudes above 60 km. The dissociation of carbon dioxide and sulphur dioxide in the upper atmosphere produces atomic oxygen that oxidizes SO_2 to SO_3, which is then hydrated as sulphuric acid (H_2SO_4). As the droplets 'rain out' and fall, they are thermally dissociated by the 100 °C temperature at about 49 km. They yield SO_3 which, upon encountering carbon monoxide, regenerates sulphur dioxide and carbon dioxide. This cycle of precipitation operates in the upper atmosphere, where the temperature is moderate, the pressure is low, and such water as remains is stable in the liquid state.

When the orbiter fell into the atmosphere in 1992, its data had shown the overall circulation system to be a single Hadley 'cell' north and south of the equator. The air rises in the equatorial zone, flows at high altitude into the polar zone, then descends. It returns to the equator in the middle atmosphere. This 'weather system' is completely isolated from the lower part of the atmosphere. Such circulation derives from the extremely slow rotation. On the rapidly rotating Earth, the Coriolis effect disrupts this simple flow pattern by inducing swirling airflows and tropical, temperate and polar components.

As regards the predominance of carbon dioxide, it isn't as if Venus has an anomalously large amount; it has a similar amount to Earth except that here carbon dioxide is mostly locked into carbonate rock. If all the carbon dioxide in the rocks were released, then Earth's atmosphere would resemble that of Venus. Maybe that is an important clue as to how Venus came to be the way that it is.

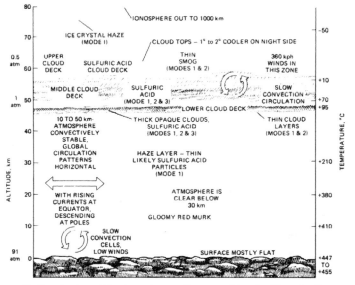

Figure 10. The structure of the atmosphere on Venus, as determined from the Pioneer entry probes. (Image courtesy of NASA/Ames.)

Figure 11. The circulation in the upper atmosphere of Venus as revealed by the Pioneer Venus Orbiter. (Image courtesy of NASA/Ames.)

The Soviets sent Venera 11 and 12. However, since the arrival velocity would be greater than in 1975 and the spacecraft could not accommodate the extra propellant that the longer orbit insertion burn would require, the carrier for the landers was downgraded to a flyby role. The two landers touched down successfully on 21 and 25 December, but neither was able to return any images because the lens caps failed to jettison – in fact, one lens cap had failed on each of the previous landers, which were supposed to have provided pairs of 180° panoramas for an all-round view.

1979

In 1979 the focus shifted to the outer Solar System. First, on 5 March, was the Voyager 1 flyby of Jupiter. It flew 270,000 km above the cloud tops on a trajectory designed to achieve a gravity-assist to send it on to Saturn. Inbound imaging documented the banded atmosphere, and in particular the Great Red Spot, in unprecedented detail. The flyby also revealed the four large satellites as individual worlds for the first time. The trajectory passed within 21,000 km of the south pole of Io. Telescopic studies had established there to be a torus of plasma occupying the moon's orbit. The spacecraft found a pair of 'flux tubes' aligned with the magnetic field of Jupiter, with an electron current of a million amperes linking the moon to the polar regions of the planet's ionosphere. Io itself appeared to have a mottled and multicoloured surface that was remarkably devoid of impact craters. On the way out of the system, the spacecraft made close passes by Ganymede, which had patches of smooth terrain separated by bright strips of ridges and grooves, and Callisto, which, being heavily cratered, resembled what the other moons had been expected to look like. Several days later, as a member of the navigation team measured the positions of stars relative to the limb of Io in order to calculate the 'sling shot' from the Jovian encounter, that moon was discovered to be volcanically active, with a number of vents sending plumes of gas and dust to high altitudes. This explained the mottled surface and the absence of impact craters – the moon was being resurfaced on an ongoing basis. It turned out that the heat required to drive this volcanism was generated by gravitational tides generated by the mass of Jupiter as the moon travels its slightly elliptical orbit, this eccentricity resulting from resonances between Io and the next two moons out from the planet, Europa and Ganymede.

On 9 July, Voyager 2 flew by Jupiter with its closest point of approach just outside the orbit of Europa. This moon had not been well placed for Voyager 1, being on the opposite side of its orbit, but this time it was a priority. The very high albedo was revealed to occur because the moon is entirely covered in ice. It was as smooth as a billiard ball and intensively cracked and stained. Only a few impact craters were suspected. If Europa suffered significant internal heating in the

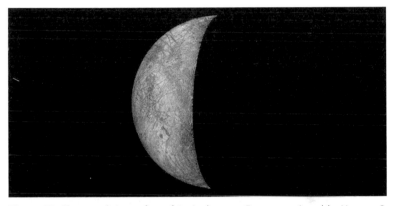

Figure 12. The smooth icy surface of Jupiter's moon Europa, as viewed by Voyager 2. (Image courtesy of NASA/JPL.)

same manner as Io, then there could well be an ocean beneath a thin crust of ice, and the insight gained from terrestrial extremophiles at volcanic vents on the ocean floor raised the prospect that there might be life in such an ocean.

After pictures taken by Voyager 1 hinted that Jupiter might possess a system of rings, Voyager 2 looked back sunward while passing through the planet's shadow and confirmed the presence of several dark rings just above the planet. Invisible to telescopic observers, they were apparent to the spacecraft by the efficient manner in which the dust forward-scattered sunlight. Like its predecessor, Voyager 2 left the system on a trajectory for Saturn.

At that time, Pioneer 11 was approaching Saturn. Its closest approach was on 1 September. The southern face of the rings was illuminated, and for terrestrial observers the plane was almost edge-on. But the spacecraft was approaching from several degrees north of the ecliptic, giving us our first view of the unilluminated face of the rings. The

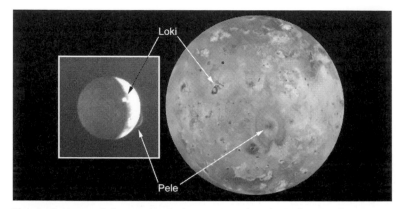

Figure 13. Voyager 1 discovered Jupiter's moon, Io, to be volcanically active. (Image courtesy of NASA/JPL.)

Figure 14. A view of Jupiter by Voyager 1 showing the planet as a crescent. (Image courtesy of NASA/JPL.)

broad classical rings were revealed to possess considerable structure. The trajectory had been designed to cross the ring plane just outside the edge of the telescopically visible system. The spacecraft found a very narrow ringlet in this region, and during its passage through the plane evidently flew very close to an unknown small moon. The large satellite Titan was known to possess an atmosphere that contained methane, but it was not known whether this was the majority constituent. As the spacecraft withdrew from the planet, it imaged Titan from 363,000 km and confirmed the hypothesis that its atmosphere was opaque.

1980

The highlight of 1980 was the flight of Voyager 1 through the Saturnian system. The inbound trajectory was designed to yield a 4,000 km flyby of Titan with a radio occultation to sound its atmosphere. The occultation data showed the atmosphere to be very deep and stratified, with layers of aerosols far above the optically opaque orange smog. The pressure at the surface was inferred to be 1.5 bars, which for such a small body equated to an order of magnitude more gas than was present in Earth's envelope. Infrared data revealed the majority constituent to be molecular nitrogen, identified a variety of hydrocarbons produced by photolysis of methane, and indicated a surface temperature of 92 K; a value that applied globally to within several degrees. It had been hoped to catch a glimpse of the surface but the smog was ubiquitous.

The ring system proved to comprise so many 'ringlets' that scientists gave up counting after reaching one thousand in just one small part of the structure! As suspected, the thread-like outer ring discovered by Pioneer 11 was constrained by a pair of 'shepherding' moonlets. One surprising finding was the presence of radial 'spokes' which rotated around the ring system at a rate comparable to the axial spin of the planet. Owing to the tilt of the equatorial plane relative to that in which

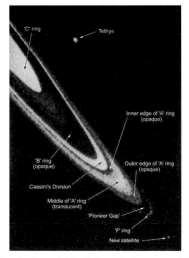

Figure 15. A view of the shadowed side of Saturn's rings, revealed by Pioneer 11. (Image courtesy of NASA/Ames.)

the spacecraft was flying, the encounter on 12 November, when Voyager 1 was 124,000 km above the planet's cloud tops, deflected its trajectory some 35° north of the ecliptic. The exit was designed to make the spacecraft pass behind the rings as viewed from Earth, and this radio occultation served to probe the fine structure of the rings. While behind the rings, it imaged Enceladus. When the pictures were later transmitted, they revealed this icy moon to have undergone a remarkable degree of resurfacing.

1981

The flyby of Saturn by Voyager 2 at a range of 101,000 km on 26 August used the ring-plane crossing which had been demonstrated by Pioneer 11, this time for a gravity-assist to reach Uranus as the next step of the Grand Tour.

1982

Venera 13 reached Venus on 1 March, followed by Venera 14 four days later. The Soviets had upgraded the landers with colour cameras and had fixed the lens cap release mechanisms. Again, each spacecraft performed a flyby and relayed the transmission from its lander. In addition to taking the desired pair of colour panoramas, each lander had a penetrometer to measure the physical characteristics of the surface. By sheer bad luck, the penetrometer of Venera 14 rotated down on to one of the jettisoned lens caps!

1983

When Venera 15 arrived at Venus on 10 October 1983, it went into orbit, as did its partner four days later. This time there were no landers. Instead the payload was a synthetic-aperture radar. Each was in a highly elliptical orbit with a period of twenty-four hours, scanning a strip of terrain approximately 120 x 7,500 km at periapsis and then transmitting this data to Earth while near apoapsis. As the planet slowly rotated beneath them, the overlapping strips built up a continuous map. However, the fact that the periapsis was at about 62°N meant that each strip began at 80°N on the inbound side of the pole and continued over the pole down to 30°N on the retreating side, with the result that only the northern hemisphere was mapped. The resolution was about

2 km. The 300-metre Arecibo radio telescope could achieve a comparable resolution when operating in radar mode, but could map only the equatorial region, so the results were complementary. It is a pity that the Soviets did not insert one of their spacecraft into an orbit that had its periapsis in the south, to map the higher latitudes of both hemispheres. The results added enticing detail to the topographic map obtained by the Pioneer orbiter, including the presence of impact craters.

1984

There were no major events in 1984. At Venus the Soviet radar mappers concluded their missions and the Pioneer orbiter continued its long-term observations of the atmosphere. Meanwhile, the two Voyager spacecraft were cruising through the vast gulfs of the outer Solar System.

1985

To wrap up their investigation of Venus, the Soviets decided to send a pair of flyby spacecraft that would release entry systems which held not only landers but also helium aerostats that were to inflate and drift at an altitude of 54 km, where the pressure was 0.5 bar, right in the middle of the cloud layers. The landers did not carry cameras because the balloons were to start their journeys on the night hemisphere. VeGa 1 delivered its package on 11 June, and VeGa 2 followed four days later. In addition to monitoring the Doppler effect, international networks of antennae were linked together to use the very long baseline interferometry technique to track how each balloon drifted with the 'superrotating' wind flow. They drifted longitudinally for 30 hours prior to crossing the dawn terminator, 8°N in the case of Vega 1 and 7°S for Vega 2, and another 16 hours in daylight. When their batteries expired they had drifted some 10,000 km, or about one-third of the way around the planet. As they penetrated further around the day side, the balloons would have succumbed to solar heating and burst their envelopes.

1986

The year 1986 began with the next stage of the Grand Tour of the outer Solar System, with Voyager 2 making a flyby of Uranus on 24 January at a range of 71,000 km. Unlike Jupiter and Saturn, whose equators are oriented near the plane of the ecliptic, that of Uranus is tipped right over, with the result that instead of crossing the orbits of the moons one by one, the approach trajectory was perpendicular to the system, compressing the encounter into a few hours, with the southern hemispheres of the planet and its satellites illuminated. But everything was new and fascinating, particularly the small moon Miranda, whose surface was a mass of faults, grooves and scarps – one forming a 16 km tall cliff that disappeared over the terminator. The juxtaposition of terrain types prompted speculation that Miranda had broken apart and reformed. In 1977 astronomers had serendipitously discovered Uranus to possess several dark rings, and the spacecraft found even more rings together with some of the shepherding moonlets. As to the planet itself, despite rotating pole-on to the Sun, the temperatures at the cloud tops were the same for both the illuminated and dark poles. Intriguingly the axis of the planetary magnetic field proved to be tilted almost 60° from the spin axis and significantly displaced from the geometric centre.

1986 was also the year of Halley's comet. Its perihelion passage was on 9 February, and an international flotilla was sent to investigate it as it withdrew from the Sun – although NASA was notably absent because of budget cuts. The Soviets extended the missions of the spacecraft that had delivered landers and balloons to Venus in 1985, using the gravity-assist there to head for the comet. VeGa 1 flew by the nucleus of the comet on 6 March at a range of 8,890 km. The Suisei probe sent by

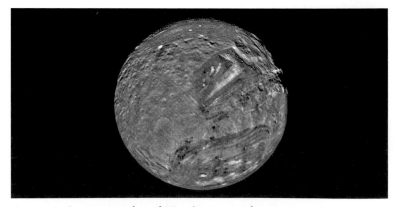

Figure 16. The strange surface of Miranda, a moon of Uranus, imaged by Voyager 2. (Image courtesy of NASA/JPL.)

Figure 17. The active nucleus of Halley's comet, as revealed by Giotto. (Image courtesy of ESA/H.U. Keller at the Max-Planck-Institut für Sonnensystemforschung, Katlenburg-Lindau, Germany.)

Japan made a 150,000 km flyby on 8 March to study the hydrogen cloud surrounding the comet by the manner in which it fluoresced in sunlight. The next day VeGa 2 flew by the nucleus at a range of 8,000 km. On 11 March Japan's Sakigake made its closest approach at a range of 7 million km, studying how the presence of the comet influenced the solar wind. Using targeting data provided by the Soviets, the European Space Agency's Giotto probe was able to manoeuvre to pass the nucleus at a range of just 600 km on 14 March.

The nucleus was peanut-shaped and measured 15 × 8 km. Only a small part of its surface was active, but the jets carried a lot of gas and dust. The prevailing 'dirty snowball' theory had suggested the nucleus would be bright, but it was as black as coal. Its temperature was around 350 K, which was too warm for water ice. Spectroscopic data indicated organic molecules, showing that the surface was a layer of non-volatile material that acted as an insulator. It was evident that this 'lag' built up on the surface as the nucleus shrank with successive perihelion passes.

1987–1988

The loss of *Challenger* in 1986 grounded the remaining space shuttles until 1989, postponing the launches of the Galileo, Magellan and Ulysses Deep Space missions.

1989

Flushed with the success of their VeGa missions to Halley's comet, the Soviets renewed their interest in Mars, and this time invited international collaboration. Not wishing to compete with the Viking landings, it was decided to make the larger of the two moons the primary target. A pair of spacecraft set off in the summer of 1988, but Fobos 1 was lost early in its interplanetary cruise when it was sent an incorrect command that made it shut down its attitude control system, and before anyone realized this it had lost its lock on the Sun and drained its battery. Fobos 2 entered orbit around Mars on 29 January 1989. By 18 February a series of manoeuvres had almost circularized the orbit at 6,270 km, a few hundred kilometres above its target. The orbit was trimmed again on 15 March and 21 March into a 5,692 x 6,276 km path that was nearly synchronous with Phobos. The plan was to produce a very low and slow pass over the surface of the moon on 9 April, during which it would release two landers. The smaller lander was to 'hop' in the weak gravity to sample a number of sites in the few hours available until its battery expired, reporting data to the parent spacecraft. The larger of the landers was to fire a harpoon to hold itself in place, and then report to Earth for a period of three months. But on 27 March the spacecraft, pursuing instructions, turned to take some more images of its target and failed to re-establish high gain contact. A weak oscillating signal implied that it was in a spin, draining its battery. One of its three redundant computers had failed early on, and another had begun to show problems. Evidently this second computer had malfunctioned and the two failed computers had 'outvoted' the functional one! To offset the terrible disappointment at Phobos was the fact that during its brief period in orbit Fobos 2 produced more data than all of its Soviet predecessors combined.

Voyager 2 completed its Grand Tour by passing within 30,000 km of Neptune on 25 August. With the planet being so far from the Sun, it had been expected that the atmosphere would be quiescent, but it hosted a number of storms, one of which was dubbed the Great Dark

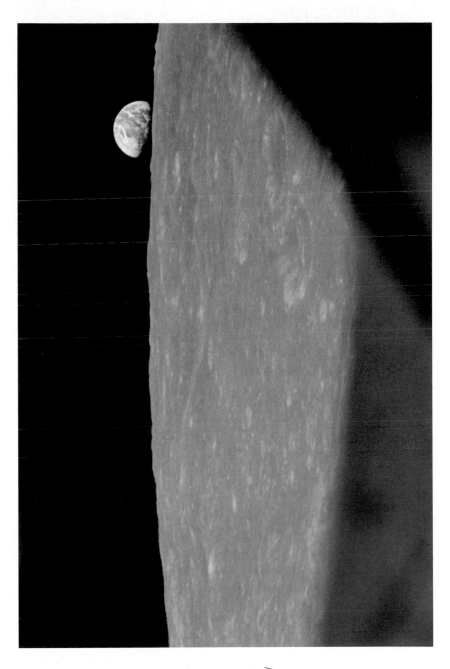

The Apollo 8 crew were the first to see Earth rising over the lunar horizon. (Image courtesy of NASA)

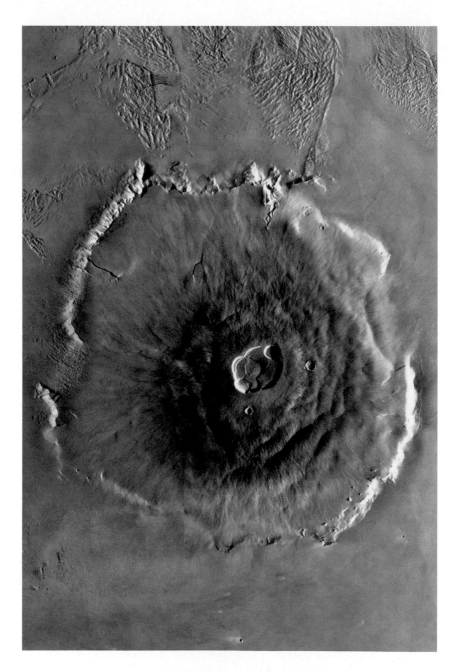

The Martian volcano Olympus Mons viewed by the Viking 1 Orbiter.
(Image courtesy of NASA/JPL)

A montage of Voyager 1 images of the Saturnian system. Dione in the forefront, Saturn rising behind, Tethys and Mimas fading in the distance to the right, Enceladus and Rhea off Saturn's rings to the left, and Titan in its distant orbit at the top.

(Image courtesy of NASA/JPL)

A Voyager 2 view of Triton, the largest of Neptune's moons.
(Image courtesy of NASA/JPL/USGS)

Jupiter viewed by Cassini during its passage through that system.
(Image courtesy of NASA/JPL/Space Science Institute)

The 'tiger stripes' at the south pole of Enceladus viewed by Cassini.
(Image courtesy of NASA/JPL/Space Science Institute)

A view of the rings of Saturn from Cassini while passing through the planet's shadow. (Image courtesy of NASA/JPL/Space Science Institute)

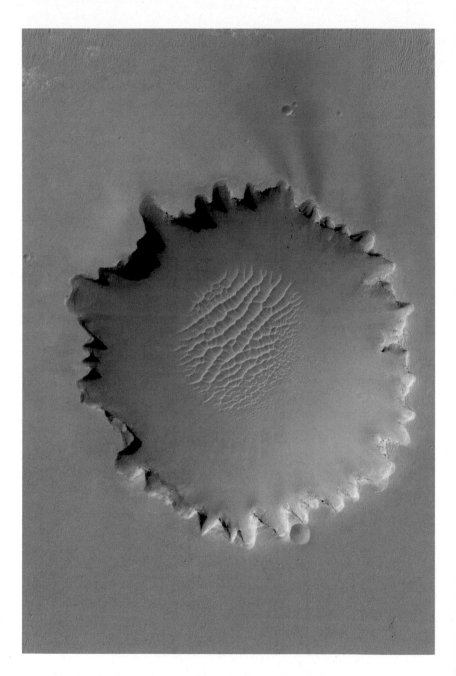

A view by Mars Reconnaissance Orbiter of Victoria crater with the Opportunity rover on its north-western rim.
(Image courtesy of NASA/JPL/University of Arizona/Cornell/Ohio State University)

Spot. Neptune also proved to possess dark rings, or at least arcuate segments. The trajectory was designed so that the planetary encounter would deflect the spacecraft towards the large moon Triton, whose orbit is inclined to the system and in a retrograde direction, suggesting that it has an interesting history. The 39,800 km flyby revealed Triton to have a subdued topography in which depressions have been filled with frozen methane ice. There were geysers of nitrogen rising from the surface, creating a tenuous atmosphere.

As the spacecraft looked back at Neptune, which had a crescent phase, team member Laurence Soderblom remarked in awe, 'Wow, what a way to leave the Solar System!' Like its partner, Voyager 2 left on a trajectory steeply inclined to the ecliptic, only in this case it was towards the south.

1990

During the decade of the 1980s, NASA had not launched any deep space missions. Only the Grand Tour by Voyager 2 conveyed the impression of activity and, ironically, when that mission was launched it was to have ended officially with the Saturn flyby. But in 1989 the shuttle *Atlantis* had deployed the Galileo spacecraft which was to pursue a circuitous route to Jupiter. On 10 February 1990, a flyby of Venus provided a gravity-assist to stretch the initial aphelion. In 1984, it had been realized that there were two narrow near-infrared 'windows' in which it should be practicable to see through the otherwise opaque clouds. This was so. Galileo was able to observe the turbulent middle atmosphere at an altitude of 50 km, and as the heat from the surface passed through these patchy clouds, it could discern their structure. A comparison of the ultraviolet and near-infrared imagery confirmed that the 'super-rotation' of the atmosphere decreased with depth. In the stagnant lower atmosphere there was a steep thermal gradient, declining by 8° per kilometre of altitude. This implied that from the top of the tallest peak to the lowest plain, a vertical range of 13 km, the surface varied by 100 °C. By processing the data to 'subtract' the atmosphere, it was even possible to gain an impression of the most elevated features.

On 14 February, Voyager 1 looked back over its shoulder and took the first image of the Solar System from outside, prompting the irrepressible Carl Sagan to describe Earth as a 'pale blue dot'.

Another mission that had been dispatched by *Atlantis* in 1989 was Magellan. Unlike its synthetic-aperture-radar-equipped Soviet predecessors, on 10 August the American spacecraft adopted an elliptical polar orbit of Venus with its 250 km periapsis near the equator and a period of just over 3 hours. It mapped 24 x 15,000 km 'noodles' that had a periapsis resolution of 100 metres. After the first planetary cycle, it had mapped 82 per cent of the surface, and by the end of the second it had increased its coverage to 98 per cent. The surface proved to be predominantly volcanic, with vast lava plains, large shield cones and fields of small domes. Channels over 6,000 km in length suggested effusive eruptions of extremely low viscosity lava. However, pancake-shaped domes implied the presence of a type of lava produced by extensive reprocessing of crustal rocks. Features typical of terrestrial plate tectonics were absent. The tectonics was dominated by a system of global rift zones and numerous broad, low domical structures which were named coronae, evidently produced by the upwelling and subsidence of magma from the mantle. Most intriguingly, the impact craters suggested that Venus suffered an almost global pulse of volcanic resurfacing about 750 million years ago, possibly because heat rising in the

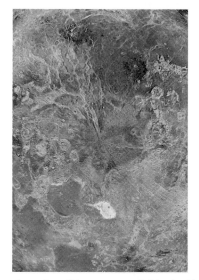

Figure 18. The north polar region of Venus, as revealed by the Magellan imaging radar. (Image courtesy of NASA/JPL/USGS.)

mantle was not being released on a continual basis by plate tectonics. It was remarkable how a planet so similar in size to Earth could have evolved so differently.

1991

On 29 October, pursuing its slow and circuitous route to Jupiter, Galileo became the first spacecraft to encounter an asteroid. It flew within 1,600 km of Gaspra, whose orbit implied it was a member of the Flora family. It was found to be 18 x 10 x 9 km with an angular shape. Its surface was smooth with some flat faces, and cratering suggested that the surface was no more than 300 million years old.

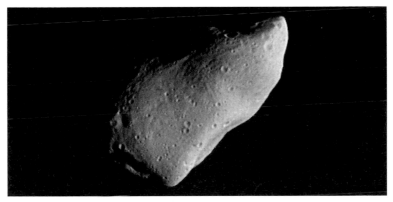

Figure 19. Gaspra, the first asteroid to be investigated close-up, as revealed by Galileo. (Image courtesy of NASA/JPL/USGS.)

1992

The shuttle *Discovery* dispatched the Ulysses spacecraft in 1990 on a mission to study the polar regions of the Sun. On 8 February 1992 a polar flyby of Jupiter produced a gravity-assist that sent Ulysses sunward on a trajectory inclined at 80° to the ecliptic. With each passing month the spacecraft penetrated further into unexplored territory. In June 1992 it crossed the transition zone that marks the boundary of the turbulent part of the solar wind confined to a disk in the equatorial zone, and penetrated the simpler mid-latitude flow. At 800 km/s this wind was twice as fast, less dense, and much smoother than in the equatorial disk. Evidently the wind left the Sun at this speed and was slowed to half that rate in the disk where the magnetic field emanating from the Sun becomes wound into a spiral. Out of the disk, the flow was radial. But the transition zone was ragged owing to the inclined axis of the Sun, and over a thirteen-month period the spacecraft flew through whorls and eddies in the solar wind; only in July 1993 did it finally leave the last trace of the equatorial zone behind. The real objective, the polar zone, was still a year away. It passed 70°S on 26 June 1994. This was the official start of the first polar passage. All this time, it was closing on the Sun. By 13 September, on reaching 80°S, it was at 2.3 AU. Although the highest latitude, this was not perihelion. It left the south polar zone on 6 November and approached the ecliptic. For a month on either side of the ecliptic crossing on 5 March 1995, Ulysses was in the disk. It reached perihelion on 12 March, at 1.34 AU. Then on 19 June it crossed 70°N to begin its second polar pass, reached 80°N on 31 July, and left the northern zone on 30 September, thereby completing its primary mission.

Its orbit carried Ulysses back to the orbit of Jupiter in April 1998, but the planet was absent and the spacecraft returned sunward to repeat its polar passes at another part of the eleven-year solar cycle of sunspots. After several such orbits and a succession of mission extensions, the spacecraft was commanded to cease operating in 2009.

Ulysses found that the most violent eruptions on the Sun, coronal mass ejections, issue plasma at 1,000 km/s. In the equatorial disk this plasma smashes into the slowly moving material and produces a shock wave, but the radial flow travels unimpeded until the solar wind is confined by the pressure of the interstellar medium. No one knew where this barrier was, but the Voyagers were seeking it.

1993

On 21 August NASA suffered a shocking loss when Mars Observer, its first Mars mission since the Vikings, fell silent when preparing to insert itself into orbit around its target. It appears that its propulsion system suffered a catastrophic failure.

Galileo encountered a second asteroid on 28 August, flying by Ida at a range of 2,390 km. At 56 x 24 x 21 km it proved to be elongated and heavily cratered. One end seemed to be a rectangular block, and the other end was irregular and dominated by a depression 28 km in diameter. The fact that two dissimilar parts were separated by a 'waist' hinted that, even although the asteroid was a member of the

Koronis family, it was itself comprised of two solid components with loose debris filling the gap. The cratering implied the surface was about 2 billion years old. The surprise was a small companion. Named Dactyl, this was more spheroidal and a mere 1.6 km on its principal axis. Its general similarity to Ida suggested that they shared a common origin.

Figure 20. When Galileo examined Ida, it found the asteroid to possess a small satellite, which was named Dactyl. (Image courtesy of NASA/JPL/USGS.)

1994

The year 1994 saw America launch its first lunar mission since the final *Apollo* landing in December 1972, only this time it was a Department of Defense spacecraft to test miniaturized technologies in a way that would avoid the accusation of militarizing space. Clementine entered polar orbit of the Moon on 19 February. In addition to mapping mineralogy using a suite of sensors sensitive across the spectrum from the ultraviolet to the near-infrared, and using a laser altimeter to measure topography, it tested a technique to detect water ice in the permanently shadowed craters at the poles. This involved the spacecraft beaming its radio across the surface at a glancing angle, and the Deep Space Network taking note of the strength and polarization of the reflection. The results suggested the presence of ice. On 3 May the spacecraft left lunar orbit to make a flyby of the near-Earth asteroid Geographos, but malfunctioned and had to be abandoned.

In July 1994 the fragments of Comet Shoemaker-Levy 9 smashed into Jupiter, demonstrating that impacts are an ongoing phenomenon, and the Galileo spacecraft was able to provide a unique vantage point.

1995

On 7 December a small probe that had been released by Galileo several months earlier entered the atmosphere of Jupiter at the boundary of the equatorial zone and the north equatorial belt, on a trajectory approaching the dusk terminator. After aerodynamic braking to reduce the entry speed of 48 km/s, it deployed a parachute to make a vertical descent and report on the ambient environment. The baseline mission was to transmit data for 38 minutes in descending from the 0.1 bar level to the 10 bar level, corresponding to a vertical column from 50 km above to 100 km below the altitude of the 1 bar level. An error in the parachute sequence delayed the start of sampling to the 0.35 bar level, which started the column some 25 km lower than intended. The probe reported until succumbing to the 425 K temperature at the 23 bar level. It would have continued to descend until vaporized, thereby adding a small measure of heavy elements to the planet.

It was self-evident that sunlight energized the upper atmosphere, and one of the objectives was to determine the contribution from internal heat by identifying the depth at which the strong 'surface' winds diminished. It turned out that the winds below the solar-energized cloud tops persisted throughout the sampling column and actually increased towards the end. This implied that the convection was extremely deep and the main source of energy was heat from the interior. There would be distinct layers of cloud having compositions appropriate to the temperatures and pressures. The probe was expected to descend through a frigid brown hydrocarbon aerosol haze immediately above the white ammonia cloud, below which there would be a thin layer of ammonium hydrosulphide and then a layer of water-ice crystals leading to a water-rich cloud. At the 8 bar level, the atmosphere was expected to become clear. In reality it was different. The delayed start meant that the sampling started in the zone expected to be ammonia cloud. However, the probe passed through the pressure range corresponding to the ammonia cloud detecting only a haze. It did detect the ammonium hydrosulphide cloud, but this was very thin. The water that was expected in the 5 to 8 bar range was absent.

Once it was realized that the convection penetrated to a great depth, it was possible to explain the 'anomalously dry' readings. A rising column of heated moist air cools with expansion and clouds of rain condense out. On Earth, this condensate is water, but on Jupiter, the rising air produces a succession of condensates that create layers of cloud. By the time the air rises above the ammonia layer it is devoid of volatiles. As this airflow 'turns over' north or south, it is deflected by the Coriolis effect, and it is this which creates the alternating latitudinal banding. As the air starts to settle it is compressed and heated. Descending volatile-free air appears visually dark, but because it is arid it is transparent at a wavelength of 5 microns and by providing a view of the hotter interior the downdraft is bright in the near-infrared. The probe fell into such an 'infrared hot spot'. Although it would have been satisfying to have confirmed the predicted layering in a more typical region, in retrospect the insight gained of this arid environment was perhaps better.

Shortly after completing relaying the data from its atmospheric probe, Galileo entered orbit around Jupiter to conduct a detailed investigation of its miniature Solar System, some of the highlights of which were: Europa almost certainly does possess an ocean beneath its icy shell; Ganymede has a differentiated interior, including a subsurface ocean; the volcanoes of Io erupt an extremely hot magma which terrestrial volcanoes have not produced for billions of years; and the dust that makes up the dark rings is blasted off the small inner moonlets by meteoroid impacts.

1996

Encouraged that Fobos 2 had almost achieved its primary objective, the Russians built a spacecraft to deploy two landers as it approached Mars. The spacecraft was then to enter orbit around the planet and deploy a pair of surface penetrators. Much of the scientific payload was provided by international partners. When launched on 16 November, the final stage of the rocket failed to perform the escape manoeuvre and the spacecraft fell back to Earth a few days later. This killed the Russian planetary programme for the foreseeable future.

1997

On its way to a rendezvous with Eros, the NEAR-Shoemaker spacecraft passed the asteroid Mathilde at a range of 1,212 km on 27 June 1997. It proved to be 66 x 48 x 46 km in size, and to have a very dark surface that was consistent with its spectroscopic classification as carbonaceous. The largest crater observed had a diameter spanning half the asteroid's major axis. The absence of any indication of layering in the wall of this crater implied that the interior of the body must be homogeneous. The inferred density of 1.3 gm/cc suggested that it was a gravitationally bound 'rubble pile', possibly with some large solid chunks inside.

The highlight of the year was the landing of Mars Pathfinder on 4 July using an airbag system in Ares Vallis, a site littered with rocks believed to have been washed down off the highlands by a catastrophic flood. The main public interest was in the Sojourner rover, which was equipped to investigate the surface chemistry of the rocks.

On 12 September Mars Global Surveyor entered a 263 x 54,026 km orbit around Mars, and then used aerobraking passes through the upper atmosphere to slowly achieve its operating orbit. The propellant saved by this technique enabled this small craft to carry several of the instruments developed for the larger Mars Observer. The Sun-synchronous orbit was designed to cross the equator in daylight north-south in the mid-afternoon in order for all the vertical imagery during the primary mission to be with the same illumination. The camera had an unprecedented resolution of several metres. One surprising discovery was made while aerobraking. Previous spacecraft had shown Mars not to possess a magnetic field. But while its periapsis was below the

Figure 21. The Sojourner rover on Mars caught the public attention. (Image courtesy of NASA/JPL.)

ionosphere Mars Global Surveyor was able to sense the surface, and discovered isolated magnetic anomalies. As data were accumulated, it became apparent that these were concentrated in the southern hemisphere. The fact that the strength of the field was not correlated with individual craters or other features meant that the sources predated the major impact basins. There were no significant anomalies in the low-lying northern plains, the Tharsis and Elysium volcanic areas, or Valles Marineris. Once the spacecraft had begun its primary mission, its high-resolution camera found ample signs of the action of liquid water in the remote past and hints of ground water recently leaking from the walls of craters. An infrared spectrometer mapped the minerals at the surface. The carbonate sediments expected in the low-lying northern plains were absent, which was puzzling since other lines of evidence suggested there had once been a shallow ocean. A laser altimeter produced a topographic map far superior to that inferred from surface pressure data measured by the Vikings. Mars Global Surveyor operated until November 2006, when it was crippled by a programming error.

1998

Lunar Prospector entered orbit around the Moon on 11 January 1998 to conduct remote sensing on a global basis. Its payload included a gamma ray detector and a neutron spectrometer. When cosmic rays hit the lunar surface, neutrons and gamma rays are produced. Most of the neutrons fly off at high speed, but if one encounters a lightweight nucleus it will slow down by yielding some energy. A water molecule contains two hydrogen atoms, the lightest of all atoms. The presence of hydrogen was inferred from the energy spectrum of the neutrons. In effect, the instrument was able to identify buried ice. On 5 March the presence of significant water-ice at both poles was announced. Although the 100 km spatial resolution of the instrument was insufficient to localize the sites of this hydrogen enrichment, there was a correlation with permanently shadowed craters at the poles.

1999

NASA's exploration of Mars suffered two setbacks in 1999. First, on 23 September Mars Climate Orbiter was lost in attempting to enter orbit around the planet, when it dipped into the atmosphere and was destroyed. The navigational error was identified as a misinterpretation of data used to model 'small forces' acting on the spacecraft during its interplanetary cruise, with the manufacturer stating the values in imperial units and the navigators presuming them to be in metric units! Then on 3 December Mars Polar Lander was lost at some point during the entry, descent and landing sequence – the specification did not call for telemetry, so it was impossible to be sure what occurred. However, one plausible explanation was that the deployment of the legs triggered the sensor that was meant to indicate contact with the surface and cut off the engine, with the result that when the altimeter indicated a height of forty metres and the computer started to monitor the contact sensor, it saw that this signal was asserted and cut off the engine, leaving the vehicle to fall to the surface. A pair of penetrator probes were also lost separately, again without the cause being known for sure.

2000

On 14 February, NEAR-Shoemaker entered orbit around asteroid Eros at a distance of about 300 km. At 34.4 x 11.2 x 11.2 km, the asteroid was similar in shape to a banana. A number of manoeuvres were performed to adjust the altitude and orientation of the orbit in order to survey the entire surface. It was cratered, with some sizeable chunks knocked out of it, but the features were smoothed off and there were rocks scattered around. To wrap up the mission the spacecraft was

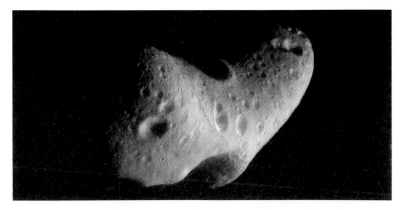

Figure 22. A view of Eros taken by NEAR-Shoemaker, in orbit around the asteroid. (Image courtesy of NASA/JPL/JHAPL.)

told to make grazing contact, and to the delight of the team it settled into place undamaged on 12 February 2001. A week later it was commanded to switch off.

2001

On 22 September, in its extended mission and having achieved its technological test objectives, the Deep Space 1 spacecraft flew by Comet Borrelly at a range of 2,200 km. Although it did not approach the nucleus as closely as Giotto had that of Halley's comet, the fact that the comet was less active enabled Deep Space 1 to obtain images of greater clarity, and showed the nucleus to be 8 x 4 x 4 km, extremely dark, and with a low density that suggested it was a rubble pile. The spacecraft did not pass through any of the sparse jets, which was fortunate because it was not equipped with debris shields.

2002

On 2 November, Stardust made a 3,080 km flyby of asteroid Annefrank, a member of the Augusta family. This was essentially an engineering test of the spacecraft and ground operations to prepare for the encounter with a comet. At 6.6 x 5.0 x 3.4 km, it was twice as large as expected, and the triangular-prism shape suggested that it was two objects in contact.

2003

In July in was announced that Mars Odyssey, which had entered orbit around Mars in 2001, had discovered a high density of ice just under the surface in both hemispheres poleward of 55° latitude. As in the case of Lunar Prospector, the observations were made by a gamma ray detector incorporating a neutron spectrometer.

On 25 December the European Space Agency's Mars Express entered orbit around Mars. Six days earlier it had released the Beagle 2 lander supplied by Britain on a trajectory to enter the atmosphere, but this failed enigmatically. Much of the orbiter's payload was carried over from the lost Mars-96 mission. At the time of writing, the orbiter is still operating with funding to the end of 2012.

2004

On 2 January the Stardust spacecraft passed through the coma of comet Wild 2 at a range of 237 km from the nucleus and used aerogel to 'non-destructively' collect dust grains. In the first sample-return with cometary material, the capsule was returned to Earth on 16 January 2006. Analysis revealed a wide variety of organic compounds and abundant amorphous and crystalline silicates. A paucity of hydrous minerals implied a lack of aqueous processing of the cometary dust. The spacecraft itself performed a deflection manoeuvre to miss Earth and pursue an extended mission.

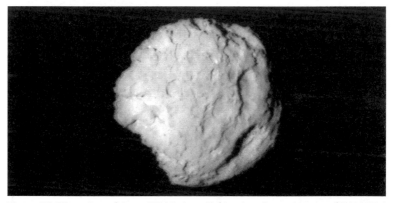

Figure 23. The nucleus of Comet Wild 2 viewed by Stardust. (Image courtesy of NASA/JPL.)

The highlight of 2004 for the public was the arrival of the two Mars Exploration Rovers, which were considerably larger and better equipped than Sojourner to undertake field geology investigations. Spirit landed in Gusev crater on 4 January, and Opportunity on Meridiani Planum on the other side of Mars on 25 January. They both greatly exceeded their three-month baseline missions. Gusev had been selected because it seemed in imagery to have been breached by a river-like channel and to have hosted a lake at some point in the remote past. Spirit found the floor of the crater to be a volcanic plain; any alluvial sediment had evidently been buried. However, by driving to a cluster of hills and ascending the tallest one, it found ample evidence of the past presence of water. Meridiani Planum was selected because of indications that its surface contained hematite, a mineral which (on Earth) forms in the presence

Figure 24. This image by Mars Reconnaissance Orbiter shows the track of Opportunity's approach to the rim of Victoria crater and, indeed, the rover itself. (Image courtesy of NASA/JPL-Caltech/University of Arizona/Cornell/Ohio State University.)

Figure 25. A view of strata exposed at Cape St Vincent, one of the promontories of Victoria crater, viewed by the Opportunity rover. (Image courtesy of NASA/JPL/Cornell.)

of water. Opportunity found hematite in abundance, as well as clear evidence of evaporites that suggested water had once stood on the surface. But that water was acidic. This offered an explanation for the absence of carbonate sediments in the northern lowlands, since dissolved carbonate will not precipitate in acidic water. On 1 May 2009, Spirit became stuck in soft soil. It was last heard from on 22 March 2010 and, being poorly oriented to collect sunlight for power during the winter, perished. As of the time of writing, Opportunity had driven an astonishing 30 km and was approaching a crater several kilometres in diameter that offered an unprecedented view of deep rock exposures.

On 11 June, entering the Saturnian system, the Cassini spacecraft flew by Phœbe at a range of 2,068 km. The fact that the orbit of this outer moon lies essentially in the ecliptic, rather than the plane of the Saturnian system, implied that it had been captured. At 230 x 220 x 210 km it was heavily cratered. Although very dark, there were indications of ice below the surface. The presence of carbon dioxide on the surface hinted that it was a captured Centaur, one of a number of icy bodies which formed in the Kuiper Belt and now orbit the Sun between Jupiter and Neptune. The spacecraft successfully manoeuvred into orbit around Saturn on 1 July to undertake a detailed investigation of the planet, its rings, magnetosphere and satellites.

After being launched in 2001 the Genesis spacecraft was manoeuvred into a 'halo' orbit at the Lagrange point 1.5 million km 'up-Sun' of Earth and collected solar wind. In April 2004 it left its station and headed back to Earth to deliver its samples so that laboratory analysis could measure isotopic abundances of the ions in the solar wind and precisely determine solar elemental abundances. When the container entered the atmosphere on 8 September, a design flaw prevented it from deploying its parachute, pre-empting the plan to snatch it in mid-air in order to avoid the shock of contact with the ground. The impact at terminal velocity broke open the container. Although the solar wind samples were contaminated, valuable scientific results were achieved.

On 16 December 2004 it was announced that at a range of 94 AU from the Sun the magnetometer on Voyager 1 was showing the strength of the ambient magnetic field to have tripled. This was a clear sign that the spacecraft had reached the 'termination shock' of the heliosphere, since the strength of the field would increase as the solar wind was compressed upon becoming subsonic. At the same time, it detected the same kind of plasma-wave oscillations as it had observed shortly prior to penetrating the bow shocks in front of planetary magnetospheres. The actual crossing of the termination shock had passed unobserved since the spacecraft was not being tracked at that time. By the time the change became apparent, it was already in the heliosheath. What had occurred was that in mid-2004 the pressure of the solar wind had declined, and the termination shock had washed over the vehicle several months later as the heliosphere was compressed by the pressure of the surrounding interstellar medium.

2005

Cassini had released the Huygens probe on 25 December 2004 on a trajectory that reached Titan on 14 January 2005, where the probe deployed a parachute and settled on the surface. The transmission was relayed to Earth by the main spacecraft making a flyby. Images taken during the descent, below the smog, showed a landscape of what appeared to be channels etched by flowing liquid which drained on to a level plain. The probe landed on the plain, and the view from the surface showed a granular 'soil' littered with small rounded lumps of ice that looked as if they had been washed down off the surrounding uplands. The temperature was 94 K, and the heat from the lamp on the probe caused a variety of organics to evaporate from the soil. The impression was that the plain occasionally hosted a shallow lake of hydrocarbons, possibly on a seasonal basis.

Figure 26. A vertical view taken by the Huygens probe as it parachuted down on to Saturn's moon, Titan, showing valley networks in rough terrain that appear to drain on to a smooth plain. The view from the surface (centre) with a sense of perspective provided by a view of an astronaut on the Moon. (Image courtesy of NASA/JPL/ESA/University of Arizona.)

Figure 27. Hydrocarbon lakes on Titan, revealed by Cassini's imaging radar. (Image courtesy of NASA/JPL/USGS.)

Over the next few years, Cassini made a series of close passes by Titan and used its radar to map swathes of the surface, revealing a landscape of elevated regions and plains, with lakes in the polar regions and vast fields of dunes aligned with the prevailing winds in the equatorial zone. It was a terrestrial landscape in a cryogenic state. The spacecraft also inspected the other satellites. The most intriguing finding was that the south pole of Enceladus was much warmer than it ought to be, and that water and other volatiles were venting from a series of long fractures, forming a tenuous ring around the planet. Iapetus, whose leading hemisphere had long been known to be anomalously dark and had been imaged at long range by the Voyagers, was revealed by Cassini in sufficient detail to indicate that the dark material was a thin veneer picked up as the moon travelled its orbit. When a ring of dark material was found just inside the somewhat larger orbit of Phœbe, it was realized that because this outer moon travels in a retrograde manner, the dust that it shed was migrating inward and being swept up by Iapetus. As regards Iapetus itself, Cassini found it to have a remarkable 'walnut ridge' running around its equator, the origin of

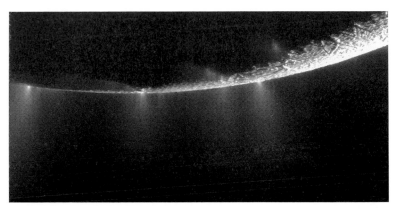

Figure 28. Cassini imaged the jets spewing from the 'tiger stripes' of Enceladus, backlit by the Sun. (Image courtesy of NASA/JPL/Space Science Institute.)

which is disputed. The spacecraft's primary mission ended on 2008 but it ran seamlessly into a two-year extension, and at the time of writing a second extension is underway with funding pencilled in through to 2017.

On 4 July 2005, the Deep Impact mission flew by the nucleus of Comet Tempel 1 at a range of 500 km, having released a 370-kg impactor which manoeuvred on to a collision course and produced surprising results. The impact created a large and bright dust cloud which obscured the view of the impact crater. The nucleus proved to be

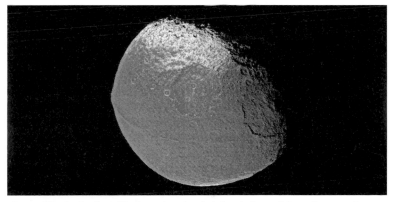

Figure 29. The 'walnut ridge' crossing the dark hemisphere of Saturn's moon, Iapetus, imaged by Cassini. (Image courtesy of NASA/JPL/Space Science Institute.)

Figure 30. The spectacular impact of a probe on the nucleus of Comet Tempel 1, imaged by the Deep Impact parent spacecraft. (Image courtesy of NASA/JPL/UMD.)

more dusty and less icy than expected – a very fluffy structure held together by gravity.

The Japanese spacecraft Hayabusa rendezvoused with Itokawa on 12 September. Since the spacecraft's heliocentric orbit was so similar to that of the asteroid, it readily maintained station at a distance of about 20 km. At 535 x 294 x 209 metres, it was a very small body, and its peanut shape implied it was several objects in contact. There was a sur-

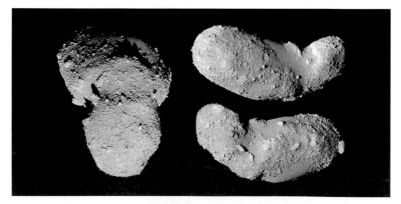

Figure 31. The asteroid Itokawa revealed by Hayabusa. (Image courtesy of ISAS/JAXA.)

prising absence of impact craters. Although the surface was studded with boulders, there were smooth patches. The low density implied it was a rubble pile. After surveying the surface to select a target, on 19 November the spacecraft touched down to collect a sample of the surface, but it did not activate its sampling mechanism. It tried again on 25 November, but with uncertain results. After many travails the sample container was returned to Earth on 13 June 2010 and, despite the failure of the sampling mechanism, proved to contain some grains.

2006

On 11 April Venus Express entered an elliptical orbit of the eponymous planet with its periapsis near the north pole and a period of twenty-four hours on a mission to undertake a long-term investigation of the atmosphere and its interaction with the solar wind. At the time of writing, it is still operating with funding to the end of 2012.

2007

On 30 August Voyager 2 reached the termination shock at 84 AU. This time, the event was monitored closely. As the location of the termination shock is constantly changing in response to the Sun's activity, the vehicle actually made five crossings, providing lots of data. The next major event for these spacecraft will be when one of them crosses the heliopause, which is expected to lie at about 125 AU, and exits into interstellar space, truly a fitting end to an extraordinary survey of the outer Solar System.

Closer to home, on 3 October the Kaguya spacecraft built by Japan entered a circular lunar polar orbit at an altitude of 100 km. In addition to a science suite, it had a high-definition camera that provided astonishing movies of the surface. And on 5 November the Chang'e spacecraft made by China entered a 200 km circular lunar orbit inclined at 64° to the lunar equator. The mission was scheduled to continue for a year, but was later extended until March 2009, when the spacecraft was deorbited. Kaguya was deorbited three months later.

2008

The Mars Phoenix successfully landed on Mars on 25 May, essentially as a follow-up to the lost Mars Polar Lander, although that had been meant for the southern polar region and Phoenix landed in the northern polar region. In addition to transmitting a 'tone' to indicate its status during the entry, descent and landing sequence, detailed planning enabled the Mars Reconnaissance Orbiter, which had arrived two years earlier, to obtain a picture of Phoenix on its parachute! The lander touched down on an open plain of Vastitas Borealis that was devoid of boulders. What had attracted the site selectors was the polygonal pattern that characterized this area, similar to patterns present in permafrost areas in polar and high latitude regions of Earth. There proved to be ice several centimetres beneath the loose soil. A robotic arm retrieved samples for onboard analysis, and the ice proved to be in the form of a brine. The soil had a moderately alkaline pH of 7.7 ± 0.5, and contained perchlorate (a highly oxidizing chemical) and calcium carbonate in amounts implying formation by the interaction of atmospheric carbon dioxide with liquid water films on the surfaces of particles. The three-month surface mission was extended by two months, but eventually the lander was obliged to shut down by the shortening of the day with the onset of local winter.

On 5 September the European Space Agency's Rosetta spacecraft flew by Šteins at a range of 800 km. At 6.7 x 5.8 x 4.5 km, the asteroid was broad at one end and tapered to a point at the other, giving a distinctive diamond shape. Scientists were astonished that the 2 km crater on the wide section had not shattered the body.

Launched by India, the Chandrayaan mission entered lunar polar

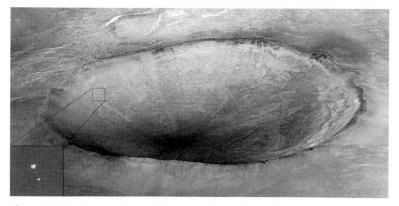

Figure 32. Mars Reconnaissance Orbiter was able to observe the Phoenix lander descending on its parachute. (Image courtesy of NASA/JPL-Caltech/University of Arizona.)

orbit on 8 November. One of its instruments was a NASA-funded imaging spectrometer to map the mineral composition of the surface. By the time the spacecraft was disabled in August 2009 by a communications failure, this had observed more than 95 per cent of the surface. It detected absorption features in the infrared indicative of water molecules in the equatorial zone. These were evidently present as a thin film of molecules on the grains at the very surface. It was belatedly realized that when Cassini made its lunar flyby in August 1999, one of its instruments had detected evidence of water at all latitudes. Scientists on the extended mission of the Deep Impact spacecraft were asked to make observations when their spacecraft flew by the Moon in June 2009 on its way to intercept Comet Hartley 2. It was decided to conduct observations over the period of a week, and the results were conclusive – there was water at all latitudes. The strength of the signature was correlated with the surface temperature, and was strong at sunrise, diminished towards noon, and returned to its original level by sunset. It was evident that there was a 'diurnal cycle' at work, operating on the lunar timescale. The fact that the strength of the signature did not increase during the fortnight-long lunar night showed the process to be associated with the solar wind. The conclusion was that protons in the solar wind react with oxides in the regolith to produce water molecules. Although these are vaporized in the midday heat, the process resumes in the afternoon. The midday temperature is hottest in the equatorial zone and decreases with latitude. In the polar regions the variation in water absorption diminishes, and in the permanently shadowed areas there is a progressive accumulation of water over time.

2009

On 23 June, Lunar Reconnaissance Orbiter entered polar lunar orbit. The Centaur stage which had dispatched it retained the LCROSS spacecraft, and exploited a polar flyby to adopt an Earth orbit that would return it to the Moon several months later. Meanwhile, as part of its research, LRO studied the shadowed craters at the south pole. The final targeting of the LCROSS mission was postponed until the final few days of the approach, and it was decided to dive into the 100 km crater Cabeus, which LRO confirmed to be hydrogen-enriched. Shortly before reaching the Moon on 9 October, the Centaur released the LCROSS spacecraft, which followed it in to observe by remote sensing and directly sample the gaseous plume raised by the impact, whose energy was equivalent to detonating 1 tonne of TNT. If the Centaur struck solid rock, then most of the energy would be converted into a bright flash which, whilst looking spectacular, would actually be bad news. It was hoped that the Centaur would strike a deep blanket of regolith. The flash would be much fainter because the energy would be expended in ejecting loosely consolidated material, hopefully with some ice mixed in. LCROSS provided data on the impact site for six minutes before itself impacting nearby. Its instruments detected the faint flash of the impact and also the presence of a plume rising into sunlight. A month later it was announced that multiple lines of evidence showed that water was present in both the high-angle plume and the lateral ejecta from the Centaur impact.

2010

On 10 July, Rosetta made a flyby of Lutetia at a range of 3,162 km. At 132 x 101 x 76 km, the asteroid was the largest inspected to date, and although it was spheroidal it had irregular features. The soft outlines of the surface features suggested the presence of a thick regolith. (Rosetta's primary mission is to rendezvous with Comet Churyumov-Gerasimenko in 2014, enter orbit around it, deploy the Philae lander to the nucleus, and accompany the comet through perihelion.)

Figure 33. Asteroid Lutetia imaged by Rosetta (left: image courtesy of ESA/MPS for OSIRIS Team MPS, UPD, LAM, IAA, RSSD, INTA, UPM, DASP, IDA) and the active nucleus of Comet Hartley 2 by the Deep Impact spacecraft pursuing its EPOXI mission (right: image courtesy of NASA/JPL/UMD).

On 4 November on its EPOXI extended mission, the Deep Impact spacecraft made a 700 km flyby of Comet Hartley 2, revealing it to have a peanut-shaped nucleus with several bright jets active.

2011

The Stardust spacecraft, completing its NExT extended mission, made a flyby of Comet Tempel 1 on 15 February at a range of 181 km in order to study the effects of the impact engineered by the Deep Impact mission. The crater was estimated to be 150 metres across, with a bright mound in the centre that probably formed when material from the impact fell back into the cavity.

One of the highlights of the year was the successful insertion of the Messenger spacecraft in orbit around Mercury on 18 March. Another was the rendezvous of the Dawn spacecraft with Vesta in July for a detailed investigation of this intriguing asteroid, which is thought to be a remnant protoplanet with a differentiated interior.

REFLECTIONS

In 1962 we achieved our first successful planetary mission with a flyby of Venus. Now we have extended our in-situ investigations to include the inner planet Mercury, our Moon, Mars and its two moonlets, Jupiter, Saturn, Uranus and Neptune and their systems of rings and many satellites, and a number of comets and asteroids. We are also about to exit the heliosphere into interstellar space. It has been an exciting fifty years. The venture began as a competition between America and the Soviet Union, but now Europe, Japan, India and China have all joined in and it is being conducted on a cooperative basis in which spacecraft carry international payloads.

The Biggest Structures in the Universe

ANTHONY P. FAIRALL

First published in the *1991 Yearbook of Astronomy*. Revised 2011.

What are the biggest physical structures in the universe? Up to around fifteen years ago, we thought them to be clusters of galaxies, at most 10 million light years in size. Today the picture is totally different; structures up to 1,000 million light years across are recognized – but that is only the minimum, our mappings still have not reached the scale where the universe can be said to be homogeneous. More remarkably, the new findings have revealed a gigantic foam-like texture to the cosmos. This seems surprising since foam-like formations more usually occur on a very small scale, such as in bath sponges or even soapsuds. Whatever the resemblance or connection, it seems as though a great new key to understanding the fabric of the cosmos is being presented to us.

Why did we not know about these structures much sooner? Because we are not able to view the three-dimensional distribution of galaxies at all readily. When we photograph the sky (Figure 1), we get a two-dimensional view. For instance, if we look at the wide-angle survey photographs from the Palomar, UK and European Schmidt telescopes, we see plenty of galaxies – if you look closely enough, you can see thousands of them on each photograph (except for photographs aimed into the thick of the Milky Way). Of course, while each photograph may show thousands of galaxies, it also shows millions of stars. However, the stars are all very much in the foreground, contained in our home Galaxy. Each galaxy we see may in itself contain towards a million million stars but, in general, the galaxies are too distant for the stars to be seen individually. It's like viewing a distant city by night; you see the collective light and not the individual street lamps.

Figure 1. Astronomers at Oxford University have completed the largest survey of galaxies ever undertaken. Their map of 2 million galaxies in an area covering 10 per cent of the sky reveals long, thin, curving lines of galaxies corresponding to sheets or 'walls' of galaxies in space, separated by 'voids'. Each photograph (originals are 14 inches square) shows about 150,000 stars and 50,000 galaxies. This picture shows a typical patch of sky represented by a single dot in the final distribution map of 2,000,000 galaxies; the image on the left is the original photograph, that on the right shows the same area of sky as stored in the survey after processing by the Automatic Plate Measurement machine. (Reproduced by courtesy of the Department of Astrophysics, University of Oxford.)

Galaxies have one big advantage over stars. Stars appear as pinprick images on the photographs, whereas galaxies show disks, although these get very small as the galaxies get more distant. Although there is a mixture of big and small galaxies, you can get a feel for the distances of galaxies by comparing their apparent sizes in the sky. It is similar to looking out over a landscape and using the apparent sizes of trees and lesser foliage to gauge distances in the landscape. When there are no trees, such as in the Apollo photos from the surface of the Moon, one has no perspective of distance (the mound behind the astronauts in the *Apollo 15* photographs is really a mountain 20,000 feet high). Anyway, this was the technique used independently, in the 1950s and 1960s, by two great pioneers of extragalactic astronomy, George Abell and Fritz Zwicky, to gauge the relative distances of the clusters of galaxies on the Palomar Sky Survey. The picture that emerged was one of occasional clusters superposed on a general random 'field component' of clusters (Figure 2). However, Zwicky and others (particularly C. P. Shane and Gerard de Vaucouleurs) suspected that much larger entities existed –

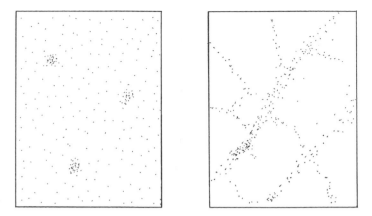

Figure 2. The dots in these diagrams represent galaxies. Left: For many years, the clusters of galaxies were thought to be superimposed on a random 'field' component of galaxies. The clusters were believed to be the largest structures in existence. However, the modern picture (Right) reveals a 'foam-like' distribution with the galaxies not scattered randomly but surrounding empty voids.

the superclusters. What nobody realized was that there were actually regions devoid of galaxies. That could only come when much more accurate distances of galaxies could be determined.

OBTAINING THE DISTANCES OF GALAXIES

Whilst it takes a fair deal of effort just to get the distance of a single galaxy, it is generally easier to get the distance of a galaxy than the distance of a star, thanks to the well-known Hubble law. Our Galaxy is not expanding, so, in general, the distances to the other stars in our Galaxy are not increasing. But the universe is expanding and the distances to the other galaxies are increasing, so they appear to be moving away from our Galaxy. It goes according to the Hubble relationship: $V = Hd$, where V is the speed of recession from our Galaxy, d is the distance and H is a constant, the so-called Hubble constant. In other words, if a neighbouring galaxy is moving away from us at 100 kilometres per second, the next-but-one neighbour would be moving away at 200 kilometres per second, the one after that at 300 kilometres per second and so on. Don't worry about everyone trying to move away from our

Galaxy. There is nothing wrong, because you get the same pictures from every other galaxy; in fact, it is not really so much that the galaxies themselves are in any way making themselves move in this way, but rather that the intervening spaces have been given a sort of Alice-in-Wonderland pill that makes them grow. The important thing is that if you measure the velocity of recession of a galaxy, then you can determine its distance. This assumes you know the value of the Hubble constant, and there is currently some controversy as to what its exact value should be, but a middle-of-the-road value is 20 kilometres per second of velocity for every million light years of distance.

So far, so good; now we just have to set up to measure the velocity of recession of the galaxy. We do this by obtaining a spectrum of the galaxy – an observation that takes a large telescope from a few minutes to much longer, according to the brightness of the galaxy. Superimposed on the colours of the spectrum are key spectral features, but, when a galaxy is receding away from us, the Döppler Shift moves these towards the red end of the spectrum. Measuring this shift gives us the velocity of recession, and from the velocity of recession we get the distance. With modern technology, this sort of observation is now very much faster than it used to be. In the early 1960s, a typical spectrogram took some hours to obtain using the largest telescopes in the world. Nowadays, with image intensifiers and photon-counting detectors, the same observation can be carried out with a much more modest telescope in only a few minutes. As a result, while there were only about a thousand redshifts available in 1960, today there are more like around forty thousand.

One more complication – while the velocity of recession of a galaxy is mainly due to the cosmological expansion, there can also be a contribution from the galaxy's own peculiar motion – perhaps up to several hundred kilometres per second. This could be one of two things. First, if the galaxy is a member of a reasonably dense cluster or if it is one of a pair, then it must have a form of orbital motion (otherwise gravity would make the cluster collapse). Second, even if the galaxy is fairly isolated from its neighbours, then it may still participate in large-scale streaming motions. Our Galaxy is such a case; apparently it is streaming at some 600 kilometres per second.

A 'FOAM-LIKE' STRUCTURE

It is very difficult to sort out all these peculiar motions, and since they are usually much smaller than the cosmological expansion, we initially overlook them and simply take velocity as representing distance. Figure 3 (southern redshift plot) is a good example. It can be seen that the distribution is anything but uniform. Where there is an adequate density of points, an astonishing foam-like structure emerges. There are regions almost void of galaxies that form the 'bubbles' in the foam. It is the scale of the structure that is remarkable – the bubbles shown are up to 100 million light years across (or, if you prefer, 1,000 million million million kilometres).

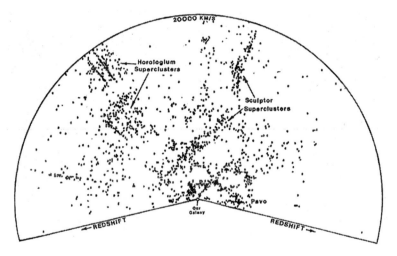

Figure 3. A 'slice' of the southern sky with individual galaxies shown as dots. Their distance from our Galaxy is indicated by the redshift (the perimeter of the diagram corresponds to a velocity of recession of 20,000 kilometres per second). This reveals a cross-section of the foam-like structure and supercluster network. The voids are typically empty, save for a few galaxies lying inward from the surrounding walls.

Foam-like structures normally occur on very much smaller scales, and usually in aqueous or organic media. It is also quite amazing to find such similar structure displayed, not by a continuous medium, but by a large number of unconnected point-like systems (on this scale, even the galaxies have shrunk to points); it would seem quite surprising if the air molecules inside a room suddenly adopted a foam-like distribution – quite against the respected laws of thermodynamics! But then, whilst the galaxies might be moving, their motions are very slow when measured against the very large-scale foam-like structure. Either this is the distribution the galaxies had when they were formed, or, since their formation, they have been drawn slowly into these foam-like formations. Whatever the case, here is a great clue as to how the cosmos evolved just before or after the formation of galaxies.

Theories abound as to how this sort of structure came about. If gravity is responsible, then there is much too little mass in the galaxies for it to have a significant effect. Thus, the most established of current cosmological hypotheses is 'Cold Dark Matter'. This theory considers that the galaxies are only a tiny portion of all the matter present. Much more mass means much more gravity – enough to draw the galaxies towards concentrations and create superclusters. It is not clear, however, that a network of superclusters will necessarily adopt a foam-like texture. If the matter were not pulled by gravity, could it have been pushed instead – by gigantic explosions, for instance? This is one of the competitive theories, which envisages a sort of 'cosmological baking powder' to give the structure of the universe the texture of a cake! Perhaps the most novel current theory concerns cosmic strings – the spaghetti-like leftovers from the very early universe. Almost immeasurably thin, yet incredibly massive, they could sweep through space with galaxies formed in their wake. At this stage there is no easy telling which theory is right and which is wrong.

But the story of large-scale structure in the universe is not yet finished, for every time we extend our surveys deeper, we find ever larger structures. The foam-like bubbles are by no means the largest entities. Rather, superclusters connect and align to make even larger structures – and the foam-like texture fills in the intervening spaces. The diagrams that accompany this article show great long structures – massive filaments, or even perhaps two-dimensional walls of galaxies.

OUR PLACE IN THE UNIVERSE

Where does all of this place our own position? Figure 4 is a schematic indication of where our Galaxy is situated relative to neighbouring structures. The band of obscuration that lies horizontally across the diagram is due to the plane of our own Milky Way Galaxy. Our Galaxy lies towards the fringe of the Virgo supercluster, so named after the Virgo cluster that lies at its centre. However, as the diagram suggests, the Virgo supercluster looks a little anæmic compared with other neighbouring structures, nor is it completely isolated from them. Filamentary links appear to stretch out towards the Hydra-Centaurus-Pavo conglomeration. In fact, our Galaxy, and presumably most of our neighbours, is streaming in the Centaurus direction at 600 kilometres per second. Not that we will ever get there – the Centaurus concentration is moving away at 4,500 kilometres a second due to the expansion of the universe.

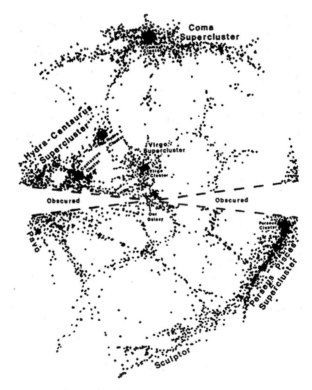

Figure 4. A schematic diagram (the dots are not real data), based on various redshift maps, to indicate neighbouring structures out to a redshift of several thousand kilometres per second. The plane of the paper conveys a general tendency for these structures to favour a 'supergalactic' plane.

This streaming motion has given rise to the idea of there being considerable mass in Centaurus, the so-called 'Great Attractor', whose gravitational pull has set our Galaxy in motion. In any case, part of the motion may be due to an imbalance in local structure, since neighbouring voids lie in the opposite direction. Also in the opposite direction is the dominant Perseus-Pisces supercluster – and it is not clear why we are not pulled that way. Perseus-Pisces has a filamentary core but links to structures in Sculptor (see Figure 3). In another direction still, and also separated from Virgo by voids, is the Coma supercluster, centred around the dense Coma Cluster. Shown in Figure 4 in cross section, it is really a thick two-dimensional sheet, often termed a 'great wall'. Its full extent is still not known, for there is still much mapping to be done. For the moment, the surrounding structures almost give a 'tree-ring' appearance, with us in the centre, but that is certain to change as the maps are extended.

It is an exciting era to live in. For the first time in mankind's history, we are seeing structures that are significant on the scale of the whole observable universe (with horizon at around 15,000 million light years). A new fabric of the cosmos has been revealed. We must now decide how it was woven.

EPILOGUE
(written by Professor Fred Watson, June 2011)

When Tony Fairall wrote this fine article for the 1991 Yearbook, he hardly could have realized that observational astronomy stood on the brink of a revolution that would transform our understanding of the large-scale structure of the universe. Or that the fundamental questions he posed would be answered unequivocally within a decade and a half.

The observational transformation took place on two main fronts. A clue to the first came a few years later in an article in the 1995 Yearbook of Astronomy entitled 'Astronomy's Multi-Fibre Revolution'. There, an author with a name very like mine described the use of fibre optics to gather detailed spectroscopic data on very large numbers of objects simultaneously, a process that was just beginning to come into its own with the introduction of robotic technology.

One of the diagrams in that 1995 fibre optics article is a slice through the local universe, similar to Figure 3 in Tony Fairall's article reprinted here. It was made with a rudimentary fibre optics spectrograph called FLAIR on the UK Schmidt Telescope using the crudest of methods, but it already eclipsed the data in Tony's diagram in depth, uniformity and velocity accuracy. The writing was clearly on the wall for the time-honoured, one-target-at-a-time method that had been used to gather the earlier data.

When FLAIR was superseded in 2001 by a whizz-bang robotic fibre system called 6dF, Tony and I became part of a large group of scientists who formed the 6dF Galaxy Survey Team. The new survey occupied the UK Schmidt Telescope for the first five years of the new decade, and produced a three-dimensional map of the local universe out to a billion light years, containing well over 100,000 southern hemisphere galaxies (Figure 5). It's worth exploring the survey's website at http://www.aao. gov.au/local/www/6df/ to see the beautiful structure it revealed. The filaments, sheets and voids that were new discoveries in Tony's 1991 article are there in profusion, confirming the foam-like structure of the universe that he had hinted at.

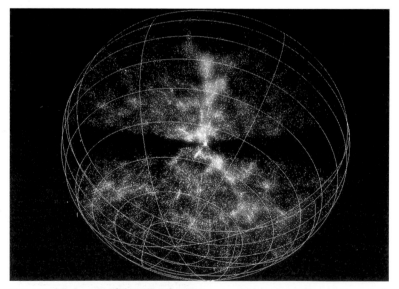

Figure 5. All-sky view of the 6dF Galaxy Survey, in which each point represents an individual galaxy. The dark band across the middle is the region hidden by the plane of our Milky Way Galaxy, but the foam-like structure of sheets and voids is clearly visible. (Image courtesy of 6dF Galaxy Survey Team.)

Tony commented in his article that, back in 1960, we knew the redshifts (i.e. distances) of about a thousand galaxies, but by 1991, that number had increased to around 40,000. Today, because of the advent of robotic fibre-optics technology, galaxy redshifts are numbered in millions, and will soon be well over one hundred times the 1991 total. No wonder I'm always harping on about the marvels of fibre optics . . .

But what of the questions Tony raised in his article? In particular, the one he ended with – how did the foam-like structure originate? Today we know the answer to that, too, and it has come from the other observational revolution I mentioned earlier. Back in 1991, a NASA spacecraft called COBE – the COsmic Background Explorer – was busy mapping the microwave sky to find tiny fluctuations in the intensity of the so-called Cosmic Microwave Background Radiation, or CMBR.

You might find CMBR an attractive acronym, but I always think it's easier to imagine this radiation as a kind of 'cosmic wallpaper' that sits behind everything else we can see in the universe. In saying 'behind everything else', what I really mean is '*before* everything else', because we are using the finite velocity of light to look back in time to an era when the universe was in its infancy. At that epoch, some 380,000 years after the Big Bang, the universe was still a fireball of intense light, and its redshifted radiation is what we see now as the cosmic wallpaper – the CMBR. It's why astronomers are so hooked on the Big Bang model of the universe – they can still see the Bang. In a sense, they can hear it, too, because the intensity fluctuations that COBE was looking for are the result of acoustic oscillations – sound waves that thundered through the early universe.

In April 1992, the COBE science team announced that they had found the intensity fluctuations, known in the trade as the anisotropy of the CMBR. And it turned out that these fluctuations formed the missing link between the Big Bang 13.7 billion years ago and today's foam-structured universe. In work published in 2005, it was demonstrated that the anisotropy (as measured by a new spacecraft called WMAP – the Wilkinson Anisotropy Probe) and the foam-like structure (as measured by a galaxy survey carried out on the 3.9-metre Anglo–Australian Telescope) are two different views of the same thing. They describe a universe in which clouds of hydrogen were triggered to collapse under their own gravity by those primordial acoustic oscillations, and then form the sheets and filaments of galaxies we see today – exactly as suggested in Tony's article. It is because of the dominance of that mysterious stuff called dark matter that this hypothesis works so well, and why it has become so important in our understanding of the universe's evolution.

Tony was well aware of these developments when his life was tragically cut short by a diving accident in November 2008. He had achieved great things in his 37-year career at the University of Cape Town and was looking forward to retirement. I believe he would have loved to have updated his article for this fiftieth anniversary *Yearbook*, and I feel honoured that the editors have asked me to do it for him.

Tony's fascination with the Big Questions of astronomy was matched by his skill in communicating them to the wider public, and one of his lectures was an event not to be missed (Figure 6). I well remember one in which the audience was asked to don red and green spectacles to see his galaxies in 3-d. When the foam-like structure of the universe leapt out in front of us from the projection screen, there was an audible gasp – followed by a beaming smile from Tony. That delight in 'switching on the lights' also shone through in his books, and for further reading on this topic, you can do no better than to track down his *Cosmology Revealed: Living Inside the Cosmic Egg*.

Figure 6. Tony Fairall's sense of humour was legendary, particularly when he was giving one of his immensely popular public lectures. (Image courtesy of University of Cape Town.)

A Universe of Darkness

IAIN NICOLSON

First published in the *2002 Yearbook of Astronomy*. Revised 2011.

Within range of large modern telescopes, which can probe out to distances in excess of 10 billion light years, lie billions of galaxies, each of which contains billions of stars. Galaxies are clumped together into groups, which contain from a few to a few dozen members, or clusters, which may contain hundreds, or even a few thousand member galaxies. Clusters themselves are loosely aggregated into huge, straggly, superclusters, with diameters of a hundred million light years or more.

According to the widely accepted Big Bang model, the universe originated some 15 billion years ago in a hot explosive event. Space has been expanding, and galaxies receding from each other, ever since. The theory has been very successful in explaining several key features of the observable universe, in particular, the recession of the galaxies, the relative abundances of the lightest chemical elements, and the fact that space is filled with a faint background of microwave radiation, radiation that was released about 300,000 years after the start of the expansion, when space, for the first time, became transparent.

WILL THE UNIVERSE EXPAND FOR EVER?

Intuitively, we would expect that gravity would cause the expansion to slow down as time goes by. If the overall mean density of the universe exceeds a particular value, called the critical density, gravity will eventually win the battle and at some time in the distant future the expansion will cease. Thereafter, slowly at first, then ever more rapidly, galaxies will begin to fall together until everything piles up into a Big Crunch. The best current estimates indicate that the critical density is about 9×10^{-27} kgm^{-3}, equivalent to an average of about five hydrogen atoms per cubic metre of space. If the mean density is less than the critical value, the expansion will slow down, but will continue at a finite rate for ever. If the mean density is precisely equal to the critical density, the universe will just, but only just, be able to expand for ever, the speed of recession of the galaxies approaching ever closer towards zero as time goes by.

A universe which expands for ever is called 'open', whereas a universe which expands to a finite volume and then collapses, is called 'closed'. A universe in which the mean density is exactly equal to the critical density, and which 'sits on the fence' between the open and closed models, is called 'flat'.

The presence of matter and energy causes space to be curved. In a closed universe, space is positively curved – wrapped round on itself rather like the surface of a sphere. In a positively curved space, rays of light travel along curved paths and parallel lines drawn at any particular location eventually intersect (just as lines of longitude – or meridians – intersect at the poles of the Earth). In an open universe, space is infinite in extent and is negatively curved – like the surface of a saddle; parallel lines diverge from each other in a space of this kind. In a flat universe, the net overall curvature is zero; rays of light travel in straight lines (apart from being deflected by localized clumps of matter), parallel lines remain always separated by the same distance, and space, again, is infinite in extent.

According to the inflationary hypothesis, first proposed in the early 1980s by Alan Guth, the universe underwent a brief but dramatic phase of accelerating expansion – called inflation – very early in its history. During this brief bout of accelerated expansion all distances in the universe increased by a huge factor (at least 10^{50}). The inflationary hypothesis has been remarkably successful in explaining why the microwave background is smooth and uniform over the whole sky (temperature variations from place to place on the background sky are only about one part in a hundred thousand) and in showing how tiny variations in density could have provided the seeds from which galaxies and clusters were formed. Inflation would have blown up the universe to such a vast size that its overall curvature would have become indistinguishably close to zero. Because the inflated universe is so huge compared to the region we can observe (the observable universe), space appears flat, just as the Earth appears flat if we look only at a tiny part of its surface. Inflation, therefore, would produce a universe with a geometry that was flat, or indistinguishably close to flat, and with a mean density indistinguishably close to the critical density.

Many theoretical cosmologists are attracted by the idea of a flat-space universe with a density equal to the critical density. What, however, do the observations show?

LUMINOUS MATTER IS NOT ENOUGH

In principle, the mean density of luminous matter – the visible stars and gas clouds that make up the galaxies – can be determined by adding up the masses of the luminous constituents of galaxies and clusters within a large enough volume of space to provide – hopefully – a representative sample of the universe as a whole. Measurements of this kind show that luminous matter contributes only about 0.5–1 per cent of the critical density. If luminous matter were the sole constituent of the universe, it would certainly be open and fated to expand for ever.

However, there is strong observational evidence to show that galaxies and clusters, and the universe as a whole, contain far more dark matter – matter which does not radiate detectable quantities of electromagnetic radiation – than visible matter. The 'dark stuff' may play a crucial role in determining the geometry and evolution of the universe.

EVIDENCE FOR DARK MATTER IN GALAXIES AND CLUSTERS

Evidence for the existence of dark matter is provided by the rotation curves of galaxies and the internal dynamics of clusters of galaxies, with further support coming from gravitational lensing.

In spiral galaxies such as our own Milky Way system, most of the visible matter is concentrated in a central bulge. If most of the mass is concentrated where most of the light is (in the central bulge), the speeds at which stars and gas clouds outside the bulge revolve around the galactic centre should decrease with increasing distance in a similar fashion to those of the planets in their orbits around the Sun. The galactic rotation curve (a plot of rotational velocity against distance from the centre) should slope downwards.

In fact, the observed orbital velocities of stars and gas clouds in spiral galaxies remain virtually constant (or even increase with distance) out to and beyond their visible boundaries. Because extra mass is needed to provide the additional gravitational pull that causes distant stars and gas clouds to move as fast as those which are closer to the centre, spiral galaxies have to contain five to ten times as much dark matter as luminous matter (Figure 1). The usual assumption is that the visible galaxy is embedded within an extensive spheroidal halo of dark matter. In the case of the Milky Way Galaxy, observations of the motions of smaller neighbouring galaxies indicate that the radius of the galactic halo is at least 150,000 light years and perhaps as great as 700,000 light years.

Figure 1. Hubble Space Telescope image of the edge-on spiral galaxy, NGC 4013. To account for the way in which spiral galaxies such as this rotate, they must contain 5–10 times as much dark matter as luminous matter. (Image courtesy of NASA and the Hubble Heritage Team (STScI/AURA).)

DARK MATTER IN CLUSTERS OF GALAXIES

Within clusters, the relative velocities of member galaxies are much higher than can be explained by the gravitational influence of luminous matter alone. In order to hold themselves together, and prevent member galaxies from escaping, groups and clusters have to contain ten to fifty times as much dark matter as luminous matter.

Further evidence for dark matter in galaxies and clusters is provided by gravitational lensing, a phenomenon which arises because light is deflected by the gravitational influence of large masses. Where a massive object or concentration of matter lies between an observer and a background object, it can act like a lens to produce a distorted image of the background source. Where light from a background cluster passes through a foreground cluster, numerous arc-like images of the background galaxies can be seen (Figure 2). Analysis of the light paths followed by these rays allows the mass of the foreground cluster to be estimated. Typically, gravitational lensing results point to cluster masses that exceed the visible masses by factors of ten or more.

Figure 2. Hubble Space Telescope image of multiple arc-like patterns within the rich galaxy cluster Abell 2218 caused by gravitational lensing. (Image courtesy of W. Crouch, University of New South Wales, R. Ellis, University of Cambridge and NASA/STScI.)

It is also possible that the universe contains galaxies that are completely dark – which emit no detectable light because they contain few, if any, stars. As possible evidence for this kind of object, Trentham and co-workers, of the Institute of Astronomy, Cambridge, have cited galaxy UGC 10214. A stream of matter appears to be flowing from this galaxy, as if it were interacting with another one, but no such galaxy is visible. The stream of matter appears to be flowing towards nothing (Figure 3).

Figure 3. The galaxy UGC 10214 showing a stream of matter heading towards 'nothing'. (Image courtesy of Simon Hodgkin and Neil Trentham, Institute of Astronomy, Cambridge.)

THE NATURE OF THE DARK MATTER

Some of the dark matter that resides in galaxies and their haloes may consist of ultra-faint 'stars', bodies with luminosities too low for them to be detected by current techniques. Potential candidates include old white dwarfs that have faded below currently detectable levels, neutron stars, black holes, brown dwarfs and bodies of planetary mass. White dwarfs, neutron stars and black holes represent different end-points for the evolution of stars. Brown dwarfs are cool dim bodies with masses of less than 0.08 solar masses which, because they are insufficiently massive, never attain temperatures sufficiently high to initiate the hydrogen fusion reactions that power normal stars. Bodies with masses below 0.013 solar masses (13 times the mass of the planet Jupiter) are more akin in nature to gas giant planets than stars and are regarded, therefore as 'planets'.

Despite their exceedingly low luminosities, significant numbers of brown dwarfs have been detected in recent years. Furthermore, recent high-sensitivity infrared studies of the star-forming region in the Orion nebula have also revealed a number of 'free-floating planets', objects with masses in the range 8–13 Jupiter masses which appear to be unattached to any stars. Although these specific results hint that our galaxy perhaps may contain as many brown dwarfs and 'planets' as stars, because these bodies are so much less massive than stars, their total contribution to the overall mass of the system is likely to be small.

MICROLENSING AND THE SEARCH FOR MACHOS

Dark bodies in the galactic halo, such as subluminous stars, black holes and planets, have come to be known as MACHOs ('MACHO' being an acronym for MAssive Compact Halo Object, or 'Massive Astrophysical Compact Halo Object'). During the past few years, several research teams have been searching for MACHOs by looking for the telltale signature of a phenomenon called gravitational microlensing (Figure 4).

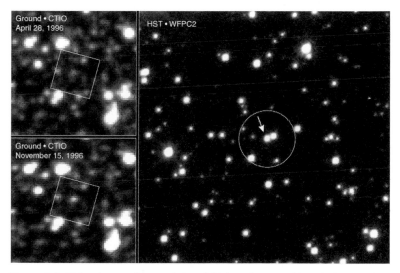

Figure 4. (Left) Two images of a crowded starfield through a ground-based telescope show subtle brightening of a star due to gravitational microlensing. (Image courtesy of NOAO, Cerro Tololo Inter-American University.) (Right) Hubble Space Telescope image of the same field resolves the lensed star and yields its true brightness. (Image courtesy of Dave Bennett, University of Notre Dame, Indiana and NASA/STScI.)

If a MACHO passes between the Earth and a distant background star, its gravitational field will act like a lens to produce a magnified image of that star. If the alignment were absolutely perfect, the observer would see the image of the star spread out into a tiny ring of light, called an Einstein ring, the radius of which depends on the mass of the lensing object. In practice, with a lensing object of mass comparable to the Sun, located in the galactic halo, the resulting Einstein ring would be far too small to be resolved by any existing telescope. However, because the gravitational lens concentrates the light of the background star, the star's apparent brightness will rise and fall in a very distinctive way as the lens passes in front of it. The duration of the event depends on the mass, distance and velocity of the lens.

During the past six or seven years, the joint US–Australian MACHO Project has recorded between thirteen and seventeen well-defined events in the direction of the Large Magellanic Cloud (LMC) – about five times as many as would be expected, statistically, on the basis of the known population of stars along the line of sight between the Earth and the LMC. The durations of these events (34–230 days) correspond to MACHO masses in the region of 0.6 solar masses, masses which match those of white dwarfs rather than those of black holes, neutron stars, brown dwarfs or planets. These results, which are consistent with those obtained by the French EROS project, imply that MACHOs are unlikely to contribute more than 10–20 per cent of the total mass of the dark matter halo.

IS THE DARK MATTER BARYONIC?

Particles such as protons and neutrons, which are acted upon by the strong nuclear force which binds together the nuclei of everyday atoms, are known collectively as baryons. Stars, planets, brown dwarfs, white dwarfs and neutron stars are composed of baryons and are, therefore, examples of baryonic matter.

According to standard theory, the present abundances of the lightest chemical elements (hydrogen, deuterium, helium and lithium) were determined by nuclear reactions that took place during the first few minutes of the hot 'fireball' phase of the Big Bang. This process, which is known as 'Big Bang nucleosynthesis' (abbreviation: BBN), depended in a very sensitive way on the mean density of baryonic matter in the universe. In order for these elements to exist in their presently observed relative proportions, the mean density of baryonic matter can be no more than about 3–5 per cent of the critical density. It is highly unlikely, therefore, that baryonic matter in any or all of its forms can account for all of the dark mass that appears to be present in galactic haloes or in galaxy clusters, let alone endow the universe with its critical density.

NON-BARYONIC DARK MATTER

If the universe has an average density equal to the critical value, and baryonic matter (luminous and dark) comprises no more than 5 per cent of the total, the remaining 95 per cent of the mass-energy of the universe has to exist in some completely different form. One possibility is non-baryonic matter – exotic elementary particles that do not respond to the strong nuclear interaction, which hardly ever interact with ordinary matter and which, therefore, are exceedingly difficult to detect. Candidate particles include neutrinos and WIMPs. Neutrinos are known to exist. The existence of WIMPs (an acronym for Weakly Interacting Massive Particles) has been predicted theoretically, but as yet none of these particles has been detected definitely.

NEUTRINOS

Neutrinos are particles which have zero electrical charge and zero, or exceedingly tiny, rest masses (if its rest-mass were zero, a stationary neutrino would weigh nothing at all). Theory suggests that a vast population of neutrinos, left over from the Big Bang, permeates the universe, there being, on average, several hundred million neutrinos in each cubic metre of space. Because they are so abundant, the average neutrino mass need be only about a ten-thousandth of the mass of an electron for there to be enough mass tied up in neutrinos to halt the expansion of the universe.

Significantly, recent experimental and observational data – in particular, some results obtained by the giant Kamiokande II neutrino detector in Japan – while not providing a direct measure of neutrino masses, do imply that their masses are finite but are probably no more than a millionth of the mass of an electron – far too small for neutrinos to provide more than a tiny fraction of the critical density.

WIMPS

WIMPs are hypothetical particles, with masses ranging from a few to thousands of times the mass of a proton, that interact only very rarely with ordinary matter. Because massive particles such as these move relatively slowly and can clump together under the action of gravity in much the same sort of way as ordinary baryonic matter does, they are known collectively as 'cold dark matter' (CDM). The existence of particles of this nature is predicted by certain theories that attempt to link together and 'unify' the fundamental forces of nature. In particular, a theory known as supersymmetry ('SUSY') predicts that at very high energies (such as would have prevailed early in the Big Bang) each ordinary particle has a heavy partner particle. For example, the supersymmetric partner to the photon is the photinos, the quark has an associated 'squark', the electron the 'selectron', and so on. The current front-runner among candidate WIMPs is the neutralino, the lightest of the supersymmetric particles, with a probable mass in the range 10–1000 times that of the proton. Although the more massive supersymmetric particles would have decayed as the universe expanded and cooled, the neutralino is expected to be stable and, if the theory is correct, to exist in large numbers in the present-day universe.

Because a WIMP is a very massive particle, when it strikes an atomic nucleus (a very rare event), it causes it to recoil. In principle, the recoil energy can be detected by measuring the heat produced by the event, by detecting the electrical charge liberated when the recoiling nucleus ionizes neighbouring atoms, or by measuring the tiny flash of light that is released when the impacted nucleus recoils. The required sensitivities are exceedingly high and any such experiments are bedevilled by spurious events caused by impacts of cosmic rays (highly energetic charged particles arriving from space) and by natural radioactivity in the laboratory and its surroundings. Although the situation can be improved in various ways, for example, by placing detectors deep underground so that the overlying rock shields the apparatus from cosmic rays while providing no impediment to WIMPs themselves, spurious events will still exceed genuine ones by a large factor.

The UK Dark Matter Consortium (a joint venture by Imperial College, London, the University of Sheffield and the Rutherford Appleton Laboratory) operates WIMP detectors at the bottom of Boulby Mine in Yorkshire (Figure 5). Their sodium iodide and xenon detectors are designed to detect light emission from WIMP-induced events. Similar types of detector are operated by the Italian–Chinese DAMA (DArk MAtter) team at the San Grasso underground laboratory. In the US, the Stanford-based CDMS (Cryogenic Dark Matter Search) uses cooled germanium detectors which are capable of measuring both the heat and the charge generated by a recoiling nucleus and so, in principle, distinguishing between WIMP-induced and neutron-induced events.

At the time of writing (January 2001) there had been no clear-cut undisputed detection of a WIMP. However, the DAMA team has claimed that a small seasonal variation in the total event rate in their detector provides statistically significant evidence for a population of

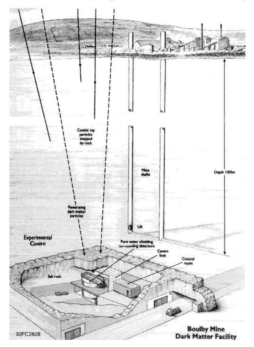

Figure 5. Schematic diagram of the UK's Dark Matter Facility at Boulby Mine. (Image courtesy of UK Dark Matter Collaboration.)

WIMPs in our Galaxy (Figure 6). The argument is as follows: if the Galaxy has a halo populated with WIMPs, then, because the Sun is orbiting around the galactic centre at a speed of around 225 km, it will appear to us as if a WIMP 'wind' is blowing past the Solar System. In addition, because the Earth is orbiting around the Sun, more WIMPs will be encountered when the Earth is heading in the same direction as the Sun (in June) as when it is heading in the opposite direction (in December). This should result in a periodic 10 per cent modulation of the WIMP event rate. From their analysis of four years of data, the DAMA team claim that there is a statistically significant 1 per cent modulation in the total event rate (which would include background events as well as WIMPs) which peaks on 2 June each year. They contend that their data is consistent with the existence of a WIMP halo consisting of WIMPs with masses in the region of 60 proton masses.

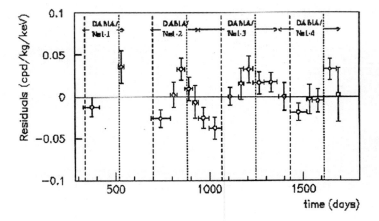

Figure 6. DArk MAtter (DAMA) count rates over a period of four years showing a possible WIMP component. (Image courtesy of R. Bernabei, Universita di Roma.)

Other researchers are sceptical. Some contend that a 1 per cent effect is not statistically significant and have suggested that the perceived fluctuations could be due to some kind of change in the background signal. The CDMS team, which by early 2000 had registered thirteen hits which mimicked WIMP events, concluded that all of their events were almost certainly caused by stray neutrons rather than WIMPs. On the basis of their data, and the sensitivity of their detectors, they are confident that the DAMA results do not reveal the presence of WIMPs.

The controversy is likely to continue until more definitive data become available and until such time as researchers can identify individual events that have the clear-cut WIMP signature.

DARK MATTER IS NOT ENOUGH

Several strands of evidence suggest that the combined mean density of baryonic and non-baryonic matter is less than the critical density. In particular, because measurements of the internal dynamics of galaxy clusters give the total mass of the cluster (baryonic plus non-baryonic matter), and the total luminosity of the cluster gives a measure of the amount of baryonic matter it contains, it is possible to calculate the ratio of baryonic mass to total mass (the baryon fraction) in clusters. If the baryon fraction in the universe as a whole is similar to that in clusters, then the overall matter density is about five to seven times greater than that of baryonic matter.

When this information is combined with Big Bang nucleosynthesis arguments – which imply that baryons (luminous and dark) provide only about 3–5 per cent of the critical density – the results imply that matter as a whole (baryonic and non-baryonic) has a mean density equivalent to about one third of the critical density.

NEW CLUES FROM THE MICROWAVE BACKGROUND

Detailed maps of the cosmic microwave background obtained by two balloon-borne experiments – BOOMERANG and MAXIMA – were published in the spring of the year 2000. Both sets of results reveal small-scale hotter and cooler patches in the microwave background (Figures 7 and 8). The predominant angular size of the strongest temperature variation (differences in temperature of about 50 millionths of a degree) was just under one degree (about twice the apparent size of the Moon in the sky). The sizes of these variation match closely what is predicted by the flat-space inflationary universe model.

In the very early universe, small differences in density between different regions of space would have arisen. In a region of enhanced density, gravity would cause particles of matter to fall inwards, but their inward motion would be halted and reversed by collisions with energetic photons. Clumps of plasma would bounce in and out, smaller clumps oscillating faster than larger ones. When the temperature everywhere in the universe dropped to about 4000 K, some 300,000 years after the start of the expansion, atomic nuclei were able to capture electrons to make complete atoms for the first time in the history of the

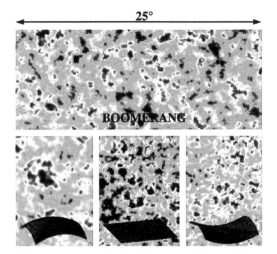

Figure 7. By observing the characteristic size of warmer and cooler spots in the BOOMERANG images, the geometry of space can be determined. They seem to indicate that space is very nearly 'flat', i.e. we live in a universe in which standard high school geometry applies. (Image courtesy of BOOMERANG Collaboration and University of California Santa Barbara.)

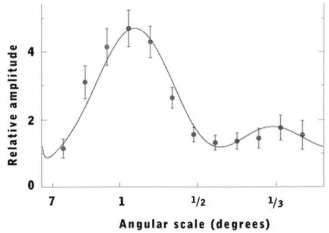

Figure 8. The power spectrum of sound waves in the primordial plasma from BOOMERANG data. These sound waves caused the temperature variations in the early universe which are seen in the BOOMERANG images. (Image courtesy of BOOMERANG Collaboration and University of California Santa Barbara.)

universe, a process that is called recombination. Space then became transparent and the radiation content of the universe was able to travel freely through the expanding volume of space with little chance of colliding with particles of matter. Diluted, stretched in wavelength and cooled by the expansion of space that has taken place since that time, this relic radiation from the Big Bang is visible now as a background of microwave radiation across the whole sky – the cosmic microwave background.

The cosmic microwave background radiation contains, in the form of hotter and cooler patches, the imprint of regions of marginally higher and lower density that existed at the time when that radiation last interacted with matter, when the universe was about 300,000 years old. The temperature variations arise because light loses energy and is shifted to a marginally longer wavelength (which corresponds to a lower observed temperature) as it 'climbs out of' regions of enhanced density. Because the drag exerted by radiation on matter reduces the amplitude of each successive 'bounce', the strongest features correspond to patches of plasma which were at maximum compression, halfway through their first cycle of oscillation, at the time when the decoupling of matter and radiation took place. Inflationary theory makes very firm predictions about the angular sizes which these temperature fluctuations should have. If space is flat, the angular size of the dominant patches should be about 1°. In a closed, positively curved, space, the bending of light rays as they travel through space would cause the patches to appear larger, whereas in an open, negatively curved space, the patches would appear smaller. The fact that the angular sizes of the dominant patches are indeed about 1° provides strong confirmation that space is indeed flat.

The maps contain temperature fluctuations on many different scales. Using mathematical techniques to extract signals of different frequencies – corresponding to different angular scales – cosmologists can plot a 'power spectrum' which shows how common patches of different angular size are. A peak in the spectrum indicates that features on that scale are common, whereas a trough indicates that features on that scale are less common. The strongest peak in the power spectrum of the BOOMERANG and MAXIMA data corresponds to features on a scale of just under 1°, and it is this peak that provides strong evidence for a flat universe.

The power spectrum ought to contain further, less prominent, peaks at angular scales of about half a degree, one third of a degree and so on. The strength of the second peak will be influenced by the relative proportions of baryonic and dark matter in the universe. Because oscillations of baryonic matter will be heavily damped down by interactions with photons, whereas oscillations of cold dark matter (CDM) will not, the second peak would be higher if the universe contains a large proportion of CDM than if CDM were a minor constituent. From the data so far, the second peak seems to be much weaker than would be expected in a universe in which CDM outweighs baryonic matter by a factor of five or ten. Some have argued that the problem would vanish if the proportion of baryonic matter in the universe were increased by a factor of between 1.5 and 2; but this seems to be in conflict with the Big Bang nucleosynthesis data. Alternatively, perhaps cold dark matter particles interact with themselves in ways that could damp down the second peak. Using a modified theory of gravity called MOND (MOdified Newtonian Dynamics), conceived in 1983 by Moti Milgrom of the Weizmann Institute as an alternative to having to invoke dark matter, Stacy McGaugh, of the University of Maryland, contends that the observed power spectrum can be matched to a universe which contains baryonic matter in the proportions given by Big Bang nucleosynthesis, and that CDM can be rejected altogether.

More data with greater sensitivity and higher resolution will be needed before any firm conclusions can be drawn about the proportion of CDM in the universe. Although under fire from several quarters, non-baryonic cold dark matter remains for the moment an important potential component of the universe. Whether or not the matter content of the universe is dominated by CDM, the BOOMERANG and MAXIMA data imply that space is flat and that some commodity other than dark matter must make up the 'missing' two-thirds of the mass-energy density of the universe.

DARK ENERGY, LAMBDA AND THE
ACCELERATING UNIVERSE

In the mid-1990s, faced with the fact that the observational evidence pointed towards a universe with a matter density only one third of that needed to make the universe flat, whereas the favoured inflation-CDM scenario requires a flat space with a mass-energy density equal to the critical density (a view which subsequently received strong observational support from the BOOMERANG/MAXIMA data), several groups of cosmologists suggested that the universe might contain an extra ingredient – dark energy. Because energy affects the overall density and curvature of the universe, dark energy – in the form of particles or fields – could provide the remaining two-thirds of the total mass-energy of the universe.

One possible form of dark energy is the cosmological constant, lambda (Λ), an extra term which Einstein added to his general relativistic cosmological equations in order to account for a static universe. At that time, which was before Hubble had shown that the universe is expanding, it was generally believed that the universe was static and that galaxies remained always at fixed distances. The cosmological constant was a property of space – a form of cosmic repulsion – which opposed gravity at very large distances. By choosing precisely the right value of lambda, Einstein could ensure that the gravitational attraction and cosmic repulsion exactly balanced and kept the galaxies at fixed distances from each other. Clearly such a balance is inherently unstable. If galaxies were nudged closer together, gravity would dominate, and they would begin to fall together, whereas if they were nudged a little further apart, cosmic repulsion would gain the upper hand and galaxies would begin to accelerate away from each other.

When Hubble discovered that the universe is expanding, Einstein called lambda his 'greatest blunder' for, had he not introduced lambda, his equations would have predicted the expansion of the universe before Hubble had discovered it. Since then, and up until very recently, cosmologists have usually assumed that the value of lambda is zero, since its existence appeared to be unnecessary to explain the observed properties of the expanding universe. In the past few years, that situation has changed.

A finite value of the cosmological constant implies that space itself (the vacuum) contains a tiny constant energy density (vacuum energy). This would affect the curvature and density of the universe and could make up the shortfall in the mass-energy that is needed to make the universe flat. For a few years, in the late 1990s, many cosmologists worked on the assumption that about one third of the mass-energy density was provided by matter (predominantly cold dark matter) and two thirds by lambda.

AN ACCELERATING UNIVERSE?

In order to act as a form of cosmic repulsion, the cosmological constant has to have negative pressure. If the pressure is sufficiently negative, it will overwhelm gravity on the large scale and cause the expansion of the universe to accelerate (rather than gently decelerating, as would be the case if it were dominated by matter).

In 1998, two research projects – the Supernova Cosmology Project and the High-redshift Supernova Search – published startling results which seem to indicate that the universe is indeed in a state of accelerating expansion at this time. Both projects have been concentrating on measuring the apparent brightnesses of Type Ia supernovae in distant galaxies. Because supernovae of this kind are highly luminous, reach similar peak magnitudes and their brightness rises and falls in a characteristic way that can be related to their peak luminosities, they provide particularly good indicators for measuring the distances of the remote galaxies within which they are embedded.

When the apparent brightnesses and redshifts of the observed supernovae were plotted on a diagram (a Hubble diagram), it emerged that the more distant supernovae were consistently fainter (and hence further away) than would be the case if the universe were expanding at a steady rate, or if the rate of expansion were slowing down. The results appear to indicate that the universe is expanding at an accelerating rate. Although it is too early to be certain that the results have been interpreted correctly, at the very least, they show that the universe is not decelerating in the way that would be expected if it consisted of matter alone.

IS LAMBDA THE DARK ENERGY?

On the face of it, the supernova results appear to be consistent with a universe with a finite (though tiny) cosmological constant, which is now in an accelerating phase of expansion. However, physicists and cosmologists are puzzled as to why lambda should have so tiny a value. Because it is a constant, its value would have been utterly microscopic compared to the energy density of matter and radiation in the very early universe. Why should it have started out with so tiny a value, and why should it have become the dominant form of energy at this particular time, 15 billion years after the beginning of the expansion? Faced with these questions, a number of cosmologists and physicists are uneasy with the concept that vacuum energy density is the dominant constituent of the present-day universe.

QUINTESSENCE

Alternative forms of dark energy – known as quintessence – were proposed in 1998 by Robert R. Caldwell (Dartmouth College, New Hampshire), Rahul Dave (University of Pennsylvania) and Paul J. Steinhardt (Princeton). There is a nice classical touch to the choice of name, 'quintessence' being the fifth element (after earth, water, air and fire), which the ancient Greek philosophers regarded as the unchanging pure element of the heavens. Quintessence may take the form of unseen particles or fields. Key features of quintessence are that it should fill all space, not clump together like baryonic or cold dark matter, and that it should possess negative pressure. If the negative pressure is high enough, quintessence, like vacuum energy density, will drive the universe into an accelerated phase of expansion.

The energy density of quintessence would have to have been feeble compared to the radiation and matter densities in the very early universe. Only recently would the energy densities of matter and radiation have dropped below the energy density of quintessence and the universe have entered a phase of accelerating expansion. One of the problems faced by any kind of dark energy (vacuum energy density or quintessence) is, why has it become the dominant component of the universe at this particular time, or in the relatively recent past?

Of particular interest, therefore, are forms of quintessence which evolve with time. Quintessence may take the form of a 'tracker field', a field of a kind that is permitted by certain of the theories that attempt to unify the fundamental forces. In the very early universe, the energy density of the tracker field would be very small compared to the energy density of radiation and matter. Initially, the universe was dominated by energetic radiation, but as it expanded and cooled, the radiation density declined more rapidly than the matter density (Figure 9). When the radiation density dropped below the matter density and the universe switched from being radiation-dominated to being matter-dominated, its rate of expansion changed, and structure (galaxies and clusters) began to form.

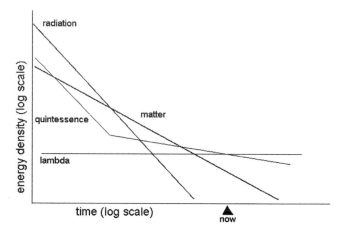

Figure 9. Schematic representation of the relative contributions to the overall density of the universe of radiation, matter, one possible form of quintessence (k-essence) and the cosmological constant (lambda). In this representation, dark energy (quintessence or lambda) has become the dominant constituent of the universe relatively recently.

Unlike vacuum energy density, tracker quintessence would change with time. It would mimic the decline in the density of radiation and matter, but in order to become the dominant form of energy, it would have to break that connection at some time in the history of the universe. One particular form of tracker quintessence, known as k-essence (kinetic energy-driven quintessence), proposed in 2000 by Armendariz, Mukhanov and Steinhardt, may have the desired properties. Its proponents suggest that k-essence would initially mimic the decline of radiation density but, when the transition from radiation to matter domination occurred, instead of tracking the matter density, it would 'freeze out' to a near constant value. Some time later, the matter density would drop below the k-essence density. Thereafter, k-essence would become the dominant component of the universe and would drive it into a phase of accelerating expansion. Because the onset of the 'freezing out' of the tracker field and the formation of structure in the universe were both triggered by the onset of matter domination, they argue that it is not surprising that dark energy should have become dominant in the relatively recent past.

INVENTORY OF THE UNIVERSE

Cosmology is in an exciting phase. Observations appear to indicate that space is flat and that the universe may be expanding at an accelerating rate. A whole range of observational evidence clearly shows that the universe contains a great deal more dark matter than luminous matter and that – even though doubts have been expressed – most of the dark matter appears to be non-baryonic cold dark matter. The overall density of dark matter in the universe appears to be no more than a third of the critical density that a flat-space inflationary universe needs to have. Dark energy, therefore, appears to be the dominant constituent of the universe at this time.

It looks as if the relative contributions to the overall density of the universe contributed by its various different constituents may be something like this: luminous matter 0.5–1 per cent, neutrinos 0.3 per cent, baryonic dark matter 3–5 per cent, non-baryonic dark matter (WIMPs or other forms of cold dark matter) 20–30 per cent and dark energy 65–70 per cent. Ground-based instruments such as the Degree Angular Scale Interferometer (DASI), ongoing balloon-borne experiments, spacecraft such as NASA's Microwave Anisotropy Probe (MAP), launched in June 2001, and ESA's Planck Mission (scheduled for launching in 2007), should greatly improve the precision of these figures within the next few years.

One thing, however, appears certain. Visible matter provides only a tiny fraction of the total mass energy content of the universe. Whether it consists of baryons, WIMPs, dark energy or something else, most of its mass-energy is dark. The splendid panorama of stars, nebulae and galaxies revealed by astronomers' telescopes is little more than thin icing on a dark and ponderous cosmic cake.

EPILOGUE

During the ten years that have elapsed since this article was first published, the observational evidence for the existence of dark matter and dark energy has increased prodigiously in quantity and quality, as has our understanding of the key rôle that these dark constituents have played, and continue to play, in determining the evolution and ultimate fate of the universe itself, and in shaping the way in which galaxies, clusters and large-scale structures have formed. The hunt for the long-sought-for, but highly elusive, dark matter particles has continued apace, while theoretical ideas about the physical nature of dark matter and, especially, of dark energy, have generated a bewildering variety of possibilities that need to be tested observationally and experimentally.

Increasingly sensitive surveys carried out by ground-based telescopes and the Hubble Space Telescope (HST) have revealed hundreds of Type Ia supernovae ('standard candles' for the measurement of distance) in distant galaxies. These observations have strongly confirmed that the present-day universe is indeed expanding at an accelerating rate and they have reached out to sufficiently high redshifts – which look back to times when the universe was less than half its present age – to reveal that the expansion of the universe switched from early deceleration (the expansion slowing under the influence of gravity) to more recent acceleration (driven by the repulsive influence of dark energy) around 5–6 billion years ago.

Precise and detailed observations of temperature fluctuations in the cosmic microwave background radiation (CMB) have yielded a treasure trove of data relating to the seeds from which galaxies, clusters and large-scale cosmic structures (such as superclusters, voids, filaments and walls) have evolved. Analysis of these data has yielded remarkably precise figures for key cosmological quantities such as the age of the universe, and the relative proportions within the cosmic mix of ordinary baryonic matter, cold dark matter and dark energy. Much of this detailed information has been provided by the Wilkinson Microwave Anisotropy Probe (WMAP), a spacecraft that was launched in 2001 and which has been studying the CMB in detail ever since (Figure 10).

A wealth of information about the distribution of luminous and dark matter, and the influence of the dark components on the evolution of structure, has been gained through comprehensive surveys (notably, the Sloan Digital Sky Survey) of the distribution of galaxies at different redshifts which reveal how structure in the universe has evolved over time. Detailed studies of gravitational lensing have enabled astronomers to measure the amount of dark matter contained in clusters and individual galaxies and to plot the ways in which it is distributed (Figure 11). Astronomers have also been able to make good use of a more subtle effect, known as weak lensing, which is the systematic distortion of the shapes of distant galaxies resulting from gravitational deflections of rays of light by the overall distribution of matter

Figure 10. Seven-Year Microwave Sky. This detailed all-sky picture of the cosmic microwave background created from WMAP data reveals 13.7 billion-year-old temperature fluctuations (shown as colour differences) that correspond to the seeds that grew to become galaxies and clusters. Analysis of this 'cosmic fingerprint' has revealed a wealth of cosmological information. Compare with Figure 7 from the original article. (Image courtesy of NASA/WMAP Science Team.)

that lies along the line of sight between these galaxies and the Earth. By measuring weak lensing at different redshifts, astronomers are beginning to build up a comprehensive picture of how the distribution of luminous and dark matter has evolved and changed over the history of the universe. All of these can be used to test theoretical models of the influence of dark matter and dark energy on the growth and evolution of cosmic structure.

Recently, a new weapon has been added to the observational armoury. During the first few hundred thousand years after the Big Bang, marginally denser clumps of matter particles would have been collapsing (under the action of gravity) and rebounding (when their inward motion was halted and reversed by collisions with energetic photons), thereby creating ripples in the overall distribution of matter and radiation, which propagated outwards at the speed of sound (which was then exceedingly high). When baryons and radiation decoupled from each other, these ripples – which are known as baryonic acoustic oscillations (BAO) – became 'frozen' into the expanding distribution of baryonic matter and therefore imposed a subtle imprint, with a characteristic size that can be calculated, on the large-

Figure 11. Dark matter in a galaxy cluster. This is a Hubble Space Telescope image of the inner part of Abell 1689, an immense cluster located 2.2 billion light years away. The distribution of dark matter within the cluster (shown in blue) has been mapped by plotting the plethora of arcs produced by the light from background galaxies that has been warped by the gravitational field of the foreground cluster. (Image courtesy of NASA, ESA, E. Jullo (Jet Propulsion Laboratory), P. Natarajan (Yale University) and J-P Kneib (Laboratoire d'Astrophysique de Marseilles, CNRS, France). Acknowledgement: H. Ford and N. Benetiz (Johns Hopkins University), and T. Broadhurst (Tel Aviv University).)

scale distribution of galaxies. This tell-tale fingerprint, which was firmly detected for the first time in 2005, provides a new and valuable 'standard ruler' – a feature of known physical size; by measuring the apparent (angular) size of baryon acoustic oscillations in the distribution of galaxies at different redshifts, astronomers have gained a new probe (which is independent of, and complementary to, supernovae 'standard candles') for investigating how the expansion rate in the universe changes over time.

Taken together, studies of the distribution of galaxies, remote supernovae, the CMB, BAO, together with other techniques, have enabled astronomers over the past decade to construct a comprehensive picture of the expansion history of the universe and the distribution of baryons and dark matter within it, and has allowed them to acquire a preliminary handle on the properties of dark energy and its influence on cosmic evolution. By combining seven years of WMAP data with BAO measurements from the Sloan Digital Sky Survey and a refined determination of the Hubble constant from Hubble Space Telescope observations, the WMAP team in 2010 pinned down the age of the universe at 13.75 billion years, the age of the universe at decoupling as 378,000 years, and the following contribution to the overall mass-energy density of the universe: baryons – 4.56 per cent; cold dark matter – 22.7 per cent; and dark energy – 72.8 per cent. The inventory of the universe now seems to be well specified.

Over the past decade or so, theoreticians have generated a plethora of possible dark energy models. The baseline hypothesis is that the dark energy which drives the accelerating expansion is Einstein's cosmological constant (Λ) in the guise of vacuum energy – a form of energy that permeates the cosmos with a density and negative (repulsive) pressure which remains constant at all times (despite the expansion of space). Other possibilities include quintessence models in which the energy density decreases with time and cosmic expansion accelerates more gently than would be the case with the cosmological constant. In so-called 'phantom energy' models the repulsive influence and concomitant acceleration increase as the universe expands, so that the stretching of space eventually becomes so extreme that all structures from galaxies to stars, planets, atoms, elementary particles and, ultimately, the very fabric of space itself, will be rent asunder in a catastrophic 'Big Rip'. Nor can theoreticians rule out the possibility that dark energy might eventually fade away, or even reverse its sign, so as to become attractive rather than repulsive, in which case – with gravity and dark energy pulling together – the universe eventually would cease to expand and would then collapse into a 'Big Crunch'.

The precise way in which the expansion of the universe varies over time depends on the properties of dark energy and how it changes (or doesn't, in the case of the cosmological constant), and the detailed fashion in which structures evolve depends on the competing influences of attractive gravity and repulsive dark energy. Astronomers are now beginning to home in on the properties of dark by utilizing more precise and detailed studies of the expansion history of the universe and the growth of structure within it. Although the data are not yet good enough to distinguish unequivocally between the cosmological constant, quintessence, phantom energy and the rest, the great majority of recently published results are consistent – within their observational errors – with the cosmological constant. For the moment, Einstein's Λ matches the evidence, and the favoured cosmological model remains ΛCDM – a universe whose principal constituents are vacuum energy (Λ) and cold dark matter (CDM), with the repulsive influence of vacuum energy holding the upper hand.

As for the nature of dark matter, the front-running hypothesis is still that it consists of weakly interacting massive particles (WIMPs), possibly weighing in at around 50–100 times the mass of a proton. The hunt for these highly elusive particles has continued with vigour over the past decade, using detectors of increasing sophistication and sensitivity in deep subterranean laboratories but, as yet (despite occasional tantalizing but inconclusive hints in the data and the continuing conundrum of the seasonal variations in the DAMA experiment's detection rate), has failed to produce any clear-cut evidence that WIMPs actually exist.

Another possibility is that the mutual annihilation of WIMPS and their antiparticles, when they collide inside massive concentrations of dark matter, may produce detectable quantities of particles (such as anti-protons and positrons) and gamma rays. Once again, despite a few intriguing hints in the data, observational searches for annihilation by-product have yet to produce definitive evidence for the existence of WIMPs. Yet another possibility is that the Large Hadron Collider, or some other high-energy particle collider, will succeed in creating and detecting dark matter particles; it remains to be seen.

The search for dark matter particles, and the quest to determine the physical nature of dark energy, are two of the pre-eminent challenges for twenty-first-century physicists and cosmologists. Despite our continuing ignorance concerning the nature of these two commodities, the evidence for their existence is compelling indeed, and now there seems to be little room for doubt that we live in an accelerating universe which is dominated by its dark side.

Choosing a Telescope

MARTIN MOBBERLEY

There is no way of saying this without sounding as though I have some kind of ego problem, so I am going to come out with it straight, and face the consequences! Frankly, apart from dealers, I reckon I answer more e-mails about buying astronomical telescopes than any other person in the UK. There, I have said it now, for better or for worse. How have I ended up in this wretched situation? Well, back in 1998 I wrote a book entitled *Astronomical Equipment for Amateurs*, which is at the root of all my problems. At the time there were no modern books on sale that described the whole range of equipment available to amateurs, despite the fact that we were obviously on the verge of an astronomical telescope revolution. Now, fourteen years later, and after numerous reprints of my first book, there are dozens of telescope books available, although they tend to be extremely specialized and some only sell in the hundreds, not thousands. From the time that first book was published I started getting letters from potential telescope purchasers and from telescope owners, too. Nowadays, the letters have been replaced by e-mails, but the volume is much higher. I suppose I did myself no favours by writing various articles and columns in the popular astronomy magazines, such as *Astronomy Now*'s 'Telescope Talk' column (formerly 'Tech Talk'). I guess I had a crazy idea that maybe if I wrote about all aspects of telescopes I would receive less mail. Wrong! I got even more.

At this point I was going to say that I have always answered every single query, but a couple of years ago I had to break that rule after the twenty-seventh e-mail from the same person in the Czech Republic, on the same sort of subject. I had to draw the line at that point! So, to cut to the chase: what, precisely, are these potential astronomers e-mailing me about? After all, surely buying a telescope and using it cannot be that mysterious? Well, the e-mails boil down into various standard questions which, as interpreted by me, roughly seem to fall into the following six categories:

1. I am thinking of buying a specific type of telescope: is this a good idea?
2. Which is best: a refractor, a reflector, or a Schmidt-Cassegrain?
3. Who makes the best telescopes?
4. I have £500/£1,000/£2,000/£5,000 pounds to spend. What should I buy?
5. If I buy a certain telescope, will I get planetary images like Damian Peach?
6. I live in a flat and want a portable telescope. What's your advice?

I receive e-mails like these on a regular basis, sometimes several per day, but none of them are that easy to reply to, simply because I do not know the mindset of the person I am dealing with, or their age, their patience threshold, or their personal circumstances. The point is, when all is said and done, using and enjoying a telescope has far more to do with the observer and his or her personal free time situation than the actual telescope. As the great observer W.H. Steavenson commented, the aperture is nowhere near as important as the man at the small end of the telescope. If you have a demanding day job, a partner, a family and a mortgage, and your spare time is full up with fixing the car, gardening and DIY, you will rarely find the time to use a telescope anyway.

WHICH INSTRUMENT WILL YOU USE THE MOST?

Every year thousands of expensive astronomical telescopes are sold in the UK. You only realize just how big the telescope business is when you visit one of the storage facilities used by the major telescope dealers. There is one not far from me, in Suffolk, and it is the size of an aircraft hangar, packed full with telescopes and equatorial mounts ranging from £500 to more than £2,000. Across the UK there are a dozen similar warehouses containing mainly imported telescopes from China, Taiwan, Mexico and Japan. Forklift trucks zip around, constantly unloading telescopes and loading them on to couriers' lorries for dispatch. It is a multimillion-pound business in the UK alone. The mystery is, though, who actually uses all these pricey telescopes every clear night? A similar question could be asked about exercise bicycles and home gym equipment; an even bigger industry. Millions of people buy sporting equipment each year in an attempt to get fit, but a lack of spare time and willpower means those dumb-bells and muscle-enhancing gizmos get used for a week and then get chucked into the garage.

So, it is obvious that, in astronomy, as in many other recreational pursuits, a lot of money is spent on equipment, but it is often used for a very short period of time, before the novelty wears off. Websites such as www.astrobuysell.com/uk/ and www.astromart.com/ are awash with used telescopes no longer required by their owners. For this reason I think that the six questions I listed earlier are irrelevant and the first two questions which the potential telescope buyer should ask are: 'If I did own a decent telescope would I enjoy using it in the cold, dark and damp British winter? Also, would I have the time to use it and am I prepared to fork out even more cash on those pricey accessories and maybe on a simple observatory enclosure too?' If the answer to these questions is no, then my best advice is to purchase a decent pair of binoculars or an inexpensive and portable refractor, rather than blow a four-figure sum on something you will rarely use.

IT'S COLD, DARK AND DAMP OUT THERE

Apart from the month of August, there are no other months in the British year when you can guarantee being able to enjoy a dark sky without being trussed up like a Christmas turkey under a layer of thermal clothing. The sky doesn't get totally dark in June and July and even by the time September comes round it can be a bit nippy after midnight. Add the ludicrous levels of humidity into the equation (with mirrors, lenses and eyepieces dripping with dew) and those kamikaze giant moths that head for any torch, and the hobby of astronomy can be pretty gruelling. The other enthusiasm killers for the telescope owner are the misery of lugging a heavy telescope outdoors and connecting all the battery-power supplies up (which rapidly die in cold weather anyway) and the typically back-breaking position of the telescope eyepiece. If you are new to astronomy, and haven't owned a decent telescope before, these energy-sapping aspects might kill your new hobby dead within weeks.

However, it is not all doom and gloom because those people who succeed in astronomy are, as in so many other aspects of life, those who patiently find ways around these hurdles and devise ingenious ways to make the hobby less arduous. Any telescope owner needs to accept that an ergonomic and user-friendly observing station, built around the telescope, is often the key to enjoying the hobby. Torches, dew heaters, power supplies, finder telescopes, eyepieces and star charts need to be within easy reach, and intuitive to use, in the dark and cold.

Now let us have a look at the telescope design to which I would award the maximum user-friendly points: the Dobsonian.

DOBSONIANS FOR A RELAXING TIME

Without a doubt the telescope that is least likely to kill your enthusiasm is the Dobsonian; a form of Newtonian reflector in which there is very little to go wrong and where the altitude and azimuth bearings are big, smooth and reliable. It is the telescope that gives you the most aperture for your money and, in astronomy, aperture is everything. Of course, huge aperture telescopes are a pain to lug around, so what you really need is a telescope that is big, but not too heavy to assemble in two parts. Personally, I'd place the dividing line, where the telescope tube weight starts to negate the aperture advantage, at around 30 cm. Any larger than this and the optical tube will weigh more than 20 kg and

Figure 1. The author with his 250 mm Dobsonian. The base was home-made from materials available at any DIY store.

become a tricky brute to lift on to the Dobsonian base without solid handles.

These days a lot is made about small and 'affordable' portable telescopes with 'Go To' drives. These telescopes, we are told, mean, in theory, that you do not need to waste time hunting for those faint, fuzzy objects. Personally, I would avoid these telescopes like the plague. You *will* waste time with them as they invariably need a tedious one-, two- or three-star alignment procedure and a power supply, and their cheap and plastic 'Go To' drives are not only inaccurate, but they also fail after a few months of use. Trust me on this one! You will be kneeling on the cold concrete of your patio with a torch trying to read the damn manual within hours of using such an instrument unless it is an observatory-mounted model with a value of £2,000 or more. A sidereal drive is only essential for imaging, and Dobsonian owners soon get used to the stars drifting in the eyepiece, requiring a tiny push to re-centre the target. Having said this, altazimuth motorized Dobsonians are also available these days.

With a decent-aperture Dobsonian, costing the same as a small 'Go To' telescope of half the aperture, you will not need to hunt for the object as just sweeping near the field will bag all the Messier objects within minutes of searching for them. However, without 'Go To' the choice of a decent finder telescope is crucial and the finder should have an aperture of at least 50 mm and preferably 60 mm. It also should be conveniently mounted on high stalks, for a straight-through unit, so that looking through it is not a neck-twisting nightmare. I would like to add one further point regarding Dobsonians. The eyepiece height is critical for observer comfort. As someone who has suffered from quite a few painful back spasms over the years I would advise minimizing bending or stooping wherever possible. My personal preference is to have a Dobsonian mounted high enough so that the draw tube is slightly higher than your standing eye height when looking at objects near the zenith, where you may need a small step to reach the eyepiece. Such a tall Dobsonian will mean that the observer can study objects at a healthy altitude without bending double. Lower altitude objects can be observed from a variable height-observing stool. In practice, for an average-height adult male, most commercial Dobsonians will force the observer to bend considerably unless the telescope is a big one. The largest Dobsonians are the most comfortable for me. For example, Meade's 16-inch f/4.5 Lightbridge Dobsonian has a focal length of 6 feet (1.83 metres) and most adults will require a step to reach the eye-piece when the telescope is pointed near to the zenith; but a small step is better than a bad back! It is easy to construct a platform on which a small Dobsonian can be raised to minimize back strain. It may seem trivial but when a telescope is always uncomfortable to use it will inevitably get used less and less.

The same argument applies to getting the telescope ready for action. If a big Dobsonian is kept indoors and has to be manhandled outside every night, enthusiasm will wane, especially when, in the UK at least, you often need to be ready to jump into action when that half-hour gap in the clouds arrives. Keeping a telescope outside in your own back garden, so you can be up and running in minutes, means you will use it. Up and down the UK there are scores of observatories owned by local astronomical societies which lie idle on most clear nights. It is easy to see why. When an observatory is a mile or two from an observer's house there is a psychological hurdle to be overcome in driving to the dome and setting up the gear, especially if it usually clouds over before you have even arrived. Then you find out you left your star atlas or eye-pieces at home! It just doesn't work out. For a user-friendly experience you need the telescope to be up and running within ten minutes of leaving the house. This may require a run-off shed or just something as simple as a tarpaulin.

If I have not yet convinced you of the joy of using a simple Dobsonian in your back garden, take the case of Gary Poyner, who lives in one of the most light-polluted areas of the country, in Birmingham. Gary is, arguably, the world's leading variable star observer and has made more than 200,000 variable star magnitude estimates since the 1970s. Some idea of Gary's level of enthusiasm and dedication can be gleaned from his observing log book for the evening of 13 January 1982, a night when all lesser mortals were indoors with the central heating turned up to the maximum, but Gary was out with his 10 inch Dobsonian:

... sky superbly clear. Began observing at 18.30 UT with the tem-perature down to −18°c already! Soon reaching mag. 14.5 with the 10 inch. Right eye 'stuck' to eyepiece whilst observing SS Aur – very painful. Conditions deteriorating by 19.30 UT. Both primary and secondary mirrors freezing over at 19.40 UT. Finder-scope useless. Closed observatory at 19.45 UT with temperature down to −21°c. Dense freezing fog by 20.15 UT.

Observing at −21°C: extraordinary! It also nicely illustrates the point that observer enthusiasm is far more important than a pricey telescope. Early in the twenty-first century Gary purchased an expensive 35 cm Schmidt-Cassegrain with, in theory, a sophisticated 'Go To' system which could slew the telescope rapidly around the sky. However, various technical hassles, drive glitches and Gary's extraordinary knowledge of the night sky convinced him to return to using a large Dobsonian which only requires a human being to power it, not dodgy electronics, plastic gears and ropey software. Like all leading observers, Gary has had his fair share of bad luck over the years but has surmounted all the hurdles placed in his way, including the destruction, by fire, of his original beloved 16 inch Dobsonian due to an electrical fault in the telescope heater. The most successful observers plough on, whatever life throws at them.

I might add that even if you can stand a bit of freezing cold weather it is, psychologically, very important just to be seconds away from a nice warm room that you can stroll back into if your fingers and toes go numb. So, having an observing station just metres from your back door

Figure 2. The tireless variable star observer Gary Poyner with his 50 cm f/4 Orion Optics Dobsonian.

will keep your enthusiasm levels high. This may sound trivial, but in reality it is crucial.

To reiterate, for the visual observer a big Dobsonian beats a small 'Go To' telescope every time and, if you want a bit of technical help for finding those faint objects, you can retrofit most Dobsonians with digital position indicators if you so wish.

NEWTONIANS

Of course, a Dobsonian is simply a Newtonian which has been mounted in a very user-friendly altazimuth style, ideal for comfortable viewing at the eyepiece, even if you have to nudge it to counteract the Earth's rotation every ten seconds or so. However, if you are choosing a telescope because you plan to do some imaging, you will want to use an equatorially mounted telescope with a polar axis carefully aligned on the north or south celestial pole, depending on whether you live in the Northern or Southern Hemisphere. The knowledgeable and pedantic reader may take issue with what I have just said, simply because it is possible to construct a Dobsonian which tracks the stars too. Various amateurs, and a few companies, make clever bases, sometimes known as Poncet platforms, in which the platform on which a Dobsonian sits slowly tilts to compensate for the Earth's rotation. In general these platforms only track safely for forty minutes or so (or the Dobsonian would tip over!) but they do enable a Dob to track the stars. However, I am assuming here that a first-time buyer will not want such an unusual instrument and therefore if he or she chooses a Newtonian, it will be equatorially mounted in the conventional way, with a polar axis and a declination axis.

Virtually all commercial Newtonians (and all other telescope designs) in the twenty-first century are sold on a so-called German equatorial mounting (GEM), in which the weight of the telescope tube is precisely balanced by a counterweight on the opposite end of the declination shaft. While this is a sturdy system, and it allows the telescope to be broken down into portable chunks, it has one main disadvantage, namely that when objects pass through the zenith the mirror end of a Newtonian tends to foul the plinth or tripod and you have to swivel the tube around to the opposite side of the mount. There are many other equatorial mount designs suitable for Newtonians but, commercially, only the GEM style is mass produced these days. This is a shame, as a well-designed fork mount can be a joy to use. Anyone who has used Patrick Moore's 15 inch f/6 fork mounted Newtonian knows how user-friendly such an instrument can be.

In recent years equatorially mounted Newtonians have, sadly, declined in popularity, due to the desire for compact portable instruments. However, Newtonians are the best instruments to modify to suit your needs and, if you can get someone to build you a custom

Figure 3. The author's 30 cm f/5.3 Newtonian. The optical tube was made by Orion Optics and the equatorial mount is a Skywatcher EQ6 Pro. Castors were added to the tripod legs to enable a smooth roll-out from a modest enclosure.

instrument, you could end up with one of the most beautiful designs of telescope it is possible to own: the long-focus Newtonian. Go back to the early 1900s and almost all Newtonian reflectors in amateur hands had f/ratios of eight or so; never faster. As very few amateurs possessed Newtonians of more than ten inches in aperture this resulted in a tube length no longer than seven feet, enabling the eyepiece to always be within reach with a modest pair of wooden steps. Apart from the fact that when the f-ratio of a Newtonian is longer than f/6 or so the secondary mirror only needs to be tiny (to catch the light cone from the primary mirror), a long, slow f-ratio has another massive advantage. The physical size of a Newtonian's 'sweet spot', the region in the focal plane where star and planet images are razor sharp, increases in proportion to the cube of the telescope's f-ratio. Let me give an example for my 25 cm f/6.3 Newtonian. At f/6.3 the central 5.5 mm of my focal plane is aberration-free as coma and other lesser aberrations are far smaller than the 0.45 arc second star images the laws of physics say will result from a 25 cm aperture. So, even allowing for the rattle of your mirrors in their mounts and the slop of the eyepieces in your draw tube, you may be able to cope with that 5.5 mm disk of perfection. At f/6 or f/7, with a collimating eyepiece, you may be able to align your optics precisely by eye, too, without having to rely on a night-time star test. If your Newtonian is f/8, that sweet spot becomes a blissful 11 mm across, making collimating, and staying collimated, a breeze. However, at f/4, the sweet spot shrinks to a painful 1.4 mm and, apart from the fact that this makes collimating the telescope a nightmare, it makes *keeping* a Newtonian in collimation a nightmare too. Without a doubt, a long-focus Newtonian of f/7 or longer is a planetary observer's dream telescope. However, commercial telescopes of this length are simply

Figure 4. Mike Harlow's home-made 35 cm Newtonian features a fork mount and a friction drive.

not mass produced any more because everyone, so we are told, wants portability. So, if you want such a dream telescope you need to have the optics specially made and build the tube yourself. Even in the twenty-first century a few amateurs do still make their own telescopes, along with the optics.

I get plenty of e-mails asking me which f-ratio is best for Newtonians and there seems to be an obsession about having a small secondary mirror obstruction; however, it is really a question of whether you want a big, semi-portable, light bucket for Deep Sky observing versus a long focus Newtonian for planetary work. In addition, whatever type of telescope the observer owns it is the observer's enthusiasm and determination to cope with all the little niggles of his or her instrument that count, and not the telescope itself. I have seen many amateur astronomers over the past few decades who seem to have decided that all they need is either a huge telescope, or an expensive telescope to make them into the next Patrick Moore, or Damian Peach, or Tom Boles. Sorry, it doesn't work like that. The best observers get good results whatever telescopes they own; they simply observe more and persevere against the hurdles that make the others give up. It's what is in the grey matter behind the observer's eyes that counts, not the telescope. If you want a perfect telescope that will do the hard work for you and make you famous, forget it: astronomy is not the hobby for you.

SCHMIDT-CASSEGRAINS

Without a shadow of a doubt the Schmidt-Cassegrain is the most popular design of telescope amongst serious amateur astronomers and astro-imagers and this has been the case for more than twenty years now. For the non-observing optical purist, though, this popularity may be hard to understand. Essentially, the Schmidt-Cassegrain (SCT) is just a traditional Cassegrain with a much shorter tube, a very fast primary mirror (f/2, amplified to f/10 by the secondary) and a Schmidt corrector plate to correct the optical aberrations created by such a fast primary mirror. The design is not, on paper at least, optimum for planetary work or for Deep Sky work and yet Schmidt-Cassegrains outsell all other types of telescope costing more than £1,000. The reason for their popularity is simple: they are just so user-friendly and versatile, for visual observing or imaging. First, the tube of an SCT is sealed against the humid atmosphere and so, even after decades, the aluminium coating on the mirrors does not degrade. At worst, the corrector plate starts to look a bit hazy after ten years. Second, because an SCT focuses by moving the f/2 primary mirror, a huge range of accessories, from digital SLRs to giant eyepieces, or CCD cameras with filter wheels, can all be focused, due to the effect the secondary mirror has on the focus position. Third, because of the highly portable compact tube and a centre of gravity near to the mirror end, the eyepiece end does not swing round by a huge amount, so a seated observer with an adjustable stool and a star diagonal can get fairly comfortable in most positions. Fourth, and this, perhaps, is the most crucial point, because Schmidt-Cassegrains exist in such large numbers and are used by so many people, there are huge amounts of accessories available for them and their design has been perfected over decades; they are a proven and tested product.

In recent years, the design has been modified to give a flatter field by adding an extra lens into the system, so that large CCD chips can be

Figure 5. A 150 mm aperture Celestron 6 Schmidt-Cassegrain is an ideal beginner's telescope with plenty of light grasp as well as a sealed rugged tube. It is also highly portable.

used, but the basic configuration and tube length is little changed. So modern Schmidt-Cassegrains may be labelled as Advanced Coma Free or, in Celestron's case, 'Edge HD', but they are still Schmidt-Cassegrains. When, in the 1960s, Celestron's Thomas J. Johnson worked out a way to cheaply mass produce Schmidt-Cassegrain corrector plates, he started a revolution.

Another way of assessing how good a telescope is for astronomy is simply to look at what instruments are used by the best and most successful observers. If you had to name one telescope brand and aperture that is the most frequently used by the world's greatest amateur CCD observers, you would probably conclude that it was the 14 inch Celestron Schmidt-Cassegrain, now also available in flatter field form as the C14 Edge HD.

The world's leading individual supernova patroller Tom Boles has three C14s, and his UK rival during the late 1990s and early 2000s, Mark Armstrong, also had three of them. They are also the instrument of choice for many planetary imagers worldwide, including the best of the lot, Damian Peach. Weighing in at only 20 kg and costing $6,000

Figure 6. The author's Celestron 14 optical tube attached to its Paramount ME equatorial mounting.

(or £6,500 in the rip-off UK) the C14 Edge HD optical tube is a compact, weatherproof and robust instrument and far cheaper than an exotic Ritchey Chrétien telescope of the same aperture. Schmidt-Cassegrain manufacturers also have numerous dealers and so, if you need accessories, they are almost always in stock.

There is one aspect of buying a Schmidt-Cassegrain that the purchaser needs to be wary of, and that is the equatorial mounting. Schmidt-Cassegrain telescopes used to be sold attached to a fork mounting which was inseparable from the optical tube. This was unfortunate as, in many cases, higher quality equatorial mountings were available if only that optical tube could be bought separately. In recent years, the SkyWatcher EQ6 Pro mount has cornered the market where a sub £1,000 mounting is required to carry a payload of up to 15 kg. While there are certainly equatorial mountings about that are better, they are all at least twice the price. So, if you are buying a Schmidt-Cassegrain telescope, you need to think long and hard about

Figure 7. Skywatcher's EQ6 Pro mounting has no rivals at the same price when a payload of 10 to 15 kg is being considered.

whether to buy one with a supplied fork, or the manufacturer's German equatorial mounting, or whether you should invest in something like an EQ6 Pro. The EQ6 Pro will handle SCT tubes up to the size of a Celestron 9.25 or Celestron 11 comfortably, but above this you really need to invest in a serious mounting such as a Vixen Atlux, or a Losmandy Titan, or even pricier mounts. By purchasing the optical tube and mount separately, you have a much more versatile system because an equatorial mount can be used for various telescopes if required. If you are planning on serious astronomical imaging, do not skimp on the equatorial mounting. It is just as important as the telescope and worth significant investment. Also, if a mounting is rated as suitable for a payload of, say, 20 kg, halve that manufacturers' figure for imaging requirements. It may well hold that weight of optical tube, but the dovetail grip, tracking and high speed slewing will be at their limit.

REFRACTORS

Refractors, with their long, sleek sealed tubes and simple design, are surely the most elegant of astronomical instruments and the ones that the general public most associate with the word telescope. In recent years, a whole host of relatively short focus refractors utilizing modern 'exotic' glass materials have entered the marketplace and attracted many purchasers, but compared to Newtonians, and even Schmidt-Cassegrains, they are very expensive instruments per inch of aperture and best used as highly portable instruments or as wide-field imaging systems. I remember that in the late 1960s, when I first became hooked on astronomy, it was often stated that a refractor was as good as a reflector of twice the aperture when observing the Moon and planets. In hindsight this advice was plainly inaccurate, although it is fairly obvious why some amateurs thought that way. There were very few Deep Sky observers about prior to the 1970s and 1980s and aperture was not as important when studying the bright Moon and planets. In addition atmospheric seeing often limits the resolving power of any telescope to little better than one arc second, so modest apertures often performed as well as large telescopes in the visual era. Another factor may have been that, throughout much of the twentieth century, the mirrors in many home-made Newtonian reflectors were of relatively low quality and were not supported well in their tubes. The thermal cool-down time of a big mirror can be a factor too, meaning that the plate-glass mirrors of the 1900s, right up to the 1960s, often changed shape until they adapted to the night air. Indeed, some were even polished so they would only hold the right shape while they were cooling! The modern glasses used today hold their shape even when cooling, although turbulence near the mirror can still be a problem. Things are a lot different now and it is generally accepted that a refractor and a reflector of the same aperture will have a very similar performance, and so if you want to look at nebulae, galaxies and comets, you will want as big an aperture reflector (or SCT) as possible.

The traditional achromatic refractor, which was the standard right up to the 1980s, had one big drawback. The two-lens design meant that only light of two specific wavelengths focused at the same point. The net result of this was that bright stars or bright crater rims ended up having a violet halo around them. This false colour could only be minimized by increasing the f-ratio of the refractor, and the bigger the telescope aperture, the longer the f-ratio. Typically, to make the false colour acceptable, a 100 mm achromatic refractor would have to be at

Figure 8. The 8 inch f/14 Thorrowgood refractor of the Institute of Astronomy at Cambridge. A magnificent Victorian instrument, but like all achromatic refractors, it suffers from chromatic aberration.

least f/10, and a 150 mm refractor, f/15. However, when you went to much larger apertures, the telescope lengths became so unwieldy that more false colour simply had to be tolerated.

In the modern era, the availability of exotic materials with ultra-low dispersion qualities (such as the large crystals of calcium fluoride used by Takahashi and the fluorite-based glass compounds made by Hoya, Ohara and CDGM) combined with triplet lenses designed by sophisticated software means that refractors with very little (or even zero) colour dispersion across the visual range can be designed. However, these materials are not cheap and so, just as during most of the twentieth century, refractors of 15 or 16 cm aperture mark the upper limit of affordability; the only difference is that whereas in the 1960s they would have been f/15 with vivid colour haloes, now they are f/7 with much smaller colour haloes.

So, is a refractor for you? Well, there is no doubt that most refractor users are also refractor lovers. They love the look of refractors and just the whole concept of a rigid, closed optical tube that does not lose collimation and gives sharp star images across a wide field. In many ways they are similar to those pricey big camera lenses made by Canon and Nikon, except that they are designed to be focused at infinity and so cannot cope with objects a few metres away. If you are keen on imaging nebulae with CCD cameras, and with digital SLRs having small pixels, then a refractor in the 60 to 150 mm aperture range, especially one fitted with a field-flattener, may be for you. They also make excellent instruments to take abroad on, for example, total solar eclipse trips, as they stay in collimation, are relatively light, and fit nicely into a standard suitcase too. Finally, they make great solar telescopes when equipped with the appropriate safe white light or H-alpha filters from a reputable dealer. Apart from those applications most observers, unless they are diehard refractor fanatics, will conclude that refractors are just a bit too pricey for the aperture.

Before leaving the subject of refractors, though, I am occasionally asked the question 'Does anyone still make big achromatic refractors for amateur astronomers?' The answer is yes, but it all seems to boil down to one company now, namely D & G Optical of Manheim, Pennsylvania USA, who make a range of standard achromatic refractor tube assemblies from a 5 inch aperture to an 11 inch; all models are available at f/12 or f/15 for anywhere between $1,700 and $10,500 at the time of writing. Before you send your order off for the 11 inch

Figure 9. Small aperture modern refractors are ideal for taking abroad, such as when travelling to solar eclipses. Nigel Evans poses near Hangzhou, China, in July 2009 with his William Optics Megrez 90 refractor.

f/15 model, though, bear in mind the tube will be 14 feet (4.3 metres) in length!

As well as refractors, there are various other expensive and exotic instruments such as Ritchey Chrétien or corrected and optimized Dall Kirkhams, but anyone willing to fork out the sort of cash required for these types of instruments will, hopefully, be a serious enough astronomer to have done their homework already.

MAKING YOUR MIND UP

Choosing which telescope to buy can be quite a traumatic business, as can the first few weeks of ownership. It would appear that nothing is ever straightforward in this life, especially where choosing and purchasing a telescope are concerned. In an ideal world the new owner would be delighted by the new instrument and, from day one, have a lifetime of pleasure with the telescope. Sadly, this is rarely the case for a variety of reasons. Telescopes can arrive months later than expected and badly packaged, so that minor or major damage has occurred. Some telescopes can be fiendishly complex to collimate and you may find that CCD cameras or DSLRs cannot be brought to focus. You may also find that you need every conceivable type of obscure Allen key to adjust the telescope and the equatorial drive may not be anywhere near as accurate as claimed. Indeed, those cheap plastic 'Go To' drives are notorious for failing within months of purchase. In addition, the fact that the optical tube's dovetail bar may not fit safely into the equatorial mounting head's dovetail channel is just as worrying. This may necessitate turning the screws on the dovetail to an insane level of tightness, causing the threads to be damaged. Then there are little niggles such as finder telescopes that won't focus on infinity, mirrors that are squeezed by the mirror cell, and discovering that no power supply is supplied with the telescope drive: you have to buy your own! It would appear that very few telescopes are 'fit for purpose' when delivered and many need days or weeks of adjustments until the new owner can actually enjoy the instrument. I am sorry to be so depressing, but my own experience and that of hundreds of other amateur astronomers confirms that telescopes are rarely perfect 'out of the box', however much they cost.

TELESCOPE REVIEWS

Of course, one source of data on telescope quality is, surely, the equipment reviews that are published in the popular astronomy magazines? Hmmm, well, call me a cynic, but as an equipment reviewer myself, I think that all these reviews need to be taken with a large pinch of salt. I know from experience that astronomy magazines never, ever, print reviews where a major advertiser in the magazine is criticized. At the editing process any harsh comments are erased or toned down to prevent the potential advertiser from throwing his toys out of the pram. Advertising revenue forms a huge part of every amateur astronomy magazine's income and so morals fly out of the window if an equipment review turns sour. On a few occasions, where I dug my heels in and objected to my criticisms being censored, a tense exchange of words would take place between dealer and magazine editor to try to decide on a compromise form of words, but, as I wouldn't budge, the review was never published and the readers were none the wiser. When the chips are down telescope dealers and magazine editors want your money in their wallets. Like everyone else they have mortgages to pay and families to feed and money comes above every other consideration. So, the worst you will see in any telescope review is a sentence such as 'we had a slight problem with the focuser but this was instantly solved by the dealers sending a replacement'. A translation would be 'the focuser was a total pile of junk and eventually we were sent a replacement which was not much better'.

For the same reason of 'not irritating our sponsors', you will never find a telescope scoring, say, 20 per cent in a group test against competing equipment. Excellent equipment will get an overall score of 99 per cent and total junk will get 91 per cent. On a similar theme, do not believe it if a well-known amateur astronomer enthusiastically endorses a very average quality product, which is quite a common occurrence in astronomy magazines in the US. They endorse the product because the dealer or manufacturer gave them a telescope free of charge! Also, be very wary of unknown dealers selling telescopes way below the market rate. They are probably selling used equipment as if it were new. If they are not authorized dealers for the product and if there is some problem over a warranty, do not touch them with a barge pole!

THE INTERNET

So, are there any independent sources of advice on telescopes? Fortunately, the answer is yes, as the Internet is a rich source of data. However, even here the amateur astronomer has to be very wary. If you go to the Yahoo! Tech Groups telescope section by entering http://groups.yahoo.com/search?query=telescopes into your web browser, you will find that there are more than 1,000 user groups devoted to telescope users. Wow! Within these groups there are some hefty forums devoted to discussing the pros and cons of the biggest-selling telescope companies, like Celestron, SkyWatcher, Meade and Takahashi. Sometimes the discussions can get quite heated as there are plenty of people in the telescope world who love the sound of their own voices and who will defend some manufacturers to the hilt, while slagging others off.

However, the bigger worry is the identity of the moderator for each group. Sometimes the user groups have been set up by someone working for the manufacturer in question and any criticism will soon be stamped upon. Even if this is not the case, manufacturers and dealers sometimes pose as customers to argue their case and calm things down. Some forum moderators have been known to pass members' address details on to dealers for their 'sponsorship', violating the data protection act.

As well as specific user groups there are also websites where all manner of telescopes are discussed. However, even here one has to be careful, as many of these sites are 'sponsored' by dealers too and they will expect any criticism of their products to be minor or a discussion thread may be terminated. One useful source of reasonably unbiased telescope advice is the Cloudy Nights site at www.cloudynights.com/ which can be a mine of useful user feedback.

However, when all is said and done the best way to make your mind up about which telescope to buy is to look at what the most successful observers and imagers use and e-mail them direct. Amateur astronomers are, in general, helpful and friendly people, who are only too keen to share their experiences. If there is a local astronomical society near to you, then join it. If it is a big one, then the chances are there will be lots of telescope owning members at their meetings, and there may even be open nights where members of the public are invited to come along and look through members' telescopes. There is no substitute for hands-on experience, on one of those cold frosty nights, which will tell you if owning a specific telescope is right for you. Even if you do not have a decent local astronomical society in your area, the biggest societies often hold star parties at weekends where astronomers can mingle and bring their own telescopes along: sometimes enormous telescopes! Walking around all the instruments at a big star party and asking questions of the telescope owners can be a mind-broadening experience.

My final suggestion is to attend the big national astronomy events, held in most countries, where the biggest telescope dealers showcase their wares every year. In the UK the annual event is called AstroFest and it is held in Kensington, London, in the first week of February, on a Friday and a Saturday. The event is always packed and hundreds of thousands of pounds' worth of telescopes are always on display. Of course, every dealer present will tell you that, without a doubt, their merchandise is the best and so it is important to ignore every word they say and just wander around the products, studying the build quality and imagining what it would be like to own one of their telescopes and to use it on a freezing cold winter's night when the tube is dripping with dew.

Whatever telescope you purchase spend a few months gathering all the data from experienced amateur astronomers before you part with any cash and bear in mind what I mentioned earlier in this article: no telescope is perfect and, however much money you spend, a telescope will not instantly turn you into the next Patrick Moore, Stephen Hawking, Rob McNaught, Damian Peach or Tom Boles. The most important items you need to succeed in amateur astronomy are determination and patience, something all the above individuals have in great abundance. A sense of humour helps too – for when things go pear-shaped!

The Story of Stellar Mass Black Holes

PAUL G. ABEL

INTRODUCTION

They are among the most exotic objects known to human beings. A place where space, time, matter – everything – cease to exist as we understand them. At their heart, matter is crushed to infinite density (zero volume) under the pull of infinite gravity at a point known as a *singularity*, where the laws of physics as we know them break down. Here, for now at least, science stops and speculation begins. I am, of course, speaking of black holes, the most intense gravitational objects in the universe, formed in the violent death throes of extremely massive stars.

The concept of the black hole is a relatively new one and, whenever I give talks on the subject, I am still surprised by the amount of confusion and misinformation there seems to be about them. This is probably partly due to science fiction writers, who quickly realized the great potential of black holes as highly exotic objects, and used them to move spaceships faster than light, make time machines bigger on the inside, or open portals to other worlds.

Part of the problem, of course, in making black holes more accessible, is the fact that the physical understanding of them depends upon Albert Einstein's General Theory of Relativity – a rather complicated theory with a complex mathematical framework. Nevertheless, the black hole has brought about some fascinating areas of research in physics, and may provide the means for us to understand how gravity works at the quantum scale – a task which has, so far, eluded even the greatest minds on the planet.

With this in mind, I thought it would be useful if my contribution to the *Yearbook* was an article which reviews what we really know about black holes, both theoretically and experimentally, and to look at what future developments are thought to be just ahead of us. Here, I have concentrated on stellar mass black holes (i.e. black holes formed by the gravitational collapse of extremely massive stars), because supermassive black holes (found in the centres of galaxies), and quantum black holes really deserve their own treatment!

I have split this article into the following sections. First, we have two sections on relativity; the first on special relativity, the second on general relativity. Since general relativity is still our best theory of gravity, it makes sense for us to look at where the theory came from, what it does and how we can be sure of its reliability. Next, we look at how stellar mass black holes form from gravitational collapse. After this we shall go on to look at one of the most startling predictions from the theoretical physics of the last century or so: Hawking Radiation. Finally, we shall look at the observational evidence we have for stellar mass black holes.

OUT WITH THE OLD – ENTER SPECIAL RELATIVITY

Although our understanding of black holes comes from Einstein's theory of general relativity, it was not Einstein who was the first to postulate the existence of such objects. In 1783, an English country parson, the Reverend John Michell, came up with the idea that there might exist stars which were so incredibly heavy that not even light could escape from their surfaces. The objects were named 'dark stars' simply because they would never emit any light. Michell's paper describing the concept of a black hole appeared in the *Philosophical Transactions of the Royal Society of London* (Figure 1). Quite independently, Pierre-Simon Laplace came up with the same suggestion in 1795. Dark stars were not given any serious consideration at the time (indeed, Laplace removed his suggestion from the later editions of his *Exposition du Système du Monde*).

These early suggestions of black holes floundered because in the eighteenth century light was thought of as a massless wave, and no one could understand how or why gravity would affect light. The other reason was, of course, our understanding of gravity at the time, i.e. Isaac Newton's theory of gravitation. Newton's theory was a great success for physics of that period; it allowed planetary motion to be predicted and enabled astronomers to calculate orbits and return periods for comets. However, as Newton himself observed, his theory did not explain what gravity actually was: it simply told us how the gravitational attraction

29. If there fhould really exift in nature any bodies, whofe denfity is not lefs than that of the fun, and whofe diameters are more than 500 times the diameter of the fun, fince their light could not arrive at us; or if there fhould exift any other bodies of a fomewhat fmaller fize, which are not naturally luminous; of the exiftence of bodies under either of thefe circumftances, we could have no information from fight: yet, if any other luminous bodies fhould happen to revolve about them we might ftill perhaps from the motions of thefe revolving bodies infer the exiftence of the central ones with fome degree of probability, as this might afford a clue to fome of the apparent irregularities of the revolving bodies, which would not be eafily explicable on any other hypothefis; but as the confequences of fuch a fuppofition are very obvious, and the confideration of them fomewhat befide my prefent purpofe, I fhall not profecute them any farther.

Figure 1. Title and excerpt from Reverend John Michell's 1783 paper which first described the concept of a black hole. The paper appeared in *Philosophical Transactions of the Royal Society of London.* (Photograph courtesy of the *Philosophical Transactions of the Royal Society of London*, vol. 74, 1783, pp. 35.)

between two bodies is proportional to the product of their masses and inversely proportional to the square of the distance between them.

Still, in the absence of anything better, Newton's theory held up until the remarkable, insightful contributions of a Swiss patent clerk, over two hundred years later. In 1916, Albert Einstein unleashed upon the world a new and most profound way of looking at the physical world, and in particular he completely replaced our notions of space, time and matter. Before we look at Einstein's theory of general relativity, let us review how physicists viewed the universe at the time.

As we have seen, it was Newton's theory of gravitation which was the fundamental theory of gravity before relativity. At the heart of this theory was the notion of fundamental space and fundamental time. Space and time were absolute, fixed and impassive. Particles and bodies in the universe moved in relation to this absolute space and the flow of time was constant, experienced at the same rate for everything and everyone in the universe, regardless of where they were. Space and time

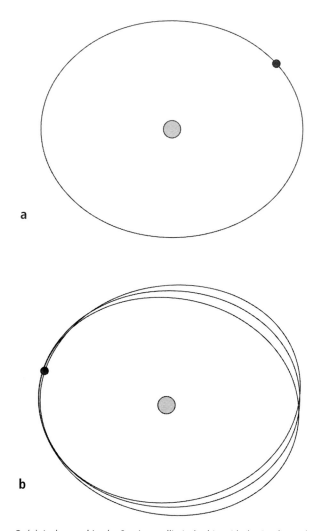

Figure 2. (**a**) A planet orbits the Sun in an elliptical orbit, with the Sun located at one focus of the ellipse. (**b**) Mercury's orbit around the Sun showing how the point of perihelion slowly precesses over time. The observed perihelion precession of Mercury is 5,600 seconds of arc per century. Taking into account all the gravitational effects of the other planets, Newtonian mechanics predicts a precession of 5,557 seconds of arc per century. Einstein's General Theory of Relativity accounts for the discrepancy of 43 arc seconds per century.

were, if you like, the two fixed pillars of the universe – set in stone – so it was thought that everything moved in relation to these two entities.

It became obvious that Newton's description of gravity was at best incomplete. Evidence of this came from observations of the planet Mercury. As we know, all planets move in elliptical orbits, with the Sun at one focus of that ellipse. Kepler's celebrated laws of planetary motion allowed one to calculate the position and orbits of astronomical bodies with very good accuracy – except in the case of Mercury. Mercury moves around the Sun in an elliptical orbit, but its orbit undergoes the phenomenon of perihelion precession. This can be explained quite simply. Look at Figure 2 a. In normal planetary orbits, a planet is at some position A and moves along its path and, after a time, returns to the point A. Now, imagine Mercury travelling on an ellipse about the Sun. Mercury starts off at a point A, and moves around in its orbit. However, because of the Sun's strong gravitational field, the orbit itself is moving, so Mercury doesn't quite return to point A, but to a point nearby: this means the orbit precesses, as you can see in Figure 2 b.

Newton's theory of gravitation could not account for Mercury's unusual behaviour. Moreover, his theory couldn't account for Galileo's observation that objects having different masses fall at the same rate. So, if we had two identical footballs, and filled one with air and the other with sand, when dropped from a tall building, both balls would fall to Earth at the same rate; and if they were released at the same time, they would hit the ground at the same time. If we were going to understand these things, we would need a better understanding of motion and gravity.

On 26 September 1905, Einstein published a paper outlining what would come to be known as his Special Theory of Relativity. It introduced two new concepts:

• The laws of physics are the same everywhere in the universe;
• Matter and energy are equivalent (i.e. $E = Mc^2$).

Special relativity also did away with the concepts of absolute space and absolute time. In their place was the notion of a frame of reference which, put simply, is a coordinate system with its own clock. All these inertial frames of reference are equal. Einstein and the mathematician Hermann Minkowski also introduced a new framework which particles and photons, indeed our entire universe, occupy: *spacetime*.

Spacetime is rather a simple and elegant concept. In relativity it is four-dimensional: it has three spatial dimensions (i.e. objects in spacetime can move left and right, up and down and backwards and forwards) and one temporal dimension. A particle moves from one point to another in spacetime along a path called a world line. Moreover, each particle has its own sense of time called *proper time*. Einstein showed that the faster we move through the spatial dimensions, the slower we move through time, and so time moves much more slowly for particles moving close to the speed of light – a phenomenon known as *time dilation*. The theory also suggested that an object possessing mass becomes shorter and more massive as it approaches the speed of light, so, in fact, nothing could travel at the speed of light because it would become infinitely massive!

Einstein had, it seems, set a speed limit for the universe. Only photons (which have zero rest mass) could travel at the speed of light, and everything else was confined to travelling below that limit. This also meant that information couldn't travel instantaneously, so if by some unfortunate mysterious event the Sun was to suddenly vanish from the Solar System, we wouldn't know about it for eight minutes and twenty seconds, because it takes light that time to travel from the Sun to Earth. So, the Sun would continue to shine in the sky for eight minutes and twenty seconds before it suddenly vanished. Interestingly, at the same time as it vanished from view, we would also notice its lack of gravitational attraction and fly off into space. Clearly, then, the action of gravity also travels at the speed of light and is not instantaneous.

These ideas revolutionized physics in ways that we have only recently come to understand. So, with new concepts of space and time, matter and energy established, Einstein embarked on modernizing the theory of gravity.

GENERAL RELATIVITY: A MATTER OF GEOMETRY

In 1916, Einstein took his ideas of special relativity further and established his theory of general relativity. Remember we said that the framework of special relativity was spacetime. In special relativity this spacetime is flat, and is called *Minkowski spacetime*. In the general theory of relativity we learn that spacetime is not flat, nor is it fixed; rather, it is a like rubber sheet and if we put heavy things like stars and planets on to the sheet, it distorts. It is this bending of the sheet (which we call spacetime curvature) which we feel locally as the effect of gravity. Thus

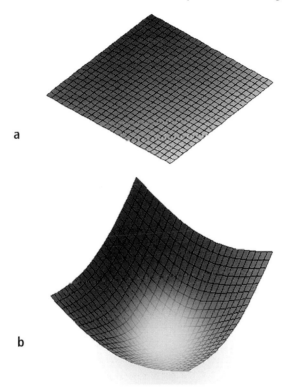

a

b

Figure 3 Diagram showing Minkowski spacetime and curved spacetime. (**a**) Flat space-time, also known as Minkowski spacetime. (**b**) When we put something heavy like a star into spacetime, it curves and stretches (like a rubber sheet would do if we were to put a heavy bowling ball on it). This curvature of spacetime is what we experience as gravity.

gravity is little more than a distortion of spacetime caused by some heavy object being there. This is illustrated in Figure 3.

General relativity is governed by a set of ten equations known as the Einstein field equations. We shall not quote them here but, simply explained, they quantify the amount of spacetime curvature (whose effect we experience as gravity) due to the presence of matter and energy. I should point out that these equations arc very difficult to solve, but by making some simple basic assumptions we have been able to solve them in some cases. When we solve the Einstein field equations, we obtain what is known in the business as a *metric*. In simple terms, a metric allows us to calculate the distance between two points in spacetime. If it is flat, this is easy – we just use Pythagoras's theorem! If it is curved, though, this is not an easy thing to calculate, and even today it is believed that there are many more solutions to the Einstein field equations yet to be found!

So what solutions do we have? Well, obviously, we have the flat spacetime of special relativity – the Minkowski spacetime of earlier. We also have spacetimes that are so curved that once anything strays too close to their intense gravitational centres, they will be pulled into the core; these are, of course, the black hole solutions of general relativity. The first ever solution of the Einstein field equations was obtained by Karl Schwarzschild – amazingly enough in the trenches of the First World War in 1915! The Schwarzschild black hole is very simple, since it is static and spherical, but more recent solutions include: charged black holes, rotating black holes and even square black holes!

We have seen that a theoretical framework for black holes was established long before our ability to look for them, and the next question to be answered was this; if black holes in the universe exist, how are they created? This leads us into the physical scenario of gravitational collapse.

BLACK HOLES FROM GRAVITATIONAL COLLAPSE

To understand how an exotic object like a black hole could come into existence, all we have to do is look at the lives and deaths of stars – the field of astrophysics known as stellar evolution. The lifetimes and deaths of stars are dictated by how massive they are, and astronomers have been able to classify various types of star and determine, in effect, how they will die.

We may think of our own star as an immensely powerful object; certainly all life on Earth is dependent upon its energy and it is the most massive object in the Solar System. However, in overall stellar terms, our Sun is a rather weak and feeble object. Astronomers now know that stars between about 0.4 and 8 solar masses end their lives in a rather quiet and unassuming manner. Stars like this eventually exhaust the supply of hydrogen fuel in their cores and expand to become a bloated red giant. The star then sheds its outer layers in a series of shells, leaving behind the core, which forms a remnant white dwarf – a hot, compact star the size of the Earth. The shells of gas that have been blown off form a beautiful object: a planetary nebula. There are many of these in the sky and M57, the Ring Nebula in Lyra, is a particularly lovely example.

The material within a white dwarf no longer undergoes nuclear fusion, so the star has no energy source, nor is it supported by the heat generated by fusion against gravitational collapse. It is supported only by something called electron degeneracy pressure. No two electrons can occupy identical states, even under the pressure of a collapsing star of several solar masses. As the core of the star contracts, all the lowest electron energy levels are filled and the electrons are forced into higher and higher energy levels, filling the lowest unoccupied energy levels. This creates an effective pressure (electron degeneracy pressure) which prevents further gravitational collapse and leads to the formation of a dense white dwarf, in which a teaspoonful of so-called electron-degenerate material would weigh several tonnes.

More massive stars of between 8 and 25 solar masses end their lives in much more spectacular fashion. The nuclear fusion process in the core of such a massive star goes through many stages. At the end of each stage, the core starts to collapse, but the collapse is halted by the ignition of a further fusion process involving more massive nuclei and higher temperatures and pressures. Eventually, when a core of iron is formed, the core starts to collapse and this time no further fusion process is available to ignite and prevent collapse. The collapse continues and the pressure continues to build, eventually forcing electrons and protons together to form neutrons. The electron degeneracy limit has been passed but now the collapse is stopped by neutron degeneracy, and the resulting collection of neutrons is called a neutron star. The outer layers of the star are blasted outwards in the most violent of stellar cataclysms: a supernova explosion. During a supernova, a single star can outshine an entire galaxy, and all that remains is the central neutron star, an object only 10–20 kilometres in diameter, but so incredibly dense that a teaspoonful would weigh about 100 million tonnes.

Stars of greater than 25 solar masses are really very massive indeed, and it is these stars – the most massive that nature can produce in her interstellar nurseries – which are thought to be the candidates for the

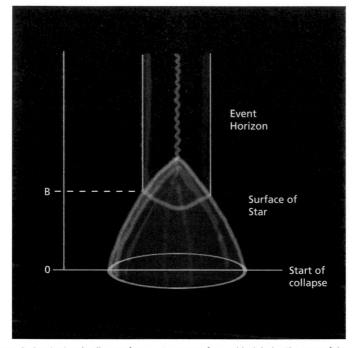

Figure 4. Gravitational collapse of a massive star to form a black hole. The start of the collapse is at t = 0 on the vertical axis. In a short time, the star collapses in on itself as can be seen by the shrinking surface as time increases. Eventually, at time t = B, a singularity forms which is surrounded by an event horizon. This is the formation of a black hole.

formation of black holes. Figure 4 shows how the collapse proceeds. Initially, the very massive star begins to collapse under its own weight. In less massive stars the collapse would be stopped by either the electron-degeneracy pressure in the case of a white dwarf, or by neutron degeneracy pressure in the case of a neutron star, but these stars are so massive that the gravitational collapse continues unopposed to form a singularity.

A singularity is an object of infinite density contained within an infinitesimal volume: it is a place of profound and total *nothing*. At this point, all the laws of physics as we know them break down and we are presented with one of the most puzzling and violent objects thought to exist. We cannot observe the singularity because even light itself cannot escape the pull of gravity here. There is a spherical region known as the *event horizon* surrounding the singularity; a boundary in spacetime beyond which events cannot affect an outside observer. It may be thought of as a 'point of no return', i.e. once an object has crossed the event horizon, the gravitational pull becomes so great that escape is impossible, and the object will fall into the hole. Light emitted from beyond the event horizon can never reach the observer.

This, then, is a black hole; a singularity surrounded by an event horizon. It is formed by the gravitational collapse of a massive star, and described by the equations of general relativity. For a long time, it was thought that this was it. Once a black hole had formed, nothing more could happen, so they were thought to be the extreme end of stellar evolution in every sense! That was until one Stephen W. Hawking had an idea – one which would radically change how we think about black holes.

BLACK HOLES AREN'T SO BLACK

How can there be anything more to black holes? We have just seen that the black hole is the final dramatic end in the life of an extremely massive star. What more could happen? Well, according to general relativity, nothing. But gravity is not the only law which has to be obeyed in the universe. There is also the second law of thermodynamics, and everything has to obey this law as well.

The second law of thermodynamics states very simply: 'entropy increases'. Entropy is simply the measure of disorder in a system, and it always increases. For example, if, like mine, your office is in a state where there are papers and books everywhere, then you have to put in some effort to tidy it up. What's more, effort and energy constantly has to be put in to *keep* it tidy. Here the entropy would be the untidy books and papers. If you keep a garden, then you know how often weeding has to be done to keep it looking neat and tidy; there is another example of entropy. Entropy always increases. Our offices don't tidy themselves, and nor do gardens weed themselves!

So it is with physical systems; they start off ordered and become disordered unless work is put in to keep them ordered, but entropy (the degree of disorder) always increases. We know from thermodynamics that if an object has entropy then it must have a temperature, and anything with a temperature radiates. So then, is there a measure of entropy for a black hole? The answer is yes! Recall the event horizon. This never decreases in size. Moreover, if we add matter to the black hole, the radius of the event horizon increases; thus the event horizon itself is a measure of the black hole's entropy! So, black holes have entropy and therefore they must radiate! To say that this result came as a shock to theoretical physicists is an understatement. Quickly it was realized that if black holes do radiate, then *how* do they radiate?

The answer comes from quantum mechanics. As we have seen, physical objects in the universe must obey all the laws of physics; we can't pick and choose which laws are satisfied. So if there is a real physical object in the universe made from matter, it is subject not only to the rules of gravity, but also to the rules of quantum mechanics. Quantum mechanics deals with the very small and tells us how atoms and their constituent particles behave. Quantum mechanics is a tried and tested theory; it has allowed us to build the computers which we all use today, so it is worth listening to what it says about empty space.

It may surprise you to learn that empty space isn't empty. Even a vacuum, which is devoid of matter, isn't empty. The theory of quantum mechanics informs us that there is an underlying background energy to the universe; we call this vacuum energy, and the effects of vacuum energy have been observed acting on atoms. This vacuum energy causes particles (and their corresponding anti-particles) to be created in pairs. Most of the time, of course, the particles annihilate each other, as a particle of matter and anti-matter do when they meet.

These particle-anti-particle pairs are coming into existence all over the universe – certainly, for example, near the surface of a black hole. So, imagine the creation of particles and anti-particles near the surface of a black hole (Figure 5). Most of the time particles and anti-particles are created and annihilate each other. However, every now and then, a pair is created on the event horizon and one particle manages to escape, taking away some of the energy of the black hole as it does so, while its partner falls into the black hole itself. These escaping photons, created by the fluctuating background energy of the universe, are the Hawking radiation, the existence of which was suggested by Stephen Hawking.

This phenomenon leads to black hole evaporation, since as a result of the black hole losing energy, it also loses mass (since we know that energy and mass are equivalent, i.e. $E = Mc^2$). Hawking went further; he managed to show in his celebrated result that the surface tempera-

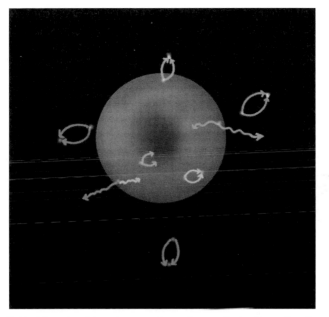

Figure 5. Particle-antiparticle pairs come into existence due to fluctuations in vacuum energy. Most annihilate each other. Some form on the surface of the event horizon, and occasionally one of the particles escapes (as a photon of Hawking radiation), while its partner falls into the black hole.

ture of a black hole is proportional to its surface gravity. This means that the larger the black hole, the less radiation is emitted.

So, our once eternal, dead black holes are actually dynamic evolving objects, although their evolution occurs over a very long period of time. However, it seems that black holes have a finite lifetime and, importantly, the smaller they get, the more Hawking radiation they emit. Eventually, as a black hole becomes smaller, it emits more and more photons until a 'runaway' process occurs and it explodes in a shower of gamma rays! Could this be the origin of the random gamma ray bursts which we see out in the distant universe? Quite possibly, but there is other observational evidence which suggests black holes exist.

OBSERVATIONAL EVIDENCE

Obviously, looking for stellar mass black holes isn't easy! As we have seen, they emit no light, and the Hawking emissions are very slow indeed. Nevertheless, there has been some indirect observational evidence suggesting that black holes do exist.

The first such candidate came from the X-ray source known as Cygnus X-1 (Figure 6), a high-mass binary system about six thousand light years from the Sun. Essentially, the system is composed of a massive blue supergiant star and an unseen companion. We know the invisible companion is there since the blue supergiant is clearly orbiting

Figure 6. An artist's impression depicting material being gravitationally sucked from the blue supergiant variable star designated HDE 226868 to form an accretion disk around the black hole known as Cygnus X-1. (Image courtesy of NASA/ESA/Space Telescope Science Institute.)

a common centre of mass. It is believed that the compact object is pulling material off the surface of the blue supergiant and creating an *accretion disk* – a disk of extremely hot stellar material which spirals into the black hole, and it is this which is the source of the X-rays. A pair of jets, directed perpendicular to the accretion disk, are carrying part of the in-falling material away into interstellar space. Measurements made of the supergiant's velocity as it orbits the unseen companion have allowed astronomers to estimate how massive this unseen object is (about 8.7 times the mass of the Sun), and it turns out this is far too massive to be something like a neutron star, so at the moment the only real candidate is a black hole. If so, the radius of its event horizon is about 26 kilometres.

At the moment, our best hope of locating stellar mass black holes is to search for their strong gravitational presence in binary star systems; i.e. to look for star systems like Cygnus X-1 whereby a visible star clearly has a massive companion, much too large to be a neutron star, and which is producing copious quantities of X-rays. It is the tell-tale X-rays that astronomers have looked for and there have been other successes.

On 22 May 1989, the Japanese Ginga satellite discovered a new X-ray source that was subsequently identified as V404 Cygni, a binary star system consisting, most probably, of a black hole of about 12 solar masses and a yellowish companion star slightly less massive than our Sun. The two bodies orbit each other every 6.5 days, and their proximity probably means that the yellow star is distorted into an egg shape by the black hole's gravitational field and that it is losing mass to the black hole. In 2009, the black hole in the V404 Cygni system became the first to have its distance accurately determined at 7,800 light years. V404 Cygni belongs to a class of X-ray binaries called soft X-ray transients. These systems are only active in X-rays for several months once every decade or so.

Another likely black hole candidate is the X-ray source known as V4641 Sagittarii, a variable X-ray binary companion located in the constellation of Sagittarius, originally thought to be about 1,600 light years from Earth, but now thought to be at least 15 times more distant. In September 1999, a short-lived, violent outburst in this system caught the attention of astronomers. R. M. Hjellming and colleagues, working at the Very Large Array (VLA) in New Mexico, discovered two superluminal jets, i.e. jets of material which *appear*, due to a small viewing angle here on Earth, to be travelling faster than the speed of light. Only something as powerful as a black hole would be capable of generating the fantastic energies required to produce such phenomena. Such X-ray binary systems are called microquasars, and they provide the most compelling evidence for black holes to date. X-ray observatories such as Chandra have been used to detect such systems and one wonders just how many black holes there might be out there within our own Galaxy.

We also have the phenomenon of *gravitational lensing*. The strong

Figure 7. A half-tone reproduction of one of the negatives taken with the 4 inch lens at Sobral, Brazil during the total solar eclipse of 29 May 1919. This shows the position of the stars, and, as far as possible in a reproduction of this kind, the character of the images, as there has been no retouching. (From F. W. Dyson, A. S. Eddington and C. Davidson, 'A Determination of the Deflection of Light by the Sun's Gravitational Field, from Observations Made at the Total Eclipse of May 29, 1919', *Philosophical Transactions of the Royal Society of London. Series A, Containing Papers of a Mathematical or Physical Character* (1920): pp. 291–333, on p. 332.)

gravitational fields around massive objects such as black holes can bend the path of light rays from a more distant background source. This lensing effect can magnify and distort the image of the background source. Indeed, this phenomenon was used in an observational test of general relativity organized by Arthur Stanley Eddington in 1919. Two expeditions were mounted – one to the island of Príncipe off the west coast of Africa and the other to the Brazilian town of Sobral – to secure photographs of the total solar eclipse of 29 May 1919. When the positions of stars recorded in photographs of the solar eclipse (Figure 7) were compared with those of the same region taken at a time when the Sun was not present, it was clear that stars near the Sun in the eclipse pictures were very slightly out of place. The Sun's gravity had apparently bent the starlight by the amount of 1.75 arc seconds, predicted by Albert

Figure 8. Simulated view of a black hole in front of the Large Magellanic Cloud (LMC). The ratio between the black hole Schwarzschild radius and the observer distance to it is 1:9. Of note is the gravitational lensing effect known as an Einstein ring, which produces a set of two fairly bright and large but highly distorted images of the LMC as compared to its actual angular size. (Image courtesy of Wikipedia Commons.)

Einstein's theory of general relativity. Although neither Eddington nor Einstein saw any practical uses for gravitational lensing, it has emerged in recent times as a most valuable technique for probing the universe.

The process of gravitational lensing would also happen if a black hole was to pass in front of a distant galaxy or the band of stars comprising the Milky Way; the intense gravity of the black hole would bend the starlight around the hole as shown in Figure 8. The effect would be quite beautiful.

'LIGHTNING IN A BOTTLE': BLACK HOLES AT CERN?

The methods described above for identifying likely black hole candidates depend on us being lucky enough to capture the effects of a black hole out there in the universe. But I couldn't finish this article without saying something about the possibilities of us creating a black hole here on Earth. I am talking, of course, about the Large Hadron Collider (LHC) at CERN and, in particular, all of the nonsense which has been talked about black holes destroying the Earth! Of course, any such black holes produced in the LHC would be tiny and would be formed by the impressive energies involved in the particle accelerator rather than by the process of gravitational collapse which we have described above.

The LHC, as you may know, is the largest, most powerful particle accelerator constructed to date. It is an extraordinary achievement and one which will allow us to recreate the physical conditions that existed shortly after the Big Bang, in the very early universe. By smashing massive particles together at speeds close to that of light, we will be able to study in detail that phase of the universe ($t \sim 10^{-11}$ s) when the cosmos consisted of a hot, dense soup: the quark-gluon plasma that existed before cooling, driven by expansion, caused quarks to form protons and neutrons. Other tasks for the LHC include the search for the elusive Higgs boson and to see if our ideas of supersymmetry (SUSY) are genuine or a false dawn.

There is also the possibility that such very high energy collisions in the LHC will produce mini black holes. There is no cause for alarm, because, as we have seen, very small black holes do not last long! Thanks to Hawking radiation, we know that they will instantaneously evaporate in a shower of gamma rays! Still, it would be quite magnificent if such an effect was observed; it would at last allow us to test our theories regarding black holes and Hawking radiation, and perhaps, too, help us achieve that holy grail of physics: a quantum theory of gravity. Whatever the future holds for black hole research, I am quite sure that it will be fascinating, and we are lucky to be alive at a time when some of the 'wild ideas' of theoretical physics finally become testable.

Lifting the Sun's Bright Veil:
Fifty Years of Solar Observing

NINIAN BOYLE

In the early 1960s, new technology in the form of telescopes, filters and even satellites, meant that we started to learn more about our local star, the Sun, than we had known in the previous few centuries. This technology, however, was largely in the hands of professional astronomers and amateurs were only able to record the passage and size of sunspots and the occasional faculae (from a Latin word meaning 'torch'), bright spots on the Sun's disk and the even rarer visible white light solar flares. More sophisticated observers would take photographs of the Sun's disk either by projecting the image from a telescope on to a large white card and photographing that, or by the careful use of heavy filtering on the telescope itself. If the optical quality of the filter was high enough, it allowed the user to see not only sunspots in all their glory, but also the solar 'granulation'. These are convection cells that cover the solar disk and make the surface take on the appearance of 'orange peel'.

EARLY DEVELOPMENTS

Over the next decade or so things really didn't change that much for the amateur solar astronomer. The quality of cameras and film improved and processing techniques also became more refined, but it would take another jump in technology to provide a significant improvement. With the advent of special filters, the professionals were able to image the Sun in narrow wavelengths of light and knowledge of the mechanisms of our star began to improve quite rapidly. These filters, though, were costly instruments and were not readily available for amateur use. This started to change in the late 1970s when a company called DayStar Filters started to produce solar filters for professional and amateur use. These instruments allowed the user to observe and image the Sun in very narrow bands of the visible spectrum and the results that they gave were, quite simply, astounding. Now the amateur solar astronomer could really observe and contribute to the rapidly accelerating wealth of knowledge that was being accumulated about our nearest star.

Sunspots were now no longer just interesting dark blotches on the otherwise bland disk of light, but dynamic, fascinating knots of activity that captured the attention and the imagination of the observer. These filters were without a doubt considerably costly devices, but as time progressed and manufacturing techniques improved, they gradually reduced in price and became more and more affordable to the ordinary but, none the less, serious amateur solar observer. White light filters, those that just reduce the amount of light being allowed through to the observer, also improved, so that amateurs could have better quality, cheaper, and most importantly, safer filters to use.

In the 1980s and early 1990s manufacturers were really beginning to tap into the potential mass market for enthusiastic amateur solar observers by improving the filters and optics that they could offer at lower and lower price tags. Other manufacturers were developing alternative types and methods of building solar filters to give the astronomer more flexibility in the type of telescope that they could use with these specialist filters, and at the same time, digital cameras were making an impression on the commercial market. Here again, rapid development of the technology meant better quality and lower cost and this quickly found a ready use with the ever technology-hungry amateur astronomy market.

One interesting development was the 'coronagraph', marketed by the German equipment specialists Baader Planetarium. This worked by using an occluding disk in conjunction with a hydrogen-alpha (H-alpha or H-α) filter. The occluding disk was designed to give an artificial solar eclipse by blanking out the disk of the Sun, leaving only the prominences visible. With the onset of digital imaging this particular piece of equipment became less popular as it was then possible to blank out the solar disk during image processing, which made the mechanical device, and the costs inherent with it, virtually redundant. The habit of processing images to give this eclipse effect has now, too, practically died out, since astronomers have realized that not only are solar prominences interesting and attractive, but so also is the activity on the solar disk, especially as improvements in the quality and design of solar filters allowed the observer to see and image more detail on the disk itself.

SOLAR SAFETY FILM

In the early years of the twenty-first century, manufacturers of specialist solar filters, such as Coronado Instruments in the US and Solarscope in the UK, were producing extremely high-quality narrow-band solar filters for the serious amateur market. Technically literate users were finding ways to adapt the now ubiquitous 'web cam' to be used with astronomical telescopes and so a new wave of affordable imaging equipment was developed in tandem with the burgeoning solar filter products. This, in turn, allowed amateur solar astronomers to produce images that were as good as, or frequently better than, pictures taken at professional observatories only a couple of decades earlier. Material such as Mylar, a very dense polymer material coated with aluminium, was found to work well as an inexpensive filter to view the Sun's photosphere, in what is known as 'white light'. This was soon improved upon with the development of 'Astro-Solar' Film from the previously mentioned German company called Baader Planetarium. This gave an improved image, was easy to use, and had a very low cost. In fact, it was so effective that it started to replace the (until then) more familiar metal-coated glass solar filters that could be attached to the front of small refracting telescopes, although this type of filter has not fallen completely out of favour. So I shall describe here in a little more detail some of these recent developments in solar filtering technology and what they now enable the modern amateur solar astronomer to observe.

Baader Astro-Solar Safety Film, to give it its correct title, is normally sold in A4-sized sheets for fitting into user-made masks or sleeves that can be made from cardboard – so the user can produce their own custom-made white light filters for use with small telescopes and binoculars. Again, the rule here must be that the finished 'filter' must be a good fit to the telescope (or binoculars) and not likely to fall off at the first puff of wind! The user should *always* check the material each time before its use for pinholes and scuffs. Should these be found, the filter should be discarded at once in such a way that it cannot be used again, even by mistake. Another point to note, while we are discussing safety, is that if you are using a telescope with a finder scope fitted, be sure that it, and any guide scopes, are securely capped off. I have heard of a case where someone, while observing the Sun through his main telescope that was properly filtered, found that his ear was burning because he had neglected to cover up the 'finder scope'! I cannot stress enough the dangers involved with solar observing, especially when using any optical device. Always seek expert advice before carrying out any such observations yourself and, if you have any doubts at all about what you are embarking upon, don't do it.

The increased popularity of Baader film is largely due to its advantage over the older-style 'Mylar' film, in that it renders the image as white, rather than with a blue cast as the previous material tended to do (Figure 1). The Sun's visible light output is, of course, white and filters that tint the image in any way are less desirable than those that do not, to any but the most casual of observers.

NARROW-BAND FILTERS

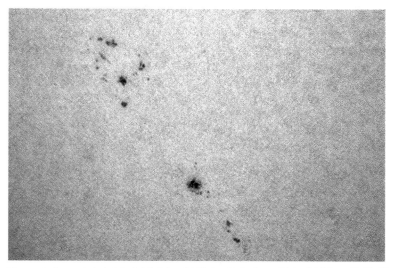

Figure 1. Sunspots in 'white light', taken by Ninian G. Boyle on 10 April 2011 at 15:18 UT using Baader Astro-Solar Safety Film on a Tele Vue TV85 Refractor Telescope and an Imaging Source DMK21 Video Camera, and processed with Registax and Photoshop.

With the advent of the 'specialist' solar filters as previously mentioned, another branch of solar astronomy opened up to the amateur observer, with the study of the Sun in a very narrow band of the spectrum located at a wavelength of 656.3 nm or 6563 Ångstroms, otherwise known as the hydrogen-alpha line. This is an area of particular interest, as it allows us to see the chromosphere of the Sun, a layer that could, until the advent of these filters, only be seen during the fleeting moments of a total solar eclipse, when the Moon's silhouette covered the bright solar disk, allowing us to glimpse the outer layers of the solar atmosphere. With this very cutting-edge technology becoming more widely available, specialist filters such as the now popular hydrogen-alpha solar filter are giving more serious solar observers the chance to view the Sun in a way that was, at one time, purely the province of professional observatories. This chromospheric layer is a region of the Sun several thousand kilometres further out from the photosphere than is observed using a 'white light' filter. It is a thin layer of rarefied gas and is a very interesting part of the solar disk. The chromospheric features of most interest to the amateur astronomer are best seen at this frequency, so that's why these filters are highly favoured.

Through an H-alpha filter you can see flares and sunspots in a way quite unlike the way you can in white light. You can also see prominences (Figure 2). These are features that tend to capture the observer's attention very quickly. Prominences are huge fountains of hot gas ejected from the Sun for thousands of kilometres into space and are

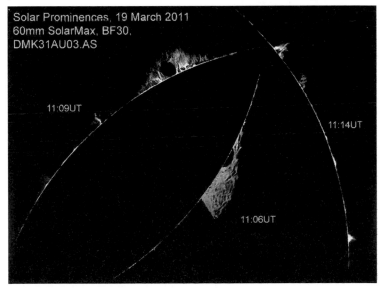

Solar Prominences, 19 March 2011
60mm SolarMax, BF30.
DMK31AU03.AS

11:09UT

11:14UT

11:06UT

Figure 2. Solar prominences taken on 19 March 2011 by David Evans using a Coronado Solarmax 60 H-Alpha Telescope and an Imaging Source DMK31 Video Camera, and processed with Registax and Photoshop.

very spectacular. They erupt from the solar disk, shooting billions of tons of matter outwards in a most awe-inspiring fashion. Prominences usually follow magnetic field lines and are connected with sunspot activity. They are frequently fast-moving events, which means that the observer can often actually see their movement, certainly over the

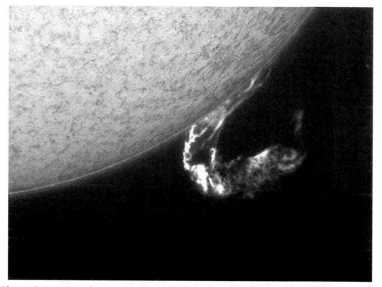

Figure 3. Massive solar prominence rising above the edge of the Sun on 19 March 2011 at 11:44 UT. Taken with a Solarscope Solarview 50 H-alpha Telescope and a Tele Vue 2x Barlow and an Imaging Source DMK21 Video Camera, and processed with Registax and Photoshop by Ninian G. Boyle.

course of a few minutes' study. Considering the considerable distance to the Sun (149.6 million kilometres), this is quite impressive. Many prominences will form features that look like 'walls of fire' or great loops and whorls (Figure 3).

Occasionally it is possible to see a prominence form and get ejected in its entirety off into space. It used to be possible only to see these prominences in particular during a total solar eclipse, but now, of course, using these filters, we can observe them practically any time. These ejection events are very violent and it is worth the observer pausing for thought at the sight of many billions of tonnes of matter being blown out into space! Should these ejections take place towards the direction of the Earth, we can experience such phenomena as auroral activity and even the disruption of communications satellites. We now get early warning of such events from the array of orbiting satellites observing the Sun, so this information is not something that we have to rely on enthusiastic amateurs to provide, but it is fascinating to watch these occasional events, nonetheless. It is also possible to observe phenomena known as spicules. The chromosphere is by no means flat or smooth and in the light of hydrogen-alpha it is possible to see 'short' jagged lines protruding upwards. They rise and fall quite rapidly, travelling at around 20,000 metres per second, and reach 1000–2000 kilometres in height. Individual spicules are quite short-lived, lasting maybe only a few minutes, but there can be a lot of them at any one time. Plages, derived from the French word meaning 'beach', are bright regions associated with sunspots and areas of intense magnetic field that appear to glow in hydrogen-alpha light.

Other features that can be seen with an H-alpha filter are filaments, which are prominences on the disk, as viewed from above. These look like dark lines that someone has 'drawn' across the face of the Sun and can appear quite jagged at times. The magnetic cellular structure of the chromosphere can also be seen using one of these filters. So owning or

Figure 4. The disk of the Sun in hydrogen-alpha, taken through a Solarscope Solarview 50 H-alpha Telescope on 10 April 2011 using an Imaging Source DMK21 Video Camera, and processed with Registax and Photoshop by Ninian G. Boyle.

having use of an H-alpha filter can be an endless source of fascination and wonder. Viewing the vibrant solar disk at this particular wavelength suddenly seems to make the Sun come alive and starts to show it off in its true glory (Figure 4). Unfortunately, the dazzling brightness of the Sun drowns out the much weaker light of the chromosphere and is therefore completely invisible to us without a narrow-band H-alpha filter. But what exactly is a hydrogen-alpha filter and how does it work?

The H-alpha line in the Sun's visible light output is a very, very narrow part of the overall spectrum that we are familiar with as the colours of the rainbow. As we have seen, this line is to be found at the red end of the spectrum, specifically at a wavelength of 656.3 nm or 6563 Ångstroms. If you were to stretch out the visible spectrum so that it could be run around the inside of an average-sized living room, and you then took a razor blade and sliced this spectrum precisely at the 656.3 nm line, the thickness of the razor cut would probably be too wide in comparison to the bandwidth at which the filter is operating! So you can see the precision with which such a filter must work. In other words, an H-alpha filter must reject every other wavelength of light in the solar spectrum. I think you will agree that this is a pretty tall order. It is probably therefore also clear that such filters are not cheap! However, modern technological improvements have helped ease the ability to manufacture such precision instruments and the price has now become within the reach of many more affluent astronomers or even societies and university departments. In fact, with such filters, these bodies are now carrying out serious scientific research and adding to our overall knowledge of our nearest star.

So how do they work? The usual design uses some form of 'energy rejection filter' at the front end of the system which, as the name suggests, removes much of the 'energy' in the form of infrared and ultraviolet light which, apart from helping to protect your eyesight, also helps to reduce the wear and tear on the filter system. This looks like a disk of very red coloured glass seen at the aperture of the filter, or held in a cell as necessary to render the correct focal length for the telescope system in certain types of instrument. The standard filter stack consists of a narrow-band blocking filter, a Fabry-Perot solid spacer crystal (better known for its use in laser technology, and known as an etalon), coupled with a broadband trimming filter. This final filter rejects any other wavelengths that are transmitted by the etalon, allowing only the particular wavelength of interest through – in this case, the hydrogen-alpha line. The quality of the quartz crystal is very important in giving the sharp- edged 'sub-Ångstrom' cut-off necessary for good detail and contrast in the solar image. Another method of creating the etalon involves the use of extremely flat and parallel glass plates with special 'interference' coatings. The flatness of the glass has to be made to tolerances of around one-hundredth of the wavelength of green light! This is a factor of at least ten times smoother than that of a good-quality telescope mirror or lens. This achieves the same goal as the crystal and is inherently more temperature stable, although the contrast is arguably not as good. These filters usually produce a light transmission that is sub-Ångstrom. In other words, narrower than the frequency of the wavelength of the light that it is centred upon. This allows for superb contrast in the image and the narrower this band-pass is, the more detail visible on the solar disk. Even narrower bandwidths can be obtained by stacking the filters or, rather, placing one of the etalons before the other. This way, it is possible to obtain a band-pass as narrow as 0.5 Ångstroms, which delivers even greater detail in the solar disk, although at some detriment to the bright prominences often visible around the rim of the Sun's disk.

The net result is an image of the Sun hitherto unknown amongst amateur astronomers that, quite apart from breathtaking views, can allow those who so desire to carry out serious scientific study of our

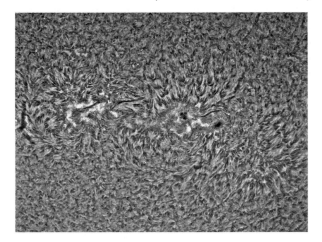

Figure 5. Active regions 1195 and 1196 imaged in H-alpha by David Evans using a Coronado Solarmax 60 H-Alpha Telescope, and processed with Registax and Photoshop.

nearest star. It is, of course, possible to take photographs (known as filtergrams) and CCD images through telescopes using these filters. With a narrow band-pass filter it is possible to study the magnetic field lines around sunspot activity (Figure 5). On the filtergrams, these tend to look rather like the lines obtained using a bar magnet and iron filings sprinkled on to a sheet of paper held over the magnet. This is an experiment that no doubt we all carried out in school physics lessons. Also now on the market are dedicated solar telescopes that have their H-alpha filters built in and so cannot be used for any other type of observing. The benefit of this is, of course, that these systems are very safe as there is no way that the filter can be removed. Personally, I prefer this type of system. These telescopes can be used as a stand-alone observing tool or can be 'piggy-backed' on a larger telescope which can have a white-light filter added that can make it possible to study the solar disk in both ordinary white light and H-alpha at virtually the same time. It is interesting to be able to make direct comparisons between the two views in this way and to be able to observe two distinct 'layers' of the Sun at the same moment.

It is possible to make a mistake with filters that you add to an existing telescope, so the need here is for extreme vigilance, especially when demonstrating solar viewing to the inexperienced or the general public. One must never become complacent! Of course, having a filter system that you can add to your own telescope is very attractive from the point of view of versatility. It comes down to personal preference. Size really is not much of an issue, unlike astronomical viewing in a dark sky where we are trying to gain as much light grasp as possible; quite the reverse. As far as resolution goes, any aperture that will resolve down to two or three arc seconds will be adequate for most purposes as, during the day, the Sun warms the atmosphere and reduces the 'seeing' (a term to describe the quality and stability of our atmosphere) considerably, so we are lucky if we manage to gain even this amount of resolution. However, this will still allow you to see prominences, filaments and the granular structure of the chromosphere in good detail. After all, the Sun is quite large!

Another area of the spectrum that is now feasible for the amateur solar astronomer to study is that of the calcium 'K' line. This is located at the opposite end of the visible spectrum to that of H-alpha, in the violet region at a wavelength of 396.8 nm. For several reasons, amateur observers use these particular filters much less. One is that the views are less interesting, as you don't see the details such as prominences and filaments that you can in H-alpha, and another is that it is much harder to see anything at all! As the calcium line is so far into the violet part of the visible spectrum it appears very dark and, to older eyes, is virtually invisible, as the sensitivity of the human retina to the shorter wave-

Figure 6. The disk of the Sun in calcium light, taken by Pete Lawrence using a Coronado 'CaK' PST Solar Telescope on 22 December 2009 at 12:02 UT, and processed using Registax and Photoshop.

lengths of the violet spectrum dramatically reduces with age. However, it is possible to capture the images on to film and CCD, thus revealing another area of solar activity not otherwise detected (Figure 6). The magnetic structures of the Sun can be appreciated in calcium light. Large-scale convective cells known as super-granulation can be seen. These large-scale cells sweep across the photosphere carrying the magnetic fields along with them and are caused by chromospheric heating. The magnetic fields themselves move towards the outer edges of the cells, where they gather and become more intense and result in the calcium emissions that can show us the solar magnetic network.

By careful tuning during the manufacture of these filters, it is possible to select other wavelengths in the spectrum for study, including that of helium and sodium, although these are less visually appealing and tend to be targets of professional, rather than amateur solar observers.

EPILOGUE

Judging by the speed of change of the technology to date, one can only guess at how it might develop over the next fifty years. Certainly the sensitivity of digital cameras seems to keep on getting better all the time and image processing techniques have kept up or even surpassed this rapid rate of improvement. This has been driven to some extent by the amateur astronomers who have used and continue to use the technology to produce superb images of the Sun and who are themselves software developers and technical pioneers. We can only marvel at what is being produced by enthusiastic solar observers and imagers right now and speculate as to what might be possible in the future. Wherever it is going, I'm sure that amateur astronomers will be at the forefront of the technology and the results that it produces.

Curious Episodes from the Observational History of the Planets

RICHARD BAUM

We have gone beyond the apparent, conceptually transforming our neighbour worlds from 'wandering stars' and objects of telescopic scrutiny, to landscapes of digital clarity. Mars shines balefully in the indigo of night, no different from how it was seen when the colourful ideas of Percival Lowell were prevalent. Nothing has really changed, except our perception of the planet, a desiccated world not dissimilar from that envisaged by Lowell, in fact, but now viewed by proxy through the cold, impersonal gaze of remote sensors. Yet, if the romance of an inhabited world has withered and its dust passed through the Ivory Gate, old fancies linger on, occasionally to trumpet their existence in odd ways. Instantly strange roads open before us and the known and familiar jostle with the hauntingly unreal, converting what began as a distraction into an absorbing adventure such as happened in August, 1989 when Voyager 2 unexpectedly revealed the existence of a tenuous ring system around the planet Neptune. Suddenly our minds switched back to Liverpool on 2 October 1846, focused on the residence of the celebrated English amateur astronomer William Lassell (1799–1880).

THE BREWER WITH A PASSION FOR ASTRONOMY

Although connected with Liverpool for much of his working life, Lassell was born in the mill town of Bolton, Lancashire. His father died in 1810 and after William had completed his education, the family moved to Liverpool, then a booming cultural and maritime centre, where, at fifteen, William was indentured to a merchant's office for seven years. Ambitious, determined and resourceful, he struck out along the road to fortune when he was twenty-six, going into business as a provincial brewer. From the venture he derived the capital and leisure to indulge his passion for machinery and astronomy, epitomized in big telescope construction. By 1820 he had built two 7-inch (17.8 cm) reflectors, one a Newtonian, the other a Gregorian, and in 1840 an exquisite 9-inch (22.9 cm) Newtonian. His most notable creations were massive reflectors of 24-inch (0.61 m) and 48-inch (1.22 m) aperture. With them he discovered faint satellites of Saturn, Uranus and Neptune, plus large numbers of galaxies and nebulae; importantly, they signalled a bright new talent on the astronomical scene (Figure 1).

Figure 1. William Lassell (1799–1880). Wealthy Liverpool brewer and amateur astronomer of note. Discovered Hyperion, the eighth satellite of Saturn, in 1848, independently of W. C. Bond in the United States. After a long and careful search, in 1851 he found two faint satellites of Uranus, Ariel and Umbriel, both interior to those found by William Herschel (1787). In 1845 Lassell completed the construction of a 24-inch reflecting telescope which for a time deceived him into believing Neptune was circled by a tenuous ring system. (Image courtesy of Gerard Gilligan and the Liverpool Astronomical Society and Michael Oates, Manchester Astronomical Society.)

The 24-inch was designed after a visit to Lord Rosse's workshops at Birr Castle, Ireland. With the help of his friend the celebrated engineer James Nasmyth (1808–1890) of Patricroft, near Manchester, he successfully installed the instrument in 1845, housing it in a 30-foot (9.14 m) dome at Starfield, his private residence near Liverpool. Almost immediately he began a study of the satellites of Saturn. Lassell's expectations were fulfilled and pride in his achievement is evident throughout his records. Doubtless the possibility of important discoveries had entered his mind.

On the morning of Thursday, 1 October 1846, the London *Times* announced that J. G. Galle (1812–1910) in Berlin, using data from the French mathematician U. J. J. Le Verrier (1811–1877) in Paris, had discovered the much-rumoured trans-Uranian planet – Neptune, as it was subsequently named; a feat described by the young astronomer John Russell Hind (1823–95), as 'one of the greatest triumphs of theoretical astronomy'.

When news of Neptune's discovery broke, Lassell's telescope had been in use for almost a year. Teething troubles had been overcome and, overall, the instrument was working perfectly, both optically and mechanically. In the meantime, Lassell's reputation had grown to international status, securing him a position in the astronomical hierarchy. His circle of friends had enlarged and included such outstanding personalities as the Rev W. R. Dawes (1799–1868) and Sir John Herschel (1792–1871).

What excited Hind instilled a sense of urgency in Sir John Herschel and he lost no time in putting pen to paper to inform Lassell where in the heavens the planet could be seen. Significantly, he added: 'Look out for satellites with all possible expedition!!' Perhaps by doing so he sought to salvage a measure of glory for the prize British astronomy had lost through its rather leisurely attitude to the search for the 'new' planet.

THE PHANTOM RING OF NEPTUNE

Lassell naturally needed no urging. A regular reader of *The Times*, he had clipped out Hind's letter and inserted the cutting in his observing diary, no doubt resolved to observe the new world at the earliest opportunity. Receipt of Herschel's letter only sharpened his anticipation;

plainly the imperative could not be ignored. That evening he directed the 24-inch (0.61 m) to the appointed spot and logged, 'Oct 2. Le Verrier's Planet.' His entry is non-committal. Perhaps conditions were not too good or moonlight too strong. Still, the bland character of the remark hints at mystery, as became clear the following night: 'I observed the planet last night the 2nd,' he confided to his observational log, 'and suspected a ring . . . but could not verify it.' On the third, in bright moonlight but in a very troubled sky, he showed the planet to all his family and was more certain of what he had previously glimpsed.

A week elapsed before he could follow up the observation; presumably bad weather had intervened. On Saturday, 10 October, in the absence of the Moon and under more favourable circumstances all doubt evaporated. Over a two-hour period the impression of 'a satellite or a suspicious-looking star . . . and a ring' persisted. Both were visible with various magnifications. 'I see the satellite [named Triton by Camille Flammarion in 1880] and suspect the ring with various powers,' he wrote. Saturn, it appeared, had a rival, for to Lassell Neptune had the look of a planet with a ring almost edge-on to the Earth (Figure 2). He even concluded that the sky could be glimpsed through the ghostly ring.

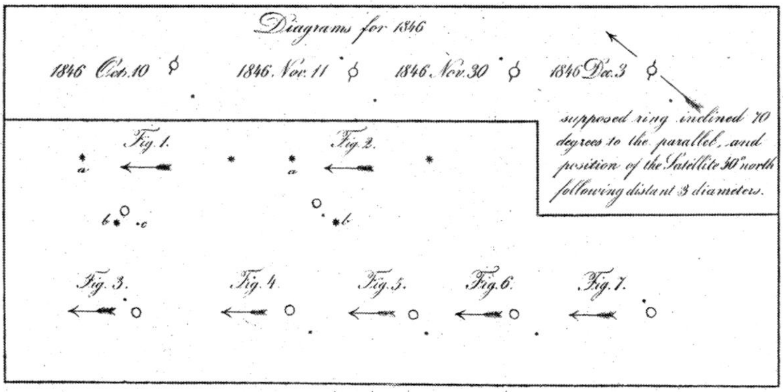

Figure 2. William Lassell's early representations of Neptune's apocryphal ring, 1846. (Image from the *Astronomische Nachrichten* 26 (1847).)

His suspicions were so strong that two days later he told John Herschel he had that very day ventured to send an account of his observations 'of a satellite and possible ring' to the editor of *The Times*.

Strangely, he thanked Sir John for 'directing [his] attention to the possible ring and satellites of Le Verrier's planet'. Herschel had not so much as hinted at a ring! It is not at all clear if Lassell expanded Herschel's suggestion about satellites before or after he made his observation, or whether it had predisposed him to imagine analogy with Saturn. However, Lassell's letter to *The Times* was made public on 14 October.

More sightings of the ring were logged before bad weather and Neptune's proximity to the Sun brought the series to a close on 15 December 1846. His communications of that time were positive and decisive; a fact reflected in the Annual Report of the Royal Astronomical Society for 1846–47, which stated that the 'existence of the ring seems almost certain'.

Other astronomers tended to agree. John Russell Hind at George Bishop's private observatory at South Villa, Regent's Park, London (Figure 3), had first resolved Neptune's disk with the South Villa telescope, a 7-inch (17.78 cm) Dollond refractor of excellent quality, on 30

Figure 3. Private observatory set up by wealthy wine merchant George Bishop at South Villa, Inner Circle of Regent's Park, London, in 1836. It housed a 7-inch Dollond equatorial refractor with which John Russell Hind made his first observation of Neptune on 30 September 1846. Initially he found the appearance of the planet perfectly normal. Once a ring was suspected he thought the planet oblong in shape. (Frontispiece of *Astronomical Observations Taken at the Observatory South Villa, Inner Circle 1839–1851*. (London, 1852).)

September, but had noticed nothing unusual about its form. After revisiting the planet on 11 October, he changed his mind and, in notes to the Royal Astronomical Society and the *Astronomische Nachrichten*, he described it as decidedly oblong in shape; its major axis tipped 30° to the meridian.

On two nights in January 1847, the Northumberland telescope at Cambridge showed James Challis and his assistant a stubby ring. Further impressions with the same instrument were obtained later that year. A rumour of sightings by De Vico in Rome drifted in, while W. C. Bond at Harvard College Observatory in the United States allegedly saw the planet less than circular in shape. James Nasmyth and the Revd W. R. Dawes also nodded their assent.

Conscious of a new feature, but cautious of illusion, Lassell hesitated. He knew even the most experienced observer could be deceived. After all, in 1787 Uranus had induced William Herschel to believe it was surrounded by two faint rings. Lassell was perplexed. Was his telescope less perfect than he imagined? Although the possibility nagged him, in writing to John Herschel on 14 August 1847, he nevertheless reaffirmed his faith in the ring and assured him that he had no intention of throwing it overboard.

Over the next six years he and his colleagues made every effort to determine its true character. In 1849, for example, he undertook a serious campaign. On 3 September, in company with Nasmyth, he easily identified the ring; a few days later it had vanished. Nor was it again seen that year. Confidence weakened and by 1851 his mood had changed as he wrote of 'the evident appendage, such as I used to take for a ring'.

The impression of 10 October 1846 continued to haunt him. The succession of strong, stable images then examined militated against illusion. That night he had closed the main dome at Starfield in the belief that 'could the planet be seen in a perfectly pure atmosphere . . . with more powerful instrumental means, it would put on the appearance of Saturn when his ring is nearly closed'.

And so it was. Frustrated by the effects of urban expansion and the turbulent air of Starfield, Lassell decamped in the autumn of 1852, and set up the 24-inch (0.61 m) under limpid Mediterranean skies on the island of Malta. On his very first night of observation, 5 October, 'the indication of the supposed ring immediately struck.' On 11 November, the impression was thought 'remarkably strong'. This was most encouraging.

As he had done at Starfield, he measured the ring's position with regard to a parallel of declination, rotated the telescope tube (and therefore the primary mirror), and repeated the measurement. A real ring, he knew, would not change its position during such a procedure; years earlier, he was unable to convince himself the one around Neptune did.

But this time there was no doubt – the ring moved. By 15 December 1852, Lassell was forced to concede that, whatever its cause, it was more 'intimately related to the telescope' than to Neptune. With that concise remark, Neptune's ghostly ring stepped aside and took its place among the 'astronomical myths of an uncritical age', to use Alexander von Humboldt's phrase. 'I suspect some illusion,' is the closest Lassell ever came to a public retraction of the ring.

And so it slipped from currency until recalled by Harlow Shapley in 1950. He was followed by Hermann Bondi in 1955 and the writer eighteen years later. This prefaced an extraordinary coincidence, for on 10 June 1982 a *New York Times* headline announced, 'Data show two rings circling Neptune.' Past and present telescoped, but investigators were sceptical – not for long. Seven years later, on 24/25 August, Voyager 2 relayed news back to Earth that Neptune does indeed possess a tenuous ring system. So what could William Lassell have seen to anticipate something visible only by sophisticated remote sensing?

Throughout the 1840s and 1850s, Neptune was poorly placed for observers in high northern latitudes. In 1846 its declination was $-13°$ and from Liverpool the planet was about 22° above the horizon when on the meridian. The problem, though, has little to do with differential refraction or, indeed, the poor quality of seeing at that site; a fact that effectively precludes serious observation of delicate features at low altitudes.

Preconception cannot be ruled out. This, of course, may have applied to Hind and Challis, who used refracting telescopes, but not to the central figures, i.e., Lassell, Dawes and Nasmyth, who each used the 24-inch (0.61 cm) reflector.

The strange case of Neptune's phantom ring must be viewed in the context of mid-nineteenth-century telescope technology. Lassell's records show he was constantly plagued by defective images. He realized that with the telescope tube at a small angle to the horizon, the speculum metal primary would inevitably bend or flex and give distorted images. Lever systems were installed to keep it from falling out of shape, but these were at an early stage of development and subject to teething troubles.

In 1848 Astronomer Royal G. B. Airy (1801–92) described some observations with Lord Rosse's 72-inch (1.83 m.) speculum, the 'Leviathan of Parsonstown': 'Upon directing the telescope to an object very near the zenith,' he wrote, 'it was seen very well defined; or at least with no discoverable fault . . . but when the telescope was directed to a star as low as the equator its image was very defective.' If the eyepiece was thrust in on the other hand, the image of the star became a well-defined straight line. When drawn out, the image was also a straight line but now at right angles to the direction of the other image. Between these two positions the image was elliptical or, at mid-point, a disk. The effect is known as astigmatism. If we mix in preconception the origin of the phantom ring seems obvious.

Even if Lassell and his colleagues were interpreting astigmatized images as evidence of a planetary ring, flexure was not the only problem they had to contend with. Imperfect temperature balance between metal mirrors and their surroundings often led to distortion. Speculum metal in particular was difficult to figure accurately and keep free of tarnish. Moreover, large telescopes such as those built by Lord Rosse and Lassell were difficult to mount and use. Taking all these factors into consideration, perhaps the real surprise is not that observers were occasionally misled, but that they achieved so much with technology that in many respects had been pushed beyond its limits.

CRATERS ON MARS AND ON MERCURY

A surprise of a different kind made headlines in 1965. As NASA investigators puzzled over the unexpected reality of Mars, following the Mariner 4 flyby in July 1965, American amateur astronomer and telescope-maker John Mellish (1886–1970) dropped a bombshell. Looking back to the period when he was a volunteer research assistant at the Yerkes Observatory, Williams Bay, in the United States, he recollected what the great 40-inch (1.02 m) refractor had revealed to him on several mornings in November and December 1915. Taken aback by the scepticism of all but one of his colleagues, and perhaps expecting similar treatment if he published, he refrained from making a formal announcement, but in subsequent years spoke readily of the experience in conversation with friends and acquaintances. Inevitably, his secret filtered out and over time developed into a brooding Homeric mystery faintly rumoured in the mythology of planetary exploration. But in 1965, with Mars and its cratered surface centre stage, Mellish finally stepped out of the shadows. Writing in the letters column of *Sky and Telescope*, he calmly announced the shock disclosure was no surprise to him, for half a century earlier he, too, had seen craters on Mars!

Critical studies by various authorities have swept away much of the mystique surrounding the story, but the truth is beyond recall because the evidence on which it rests was lost when fire destroyed Mellish's workshop the year before Mariner 4 arrived at Mars.

Some credibility attaches to the parallel case of the American astronomer T. J. J. See (1866–1962) and the craters of Mercury (Figure 4). The problem here is that the claim is bedevilled by See's controversial psychological motivations, philosophical beliefs and sociological milieux; factors which advise caution in any discussion about the man and his work. He was difficult, complex, contentious, and frequently at

Figure 4. Thomas Jackson Jefferson See (1866–1962) who, in the period 1900–1901, claimed to have glimpsed the craters of Mercury with the 26-inch refracting telescope of the United States Naval Observatory, Washington DC., over seven decades before they were imaged by the Mariner 10 spacecraft. (Image from W. L. Webb, *Brief Biography and Popular Account of the Unparalleled Discoveries of T. J. J. See* (Lynn., Mass., 1913).)

war with the astronomical fraternity in the United States. Not surprisingly, his was 'a career of controversy' as a recent biographer has termed it. Even so, British astronomers always accorded him a polite hearing, but that is not to say his ideas were any more acceptable to them.

'If we consider a small planet such as *Mercury*,' See wrote in his magnum opus *Researches on the Evolution of Stellar Systems, Vol. II*, (1910), 'which cannot retain or build up an atmosphere of much density . . . we shall perceive that its surface ought still to be similar to that of the Moon. Now in 1901, while observing under the best atmospheric conditions with the twenty-six-inch [66.04 cm] Equatorial at Washington, the author obtained the impression that the planet . . . actually has a surface similar to that of our Moon. In view of our present knowledge of the causes which have produced the craters and larger markings on the lunar surface, it is impossible to doubt that the impression gotten at Washington rests on a real foundation.'

In 1989, William Graves Hoyt (1921–85) noted that although the broad, theoretical ideas See advanced in his *Researches Vol. II* and summarized in contemporary journals were reviewed and criticized extensively at the time, none of the reviewers appeared to pay any attention to his remarks about Mercury or his drawing of the planet.

Why is this? Quite simply, other than commentators and specialists, few bothered to examine his ponderous tome in depth; his cosmogony was flawed and rejected, his standing in the scientific community in decline. It was not until 1978, sixty-eight years after its publication, that his extraordinary claim received serious scrutiny, and then only by chance.

There can be no doubt of its novelty. That is its principal attribute, not what it tells of Mercury's physical appearance. For by the late nineteenth century photometric investigations had demonstrated that the optical properties of Mercury were virtually identical with those of the Moon.

Leading astronomers of the day such as G. V. Schiaparelli (1835–1910), E. M. Antoniadi (1870–1944) and W. F. Denning (1848–1931) were more liberal and claimed traces of a thin atmosphere had been detected. John Browning (1835–1925), the well-known English instrument-maker and amateur astronomer, accurately summed up what was then known of the planet's physical state in *The Student and Intellectual Observer* of February 1870. 'Mercury, midway between the

Moon and Mars in size,' he said, 'presents, as might be anticipated, some resemblance to the Moon. Messrs. De La Rue, Huggins, and other observers, having made out markings, like the lunar craters, of a dazzling whiteness. These were seen very indistinctly as through a veil of mist.'

More exciting still, Charles Leeson Prince (1821–99) in 1867, and W. F. Denning in 1882, claimed to have seen white spots surrounded by bright rays. The most likely explanation is that they glimpsed the bright ejecta blanket surrounding impact craters of which the planet has an abundance. Two modern observers, Mario Frassati and Frank Mellilo, both affirm that such features are visible in modest apertures.

All things considered, there is no significant mystery about See's observation, only a lingering sense of puzzlement about its true nature. Quite possibly it owes its origin to See's prescience about impact theory: 'We may dismiss the old volcanic theory once and for all, as false and misleading,' he proposed in his *Researches*, 'and may look upon our satellite as a growing mass of cosmical dust, which presents to us the most lasting and convincing evidence of the processes of capture, collision and accretion by which the heavenly bodies were formed.' And in extension of the argument: 'No doubt all the planetary bodies, both large and small, have gone through a similar experience, and we may conceive all of them, the Earth included, to have suffered countless collisions with satellites of considerable size.'

Given that See was fully aware of what photometry had disclosed about the surface of the planet, his claim in general assumes the character of an ambiguous truth; something that owes as much to theoretical considerations, in this instance impact theory, as it does to actual observation. Analogy with the old Martian problem is self-evident.

Is there evidence other than his verbal description to substantiate the observation? In part yes; a fragment exists in his official observing diary. It takes the form of a rough sketch showing a half-illuminated Mercury inhabited by a circular feature. The entry is dated 24 August 1900 and annotated: 'Markings of this nature strongly suspected.' It is a telling statement, pregnant with hidden meaning; an index of immediate perception, an aide memoire, of something seen that departs from the ordinary. Still the exact truth will never be known.

Unlike Mellish, See did not claim to have seen craters, only that the surface of Mercury appears similar to that of the Moon. He backed this up with a finished drawing published as an illustration in his magnum opus. It shows the planet half illuminated and heavily pockmarked with circular features – self-evidently craters (Figure 5). In effect, it tells us this is how he believed Mercury would look like if we could examine it under good conditions at high resolution. Straightforward enough, it may be said. Yet it begs the question. Why did Dr See prefer to describe his observation in terms of the Moon rather than openly state that he had seen craters? Why did he avoid the use of the word craters? Did he seek to exemplify scientific caution? Perhaps, but his graphic illustration in *Researches Vol. II* contradicts that assumption. Possibly it evinces the hope his observation would thus make a greater impact? If so, he failed, because he succeeded only in turning what is otherwise a most remarkable observation into an ambiguity whose veracity has long been questioned.

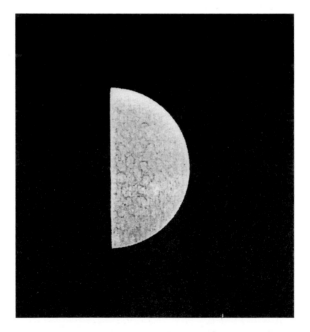

Figure 5. How T. J. J. See imagined the surface of Mercury would appear if viewed under good conditions at high resolution. Illustration shows the planet to be heavily cratered. It is probably based on impressions obtained with the 26-inch equatorial refracting telescope of the U.S. Naval Observatory, Washington DC in 1900 and 1901. (Image from T. J. J. See, *Researches on the Evolution of Stellar Systems, Vol. II*, Plate 20, Figure a. (Lynn., Mass., 1910).)

EPILOGUE

What is to be made of such episodes? The archives are full of surprises – some fanciful, some astonishing, others prescient, as in the case of the three instances above. Whatever their truth, no matter how they are construed, each has a story to tell, and something relevant to impart. Significance is to be found in their historical, philosophical and sociological context. We may exemplify this by reference to the famous lines of Samuel Taylor Coleridge in his ballad *The Rime of the Ancient Mariner*:

> *While clome above the Eastern bar*
> *The horned Moon, with one bright Star*
> *Almost atween the tips.*

Poetic licence bordering on the absurd, the cynic may observe. Yet poetic faith is vindicated. Evidence of the phenomenon abounds in the archives. The star-dogged Moon may indeed be identified as a denizen of a visionary sky. But to the astronomer there is no abrogation of physical law, only a colourful representation of what is known as an anomalous occultation, where a bright object such as Venus, or a first-magnitude star at first contact with the limb of the Moon appears to jump on to the hemisphere facing the observer. It is in this context that examination of the presumed objectivity of science may importantly influence its future development.

FURTHER READING

The Apocryphal Ring of Neptune

Richard Baum, *The Haunted Observatory: Curiosities from the Astronomer's Cabinet* (New York, Prometheus Books, 2007).

Richard Baum, *The Planets: Some Myths and Realities* (Newton Abbot, David and Charles, 1973).

Norris S. Hetherington, *Science and Objectivity: Episodes in the History of Astronomy* (Ames, Iowa, Iowa State University Press, 1988).

William G. Hoyt, 'Reflections Concerning Neptune's Ring,' *Sky and Telescope* 55 (1978), 284–5.

Robert W. Smith and Richard Baum, 'William Lassell and the Ring of Neptune: A Case Study in Instrumental Failure', *Journal for the History of Astronomy* 15 (1984), 1–17.

Mellish and the Craters of Mars

Rodger W. Gordon, *Observing the Craters of Mars, Part 1* (Middlesex, New Jersey: Typographica Publishing, 2003).

Rodger W. Gordon, 'Martian Craters from Earth', *Icarus* 29 (1976), 153–4.

Richard McKim and William Sheehan, 'Visibility of Martian craters from Earth', *Journal of the British Astronomical Association* 105 (1995), 137.

John E. Mellish, 'Letter to Editor', *Sky and Telescope* 31 (1966), 339.

Thomas R. Williams, 'Telescopes, Marriages, and Mars: The Life of John E. Mellish', *Sky and Telescope* 98 (1999), 84–8.

T. J. J. See and the Craters of Mercury

Richard Baum, 'Historical Sighting of the Craters of Mercury', *Journal of the Association of Lunar and Planetary Observers* 28 (1979),17–22.

William Graves Hoyt, 'T. J. J. See and Mercurian Craters', *Journal for the History of Astronomy* 12 (2) (1981), 139–42.

John Lankford, 'A Note on T. J. J. See's Observations of Craters on Mercury', *Journal for the History of Astronomy* 11 (2) (1980), 129–32.

Thomas J. Sherrill, 'A Career in Controversy: The Anomaly of T. J. J. See,' *Journal for the History of Astronomy* 30 (1) (1999), 25–50.

T. J. J. See, *Researches on the Evolution of the Stellar Systems, Vol. II. The Capture Theory of Cosmical Evolution* (Lynn., Mass., USA, Thos. P. Nichols & Sons, 1910).

Andrew T. Young, 'Mercury's Craters from Earth', *Icarus* 34 (1978), 208–9.

Fifty Years of Hi-Tech Magic: From Astrophotography to Astrophotonics

FRED WATSON

Astronomers are used to dealing with tricks of time and distance. Our old friend and ally, general relativity, tells us that on very large scales, space and time are ridiculously bendy, and we are all quite comfortable with that – cosy, even. Yet, on scales a lot smaller than our cosmic comfort zone, time and distance can take us completely by surprise – bamboozle us at times. Now, however, it's not relativity calling the shots, but the foibles of the human mind . . .

Sitting here in comfort with the United Kingdom Schmidt Telescope on its mountain-top in rural New South Wales, I am brought up short by thoughts of time and distance. It's a pleasant late-summer day outside, and through the window I can see the shadows of clouds dappling eucalyptus-green foothills in the afternoon sunshine. The clouds don't worry me – they are fair-weather cumulus, and will evaporate as soon as the sun goes down. The telescope upstairs – itself a survivor from a bygone age – will once again swing into action and, with a good deal of coaxing from me and my robotic assistant, will resume its task of measuring the intimate details of stars a hundred or so at a time. How can an elderly 1.2-metre telescope, which, by modern professional standards, is pretty miniscule, continue to produce leading-edge data of this kind? As we shall see, the answer lies in technology – and in scientific imperative.

But back to time and distance. Often, when I stroll around the telescope's outside catwalk (think moggies checking the weather for night-time prowling, rather than astronomers modelling observatory chic), I'm struck by the beauty of my surroundings. Australia is a land of extremes, but this range of rolling mountains is a peculiarly gentle environment – almost English in its rural charm. The England I grew up in was very different: I was surrounded by the soot-blackened stonework of a northern industrial city. Until my first visit to the south, I thought all stone buildings were black. Crumbs. Small wonder that, as a youngster, I turned to the stars for comfort – not that they were visible all that often through the murk.

One day, into the grimy world of this star-struck Yorkshire schoolboy, dropped the very first edition of the *Yearbook* you now hold in your hands. Regrettably, I no longer have a copy of that 1962 volume – it belonged to the school library, and I was still far too honest to nab it. But in my mind's eye, I can see it today, with its dramatic cover photo of the southern end of the Moon's last-quarter terminator.

For me, it was 200-odd pages of pure inspiration. Everything you could want as a would-be astronomer, all between one set of covers. Not just monthly star charts and sky notes, but up-to-date, informative articles on everything astronomical, from the moons of Jupiter to the structure of our Galaxy – not to mention Patrick Moore's brave stab at explaining the origin of the universe.

If that all sounds familiar, you'll be starting to understand my problem with time and distance. It will not have escaped your attention that this *2012 Yearbook of Astronomy* contains the self-same blend of monthly star charts and sky notes, and, yes, up-to-date, informative articles on everything astronomical. So, what does that prove? Well, for a start, it tells you that the original editors – Patrick, and his senior colleague J. G. Porter – got the recipe right first time in their fledgling *Yearbook*. Perhaps it also tells you that astronomers are a conservative bunch, and like to see the same sort of thing year after year.

Paradoxically, though, what it brings home to me is just how much has changed. The fact that, year after year, the latest developments in astronomy can fill a sizeable fraction of a chunky paperback, speaks of a dynamic science that is vigorously evolving. If you compare what we know today with what we knew fifty years ago, the difference is staggering. And there you have it. When I think back with affection to that familiar first edition, it hardly seems possible that half a century has gone by. We now live not only in a different world, but also in a radically different Solar System, a spectacularly different and more colourful galaxy, and an unbelievably more complex universe.

If you'll forgive one last bit of self-indulgence, there's a further aspect of this that plays tricks with my mind. In the second *Yearbook of Astronomy*, the 1963 edition (a copy of which I did obtain – by completely honest means – and still have), there's a useful article entitled 'Astronomy as a Career', which I remember reading with interest. Perhaps you will be surprised to know that the present-day Astronomer-in-Charge of the Australian Astronomical Observatory looks back on that article, and wonders quite how he got from there to here. From nerdy schoolkid to astronomy dream job. From grimy factories to gleaming telescope domes.

Yes, yes, I know. This is starting to sound uncannily like Monty Python's 'Four Yorkshiremen' sketch:

"Oo'd a' thought, forty year ago, that we'd be sittin' 'ere drinkin'
Chateau de Chasselais?'
'Aye. Looxury.'

It's all about time and distance, you see. . .

BRING ON THE TECHNOPHILES

I have one more factoid to relate about the first *Yearbook of Astronomy*, and then I promise I will stop harping on about it. One of its articles was an extraordinarily prescient piece by Cambridge's Roger Griffin entitled 'Automation in Astronomy'. Now, you have to remember that, in those days, astronomical observation was almost entirely a manual process. The most highly automated piece of equipment in most observatories was the telescope itself, which, by virtue of having one of its rotating axes tilted parallel to the Earth's own axis, could be automatically driven to follow the stars by means of a single electric motor. Thus, the inconvenient fact that the Earth's rotation causes the stars to move slowly across the sky could be eliminated.

That so-called 'equatorial' style of mounting had been the norm in observatory instruments for a century and a half, but is today extinct in new professional telescopes. Now that we have computers to do all the calculations needed to move both axes simultaneously by exactly the right amount, a straightforward up-and-down and side-to-side mounting (the so-called 'altazimuth') is much cheaper and easier to build – and is more compact.

The old-fashioned equatorial reached its apotheosis in the era of 4-metre class telescopes in the 1970s, of which the 3.9-metre

Anglo–Australian Telescope (AAT) of the Australian Astronomical Observatory (the larger sibling of the United Kingdom Schmidt Telescope) is a prime example. Paradoxically, despite its equatorial mounting, the AAT is computer-controlled to compensate for subtleties such as the bending of starlight in the Earth's atmosphere, and the slight flexing of the telescope as it points around the sky. It was the first large telescope to use a computer for this purpose, and, when it was new in 1974, amazed visiting astronomers with the precision by which it could be set on target objects. In fact, the mainframe Interdata 70 machine, with which the AAT was equipped when new, worked so well that it was only replaced, after thirty-five years' service in 2009 – with something the size of an iPad.

But, to return to 1962. Not only was automation virtually unheard of in astronomy, but so also was electronics. True, a few pioneering observatories had experimented with photoelectric photometry (the electronic measurement of star brightness) before the Second World War, and the technique had been more widely adopted since 1945 using sensitive photomultiplier tubes. But the workhorse technology of observatories was still photography with emulsion-coated glass plates. Every observatory had facilities for preparing and storing the plates (usually under refrigeration), plus its own darkroom – and sometimes a whole suite of darkrooms – for processing them.

If you wanted to make brightness measurements of stars, for example, you put your developed photographic plates into an ungainly device called an iris diaphragm photometer, which measured the sizes of images on the plate. Likewise, for accurate measurement of star positions in the sky, you would peer into a measuring machine, which was a fancy microscope able to scan the whole area of the plate by being driven along two perpendicular axes.

The ubiquity of photography extended beyond the straightforward imaging of stars. For a century, astronomers had been obsessed with recording the rainbow spectrum of stars to explore the bar-code of bright and dark lines that yields so much information about its distant source. Even with a large telescope, the spectra of only the brightest stars are visible to the naked eye. The light is highly diluted as it is spread out into its rainbow, and is very faint.

Thus, the recourse of spectroscopists had always been to use photographic plates, often chemically hypersensitized to enhance their responsiveness to faint light. If you then wanted to measure your spectrum – for example, to determine the exact position of spectrum lines to calculate the line-of-sight speed of your star using the Doppler effect (the so-called radial velocity) – you had to put the developed plate into another machine. Although such a 'measuring engine' would have only one calibrated axis rather than two, it was again equipped with a microscope for eyeballing the spectrum. And it was all done manually.

The long process of automating these techniques began to take off in the wake of Griffin's contribution to the first *Yearbook*. Scientists and engineers realized they could fairly easily automate the measuring machines by replacing hand-wheels with motors, eycballs with photo-electric cells, and human operators with logic circuits. I remember, in the late 1960s at St Andrews, where I did my MSc, a well-known type of measuring machine called a Joyce-Loebl Microdensitometer being automated by technicians at the University Observatory. No one was more surprised than they were when it actually worked.

MARRIAGES OF CONVENIENCE

In fact, it was just down the road, in Edinburgh, where the marriage of astronomical photography and automated measurement techniques was truly consummated. A new generation of measuring machines began with the development of GALAXY (an acronym for General Automatic Luminosity And XY) by Peter Fellgett and others at the Royal Observatory, Edinburgh, again in the late 1960s. This clever machine not only made rapid measurements of images but it also found them for you, too, and stored the data, all under its own computer control.

Photographic plates were scanned by a moving spot of light from a cathode-ray tube, projected on to the plate by a backwards-working microscope. GALAXY was the forerunner of a later Edinburgh machine, COSMOS (Co-Ordinates, Sizes, Magnitudes, Orientations, Shapes – these were truly awful acronyms), which operated from 1978 until 1993, and produced huge quantities of digitized data from photographs obtained with the United Kingdom Schmidt Telescope (UKST) and others.

The UKST itself, originally another Edinburgh initiative, was opened in 1973 specifically to provide a photographic atlas of the southern sky and to scout out the most interesting objects for the AAT. It took glass plates 1 mm thick and 356 mm (14 inches) square (Figure 1). Covering 6.5° square on the sky, each of these photographs would contain at least half a million images, which were clearly impossible to measure by manual methods. Thus, machines ruled supreme. COSMOS was eventually supplanted at Edinburgh by the more advanced SuperCOSMOS, while a radically different machine, APM (Automatic Plate Measurement), was developed further still down the road, in Cambridge, using laser-beam scanning.

Figure 1. Malcolm Hartley, discoverer of Comet Hartley 2, which was visited by NASA's EPOXI spacecraft in November 2010. This picture dates from the 1990s and shows Malcolm checking a film copy of one of the UK Schmidt Telescope's 356-mm square photographic plates. Many comets were accidentally discovered during this routine process. (Image courtesy of Duncan Waldron.)

As an aside, the digital techniques used in these machines contrasted dramatically with the control technology of the telescope itself. Although it was built only a year before the AAT, the UKST's control system could not have been more different. It used essentially the same technology as had been developed to aim large-calibre naval guns during the Second World War. Entirely electromechanical, it even boasted an analogue computer containing a tiny replica of the telescope and its equatorial mounting that was used to control the alignment of the rotating dome. That little model always raises a smile among visitors to the telescope – because, believe it or not, it's still in use.

Primitive control technology notwithstanding, the astronomy that came from the marriage of the UKST and the automatic measuring machines was truly spectacular in its day. Since the data they produced were high-resolution digital images of entire large-format plates, it was possible to run computer routines to separate stars from galaxies, and then perform statistical analyses on relatively pure samples of either type of object. That told us much about the distribution of stars in our galaxy, and the distribution of galaxies in the universe. It also helped us discover such subtleties as faint companions of our own galaxy, revealed by a localized increase in star numbers quite imperceptible to the human eye.

The technique became even more powerful if the data from two plates of the same area of sky taken with different coloured filters, or at different times, were compared. This allowed the classification of objects by colour, variability or motion across the sky for statistical studies. It also showed up the rare objects – for example, those that are very blue (like many quasars), very red (like cool 'brown-dwarf' stars) or are moving at high speed (like some faint nearby stars). The discovery of rare or moving objects from pairs of plates used to be carried out with a specialized microscope called a blink comparator, which presented images from the two plates to the eye in rapid succession. That's how Clyde Tombaugh famously discovered the dwarf planet Pluto in 1930. But the technique of comparing the digital images in a computer is far more effective at showing up small differences.

Despite the scientific success of the combined photographic/digital measurement method, the technology itself eventually ground to a halt in an evolutionary dead end. Faced with newer, more effective image detection techniques, astronomical photography suffered the same fate as its everyday counterpart. Kodak stopped manufacturing their specialized plates as well as the large-format films that were used towards the end of photographic astronomy. The great automated measuring machines closed down. Apart from tying up a few loose ends, the UKST stopped taking photographs in 2001. By then it had some 19,500 plates and films to its credit. Most of this original material is still stored in the vaults of the Royal Observatory in Edinburgh.

Far more useful than the plates themselves, however, is their digital legacy. The images are easy to find on the World Wide Web, whether they be in the Digital Sky Survey or the starry backdrop of Google Sky. Even more significant are the analysis techniques originally developed for digitized photographic images, which have also migrated way beyond the confines of professional observatories. Most amateur astronomers involved in digital imaging will use these techniques without even thinking, as they unleash their favourite proprietary software packages on the stunning images from their computer-controlled telescopes.

As most readers of this *Yearbook* will know, the imaging technology that eventually supplanted photography used solid-state electronic detectors, most notably charge-coupled devices, or CCDs. Many readers, too, will know that one of the main reasons they found such favour with astronomers was their remarkable efficiency in turning particles of light – photons – into a measureable signal. Even at the peak of their evolution, with the most effective possible chemical hypersensitization, photographic plates could record only three or four of every hundred photons falling on their surface. But some modern CCDs can record more than ninety. With light intensity always at a premium in astronomical imaging, no wonder the CCD is now ubiquitous.

It is also the reason that today's amateur astronomers can produce beautiful colour images comparable in quality (though not in fineness of detail) with the pioneering images that my colleague, David Malin, made photographically with the AAT, and with plates from the UKST, during the 1980s and 1990s. David showed the way with photographic three-colour image processing – he, too, is now doing the same sort of thing digitally.

Back in the 1960s, however, it wasn't at all clear in what direction electronic imaging would go. The first mention of such techniques in the *Yearbook of Astronomy* came in 1967, in an article by R. C. Maddison entitled 'Image Converters in Astronomy'. Even the term was a carry-over from the Second World War, when the British Army's 'Tabby' infrared imaging device used a crude electronic converter tube to render infrared radiation visible on a phosphor screen. Maddison's article highlighted a number of pioneering devices, such as electronic image intensifiers and slow-scan TV cameras, that might have applications in astronomy.

Indeed, in the mid-1970s, it was a marriage of exactly those two components that Alec Boksenberg and his team at University College, London, used to build the legendary IPCS – the Image Photon Counting System. This detector effectively filled the gap between photography's demise and the introduction of CCDs for recording the spectra of stars and galaxies. By lifting the quantum efficiency (success rate in recording photons) from the 3–4 per cent of photography to somewhere around 20 per cent and, additionally, by allowing the astronomer to watch their image building up in real time, the IPCS was an instant success. One of these machines found a permanent home on the AAT, and dominated astronomical spectroscopy throughout the late 1970s and early 1980s. Fittingly, it is now part of the collection of the Powerhouse Museum at Sydney Observatory.

Both the IPCS and the rapidly improving CCDs that supplanted it during the 1980s were relatively small-format devices. This made them ideal for spectroscopy, but less useful for direct imaging. By the 1990s, the AAO was a two-telescope institution, having absorbed the UKST in 1988, and both its telescopes had physically large focal surfaces. In 2000, a mosaic CCD camera called WFI (Wide-Field Imager) was built for the AAT. Another marriage of convenience, it was shared with the 1-metre telescope of the Australian National University (ANU), located on the same Siding Spring mountain-top. Using no less than eight CCDs to image an area 125 mm square (but still only covering a quarter of the area of the old AAT photographic plates), WFI was operational for the first five years of the new century, before the AAT ceased visible-light imaging altogether.

The UKST's image area was even bigger than the AAT's, and there was no possibility of obtaining very large-format CCDs to cover it. That's the only reason photography lasted into the twenty-first century at the UKST. Now, however, a facility of comparable size but using modern imagers is opening up on a brand new telescope at Siding Spring Observatory. Skymapper, a 1.35-metre dedicated robotic imaging telescope operated by the ANU, has a detector mosaic four times larger than WFI. By the time this article appears in print, Skymapper will be fully operational, producing 1 Gbyte of data every ten seconds as it undertakes its five-year project to image each part of the southern sky thirty-six times.

The kind of science that Skymapper will accomplish will be similar in principle to what the UKST and the automated measuring machines achieved in the 1980s and 1990s. But we now know vastly more about the universe than we did then. Dark energy, dark matter, dwarf planets beyond Neptune – all these and much more will begin to yield their secrets to the advanced technology that is embodied in Skymapper.

LOOKING FOR GUIDANCE

I have elaborated on the development of digital imaging in the half-century since the first *Yearbook* because it provides a great example of the way technology has changed the way astronomers work. But it's only one aspect of the technological revolution that has swept through astronomy in recent years.

The revolution has many facets, ranging from asteroseismology to Zeeman Doppler imaging. Many of those techniques have been pioneered at the Australian Astronomical Observatory, but I'd like to devote the remainder of this article to just one of them. It's an area that has transformed astronomy at least as much as electronic imaging – and promises even more spectacular gains.

I admit this will not be the first time I have waxed lyrical about fibre optics in the pages of the *Yearbook*, but those slender glassy filaments have allowed astronomers to perform feats of science that would have been undreamed of back in 1962. Who'd have thought that the simple idea of guiding light along a flexible strand of quartz would trigger such advances? Those advances are only part of the story, however. What is completely new is the fledgling science of astrophotonics, which has evolved from fibre optics, and here the possibilities become even more amazing.

But to the beginning. Who dreamed up fibre optics? As with so many technological breakthroughs, no single person can be identified as the originator. The idea of light being guided inside a glass rod by repeated reflections from its surface (technically known as total internal reflection) would probably have been familiar to glassblowers in ancient times. The first recorded scientific demonstration, however, was by John Tyndall in a popular lecture at the Royal Institution in 1854. It's an experiment you can try yourself – but you're sure to wind up needing a towel. Tyndall drained water from a horizontal pipe at the bottom of an illuminated container, and showed that a beam of light was constrained to follow the parabolic arc of the water – no doubt to the astonishment of the crowd.

Light guides aren't much use if they're made of water, however, and it wasn't until 1887 that the first slender quartz fibres were successfully demonstrated by the physicist Charles Vernon Boys. He used them not for transmitting light, but for suspending components in sensitive measuring instruments, thereby losing the opportunity to become the inventor of fibre optics. Rather carelessly, he also lost his wife – to a colleague, in one of the scandals of the age. But that's another story . . .

It took until the 1950s for fibres to become truly useful as light-guides with the application of an outer layer of glassy material to act as a so-called optical cladding. This keeps the light inside the fibre core. Then, a 1966 study predicted that fused silica (quartz) fibres made from high-purity materials would have very low light-losses, producing an upsurge of interest in their potential for optical communications. The first practical low-loss fibres followed in 1970, produced by the Corning Glass Works, and, well, the communications industry has never looked back.

Until a handful of excited astronomers seized on these efficient, flexible light-guides, observations of the rainbow spectra of celestial objects had almost invariably been carried out one at a time. There were exceptions, but nothing that could really be called 'multi-object spectroscopy'. But that was all about to change.

Some seven years after Corning's introduction of low-loss fibres, Roger Angel and his colleagues in Arizona suggested that they might be used to connect several independent telescopes to make a 'Fibre-Linked Optical Array Telescope', or FLOAT. At a stroke, Angel's group had both recognized the potential of a revolutionary new technology, and set a trend for daft acronyms that persists among instrument-builders to this day.

As it turned out, FLOAT never materialized, but its inverse did. Instead of using multiple fibres to link several telescopes looking at one celestial object, why not use them to deliver light from several target objects in the field of view of one telescope to its spectrograph – the instrument that records the spectra of stars and galaxies? That way, the random distribution of the objects could be tidied up into a neat straight line for the spectrograph's input slit, taking full advantage of the flexibility of the fibres. Like rearranging the icons on your desktop, the spectra could be made to fill the available detector space exactly.

This possibility was not lost on Angel's group, and by December 1979, they had used a primitive multiple-fibre system called 'Medusa' to observe galaxies twenty or so at a time using the Steward Observatory 2.3-metre telescope. Multi-fibre astronomical spectroscopy had arrived.

ROBOTS TO THE RESCUE

It didn't take long before astronomers down under got wind of these developments. A memorandum dated 16 January 1981 and simply entitled 'Fibre Optics' seems to have marked the beginning of interest at the AAT. It proposed an investigation into the practical use of fibres, nominating a young engineer called Peter Gray as the project manager. While the AAT was certainly well equipped to carry out fibre-coupled multi-object spectroscopy, it was Peter's involvement that guaranteed success, since he had the ideal mix of precision and pragmatism. (Today, it is the Thirty Metre Telescope project in California that is the fortunate beneficiary of Peter's expertise.)

Peter's first essay in fibre instrumentation was FOCAP (Fibre-Optically Coupled Aperture Plate), which allowed up to 25 targets within a small area of the AAT's field of view to be coupled with 2.5-metre long fibres to the spectrograph (Figure 2). It was first used on the telescope in December 1981, allowing the AAT to record the spectra of a couple of dozen objects simultaneously. FOCAP soon under-

Figure 2. The world's first fully-reusable fibre-optic system was Peter Gray's FOCAP, seen here at the Cassegrain auxiliary focus of the Anglo–Australian Telescope in 1983. Each fibre aligns with a particular star or galaxy, having previously been 'plugged' into an accurately drilled brass plate, and carries the light to a spectrograph for analysis. (Image courtesy of David Malin.)

went a process of improvement, and, over a lifetime of some six years, produced excellent results in a wide range of research programmes. It was eventually operating on the AAT for no less than 30 per cent of all observing time.

Meanwhile, in June 1982, two visiting astronomers from Edinburgh came to use some observing time with FOCAP. My colleague, Victor Clube, and I planned to observe variable stars in the galactic centre. The observing didn't go particularly well due to poor weather, and we eventually retreated to the Common Room of the nearby UKST, where we drowned our sorrows with the then Astronomer in Charge, John Dawe.

The UKST was, of course, an outpost of Scottish culture and tradition, but the whisky-fuelled conversation that followed resulted in an outrageous suggestion. Perhaps these new-fangled optical fibres could find an application on the UKST? Since the telescope was then fully occupied with survey photography, we all roared with laughter at such a crazy idea, and almost forgot about it. It was John Dawe who, a few days later, realized that the telescope's huge field of view – 6.5 degrees square – made it a perfect candidate for multi-object spectroscopy with fibre optics.

Subsequently, John and I worked together with Peter Gray in kicking off a scheme to use fibres on the UKST that eventually became my project. During the 1980s, I was responsible for three fibre optics instruments for the telescope, with, I'm afraid, steadily more ridiculous names. The prototype was FLAIR, the 'Fibre-Linked Array Image Reformatter' – although David Malin tended to refer to it (rather unkindly, I thought) as 'Watson's Folly'. It did show promise, however, so in 1988 we built an improved version, which became known as PANACHE, for 'PANoramic Area Coverage with Higher Efficiency'. What else?

When further improvements were suggested, I thought FINESSE might be rather a good name, but another colleague put me right by suggesting that it would have to mean 'Fails to Interest Nearly Everyone Save Spectrograph Engineers'. In the end, duly chastened, we simply called it FLAIR II.

All these instruments were characterized by the fact that the individual fibres carrying the light from each target object were bundled together in a protective cable and brought out of the telescope to a spectrograph that remained motionless in the dome. This made for an extremely stable system, and was perhaps the most important innovation in our work. But I also used to find a certain satisfaction in picking up the cable from the floor when observing was in progress, and imagining those ancient photons from galaxies hundreds of millions of light years away pulsing silently through my fingers . . .

The spectrograph itself was mounted on a vibration-free optical table in the dome – the most expensive component of the whole system (Figure 3). This thick metal bench-top, with its chunky pneumatic legs, actually cost more than it should have done when it was imported from the US. Being a table, it was deemed to be furniture rather than scientific equipment by the Australian customs authorities, who promptly levied the highest rate of duty on it. Nothing – not even a letter from the Astronomer Royal for Scotland – would change their minds. Eventually, Customs decided it was not just 'furniture',

Figure 3. The UK Schmidt Telescope in 1988, linked by the PANACHE optical fibre system to its spectrograph on an optical table in the dome. When it was imported into Australia, the table was deemed by customs authorities to be 'furniture', which attracted a much higher duty than scientific equipment. It thus became the most expensive of PANACHE's many components. (Image courtesy of Duncan Waldron.)

but was actually 'kitchen furniture'. It still attracted the highest rate of duty, however.

The drawback of the early fibre-optics systems on both the AAT and the UKST was that they relied on each fibre being manually positioned in exact alignment with its target object. This was not only a time-consuming process, but, in the case of the AAT, it also had financial consequences, with the need for expensive pre-drilled brass plates into which the fibres were plugged prior to observing each set of targets.

Some sort of robotic positioning of the fibres was clearly necessary, and during the 1980s, initially in collaboration with the University of Durham, engineers at the AAT began experimenting with intelligent machines that would pick up, move, and put down tiny magnets on a large steel plate – so-called 'pick-place' systems. With a fibre attached to each magnet, and a right-angled prism stuck on the end of each fibre, the stage was set for a revolution in the way multi-fibre spectroscopy was carried out.

That early work culminated in what are still today's workhorse fibre-positioning robots on the AAO's two telescopes (Figure 4). The

Figure 4. When not in use on the telescope, the 2dF robotic fibre positioner resides on the AAT's second floor, where it can be inspected and serviced. Here, the Australian Astronomical Observatory's Director, Matthew Colless (left) and a past director, Don Morton, watch with evident approval as the robot positions its fibres in alignment with 400 target galaxies. (Image courtesy of Jonathan Pogson.)

2dF, or two-degree field system, was commissioned on the AAT during the mid-1990s, while 6dF on the UKST (yes, you've guessed it – six-degree field) saw first light in 2001. Although the constructional details of these machines could hardly be more different, they both do essentially the same job. Their task is to place a sequence of 0.1-mm-diameter optical fibres, equipped with magnets and right-angled prisms, on to a metal plate with a positioning accuracy of 0.01 mm, taking no more than a few seconds per fibre – and to do it entirely automatically. Since the fibres themselves are rather delicate, another important requirement for the robots is that they try not to break them.

The numbers of fibres that these two machines have to deal with in their respective telescopes is related to the expected density on the sky of the targets that will be observed (in objects per square degree). In the case of 2dF, it is 400 fibres, while 6dF has a maximum capacity of 150 fibres, and typically uses 100. Being able to obtain the spectra of even only 100 objects simultaneously is a staggering improvement on the one-at-a-time methods of half a century ago.

That overwhelming advantage is reflected in the science that these two systems have accomplished. 2dF's first task was a survey of the positions of galaxies out to about two and a half billion light years to provide a detailed cross-section of the local universe. By measuring the amount by which a galaxy's light is shifted to the red end of the spectrum (its redshift), the galaxy's distance can be determined. The project was therefore known as the 2dF Galaxy Redshift Survey. It was completed in 2002, and catalogued 220,000 galaxies, resulting in a deluge of scientific papers. In 2005, it was used to find the 'missing link' between the ripples in the baby universe some 13.7 billion years ago (as revealed by the background of fossilized Big Bang radiation that covers the entire sky), and the way matter is distributed today.

This remarkable work highlighted the importance of large-scale astronomical surveys, and led to other major galaxy surveys with 2dF. 'WiggleZ', now completed, amassed another 200,000 galaxies with look-back times of up to half the age of the universe. Aimed at gaining insights into the mysterious dark energy (which is causing an acceleration in the expansion of the universe), the survey's whimsical name is a play on the ripples in the distribution of matter, and the scientific symbol for redshift, z. Still ongoing is the Galaxy And Mass Assembly (GAMA) project, which aims to measure 400,000 galaxies to address fundamental questions concerning gravity, dark energy and the mechanisms by which galaxies evolve.

In contrast with 2dF's ability to look very deep into space in narrow pencil-beams, the strength of 6dF is in its ability to perform surveys over the entire southern sky. Thus its first major task was a survey of some 136,000 galaxies in the 6dF Galaxy Survey (6dFGS). You might notice, by comparison with its 2dF equivalent, that the word 'Redshift' is missing from the survey's name. While redshifts were certainly determined for all these galaxies in order to measure their distances, the survey had another important side-line in the measurement of 'peculiar velocities', the individual speeds of the galaxies as distinct from their motion due to the expanding universe. Peculiar velocities tell you about the way galaxies are being pulled by gravitational forces, and the 6dFGS will eventually give us insights into the mysterious dark matter that makes up 80 per cent of all the matter in the universe, but only reveals itself by its gravity.

The 6dFGS was completed in 2005, and was superseded by RAVE – the RAdial Velocity Experiment. Emerging phoenix-like from the ashes of a proposed European Space Agency mission that failed to get funding, RAVE has turned the attention of the UKST from galaxies back to stars. It aims to measure the speeds and physical characteristics of more than half a million stars to give us insights into the way our Milky Way Galaxy has been put together over its 12-billion-year history. It forms part of a new field of study called galactic archaeology, which I wrote about in the *2011 Yearbook of Astronomy*.

By the time these words appear in print, RAVE will have collected more than half a million spectra of well over 400,000 stars, making it the biggest survey of its kind yet undertaken. And, yes, RAVE is the answer to the question I posed at the very beginning of this article. How can an elderly 1.2-metre telescope continue to produce leading-edge data? By using 6dF to provide a large group of astronomers worldwide with the raw material that will give us the most detailed snapshot yet of our home in the universe – the Milky Way Galaxy.

FIBRES TO PHOTONICS

Back in the early 1990s, a few of us in the astronomy world did some thought experiments on how to get the very best from fibre optics. The thinking went a bit like this. One of the things you have to get absolutely right in fibre-optics astronomy is shoehorning the light from each target image into its fibre. Whether the target is a star or a galaxy, it is inherently a lossy process. The risk is that the light from the telescope won't enter the fibre at the right angle, or that it will spill around the edges, or be reflected away. There's a certain contrariness, anyway, in the fact that your telescope might be focusing the light from a distant galaxy containing 100 billion stars on to a polished fibre face which is only 0.1 mm in diameter.

But let's assume we've got the star or galaxy's light successfully into the fibre. What do we do next? Well, the fibre does its job of transferring it from one place to another – but then we let the light out again. That's so the spectrograph (which is just a collection of components like lenses, mirrors and other optical hardware) can do its job properly. But it turns out that light leaving the fibre is also a lossy process. So, we wondered whether there might be useful things you can do with the light by keeping it locked up within the fibre, or some sort of similar device. To use the trade term, by keeping it within a waveguide.

In fact, the answer is yes, there are. But there is a problem. When light bounces along inside a fibre from one end to the other, it follows a zig-zag path as it reflects off the sides. Despite its small diameter, a conventional optical fibre allows light to take many different zig-zag paths inside it, and the different paths are called 'modes'. Hence, such a fibre is called a 'multi-mode' fibre. It turns out, however, that to do anything *really* clever, you must use a 'single-mode' fibre, in which the light path simply goes straight down the middle of the fibre. But the problem with single-mode fibres is that they are only a few thousandths of a millimetre in diameter, far too small to get the light from a typical ground-based telescope into them. So it looked as though our wild ideas of fancy optical components glued on to the ends of our fibres would be doomed to failure.

Then, in 2004, a smart AAO astronomer called Joss Hawthorn perfected a device he called a 'photonic lantern', which would allow light coming down a multi-mode fibre from a telescope to be separated into many single-mode fibres. Here, the light could be manipulated in novel ways. Suddenly, a whole range of possibilities was opened up, effectively creating the new field of astrophotonics. This is now being actively pursued in Australia at the AAO, the University of Sydney (where Joss is currently a Federation Fellow) and at

Figure 5. Joint AAO/Macquarie University MSc student, Dionne Haynes, checks the alignment of a microlens assembly in the astrophotonics laboratory at the Australian Astronomical Observatory in Sydney. (Image courtesy of Tim Wheeler.)

Macquarie University (Figure 5). In addition, there are links with the InnoFSPEC research centre at the Astrophysikalisches Institut Potsdam in Germany.

The reason this breakthrough is so exciting is that over the past couple of decades, the technology of photonics has blossomed in the commercial and technical world. Photonics allows single-mode beams of light to be manipulated in much the same way as electricity is handled in electronic circuits. Eventually it is sure to revolutionize the way we build computers, mobile phones, remote sensing devices, medical equipment, and so on.

The first task which fell to astrophotonics was something of a holy grail for astronomers working in the near-infrared. This region of the spectrum is plagued by what are called telluric emission lines – unwanted light at many discrete wavelengths that come from atoms in the Earth's upper atmosphere relaxing after a day in the Sun. If these narrow spectrum lines could be selectively filtered out, ground-based astronomers would have a completely new window through which to explore the universe. And it turns out that with single-mode fibres, they can be – if the fibre cores are imprinted with a precisely undulating pattern of density fluctuations. Such esoteric devices are called fibre Bragg gratings (named after the pioneering physicist) and, over the past few years, they have been perfected by Joss Hawthorn's research group and its industry partners. An operational version of this technology is soon to make its debut on the AAT.

Another photonics technology that is being developed at the AAO is a spin-off from one of those thought experiments I mentioned earlier. I was always particularly fond of this one. Imagine a wafer of transparent material about the size and shape of a playing card, with an edge to which a fibre could be cemented. This would allow the light to be efficiently coupled into the wafer, which itself might have waveguides and other optical components etched into it. You could then mimic the function of a spectrograph within the wafer, avoiding the losses incurred in releasing the light from its fibre. *Voilà* – a spectrograph on a chip!

It could have huge advantages. Compactness is just the first, and these wafers, or 'integrated photonics spectrographs', could be stacked one on top of another, just like a deck of cards, each with its own fibre attached, to give a multiple-object capability. AAO is already experimenting with this. Another advantage is that being solid-state devices, they would be ideal for operating on spacecraft, where all onboard components have to withstand high launch vibration. And, if you happened to be working in the infrared waveband, measuring heat radiation, it would be very easy to cool the tiny spectrograph to the low temperatures required. Finally, as telescopes get bigger, conventional spectrographs get bigger too – and on the horizon is a new generation of extremely large telescopes which will require extremely large spectrographs. But it's possible that tomorrow's integrated photonics spectrographs will be engineered to break that size dependency.

A TOAST TO THE FUTURE – AND THE PAST

At AAO and elsewhere, scientists and engineers are just beginning to scratch the surface of astrophotonics. I don't have the space to mention photonic bandgap fibres, micro-ring resonators or direct-write 3D waveguides. Coupled with novel robotic technologies like the AAO's 'Starbugs' (autonomous micro-robots that can move over a telescope's focal surface to position fibres or photonic devices), we are about to see a new revolution in astronomy (Figure 6).

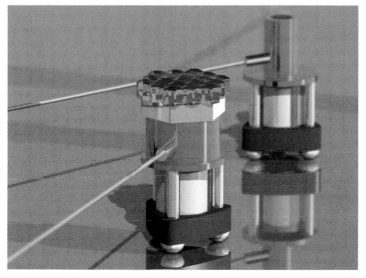

Figure 6. 'Starbugs' are miniature robots that can carry tiny payloads to precise locations in the focal surface of a telescope. The payloads can include single optical fibres or microlens arrays to dissect a star or galaxy image, as shown in this early schematic. (Image courtesy of Andrew McGrath/Australian Astronomical Observatory.)

And beyond that, there is even more remarkable technology on the horizon. You might have heard of so-called 'metamaterials' – materials with structures on the nanometre scale, well below the wavelength of light. Metamaterials can manipulate light in apparently impossible ways. One well-publicized application, still some years down the track, is the real-life invisibility cloak. Move over, Harry Potter – soon, we will all be able to hide . . .

But more seriously, metamaterials will have applications in astrophotonics that might begin to remove some of the fundamental barriers to observing which we currently face. Literally, the sky is the limit. It's very hard to imagine what sort of instrumentation could be commonplace in fifty years' time, when the editors of the *2062 Yearbook of Astronomy* are collecting together the articles for their centenary volume.

Talking of which, there is one more task that remains for me in this tour through half a century of high-tech astro-magic. That is to go back to the beginning of the article, pick up that glass of 'Chateau de

Chasselais', and drink a toast to the *Yearbook of Astronomy*. Its illustrious editors, Patrick Moore and John Mason, have much to be proud of, as have all the contributors – not to mention their readers – throughout its fifty-year history.

Cheers, everyone, and congratulations!

'Aye. Looxury.'

ACKNOWLEDGEMENTS

As always, it is a pleasure to acknowledge the support of the director and staff of the Australian Astronomical Observatory, which is a division of Australia's Department of Innovation, Industry, Science and Research. To my many colleagues in the field of astronomical instrumentation, I send a big thank-you for support and inspiration over the years. Finally, I thank John Mason for his invitation to contribute this article.

FURTHER READING (AND LISTENING)

Australian Astronomical Observatory Newsletter. Current and back issues of the AAO Newsletter (now called AAO Observer) contain up-to-date reports on astronomical technology and can be found online at http://www.aao.gov.au/library/news.html.

Australian National University, Research School of Astronomy and Astrophysics, 'Skymapper – Mapping the Southern Skies', 2011, http://rsaa.anu.edu.au/skymapper/index.php.

J. Cleese, et al., 'The Four Yorkshiremen', in *The Secret Policeman's Ball*, The 1979 Amnesty International Comedy Gala at Her Majesty's Theatre, London, audio cassette, Festival C 37253.

J. B. Hearnshaw, *The Analysis of Starlight: One Hundred and Fifty Years of Astronomical Spectroscopy*, (Cambridge, 1986).

J. B. Hearnshaw, *The Measurement of Starlight: Two Centuries of Astronomical Photometry* (Cambridge, 1996).

Monash University, 'Data Acquisition upgrade to a Joyce Loebl Microdensitometer', 2009. http://www.physics.monash.edu.au/services/eis/daq.html.

The Moon and the Amateur Astronomer

BILL LEATHERBARROW

More often than not the Moon is the first target to be viewed by the novice astronomer armed with a new telescope. And what a target it proves to be! Readily and regularly visible even in the most light-polluted of skies, it reveals a world of wonders to the user of even the most modest of instruments. Huge craters, towering mountains and vast lava plains constantly alter their appearance under the Sun's ever-changing light as the lunar day progresses. There is always something new to see, and the Moon's spectacular face is capable of impressing and delighting even the most sanguine of telescopic explorers. But not all are readily enthused by the spectacle: those observers whose pleasure is to chase down elusive galaxies and other faint deep-sky objects tend to regard the Moon as an intrusive interloper, whose bright glare frustrates them in their quest. And then there are those who would argue that lunar observation, although capable of providing dramatic views, offers little to the amateur astronomer who wishes to go beyond mere celestial sightseeing by undertaking a more systematic and science-based study of the sky. Such people would argue that, in this age of regular spacecraft investigation of the Moon, the amateur and his telescope can no longer contribute usefully to lunar study; it has all been done, the Moon is thoroughly charted in the sort of detail that the Earth-based observer cannot match, and our satellite itself is a long-dead world where nothing really changes, apart from the visible appearance of features under different conditions of lighting.

It might at first seem difficult to resist this view, especially given the regular publication of ever more detailed pictures of the lunar surface taken from probes such as NASA's Lunar Reconnaissance Orbiter (LRO), some of which show features only a few centimetres in size. The great European tradition of selenography, starting with the rudimentary sketches of Galileo and Thomas Harriot in 1609 and ending with

H. P. Wilkins's 300-inch map, published in the 1950s, was rooted in the charting of the lunar surface in ever-increasing detail. Such work is indeed now complete, and further lunar cartographical work will be done not by the amateur telescopist, but by spacecraft with a view to producing specialized geological charts of the Moon's surface. But this does not mean that the amateur observer is now redundant. We have been here before, in the 1960s, when the American Ranger probes took close-up photographs before smashing into the Moon and revealed in the process that the amount of fine lunar detail was unlimited and its charting an endless pursuit. Later in that decade, Orbiter and Apollo missions photographed most of the Moon's surface in high resolution. Moreover, the professionalization of lunar science in the run-up to the Apollo programme saw a seismic shift away from the amateur's pursuit of finer and finer surface detail towards a realization that true understanding of the Moon was to be gained more from studying the big picture, rather than the minutiae of surface topography. The recognition by W. K. Hartmann and others of the true nature of multi-ring impact basins and the lunar maria did far more for lunar science than the addition of a few more craterlets to the latest lunar map.

In the late 1960s all of this led to a feeling among some that the glory days of the amateur lunar observer lay firmly in the past – a feeling very similar to that currently experienced by many in the wake of recent spacecraft missions such as Japan's Kaguya, India's Chandrayaan-1 probe and the American LRO. In 1971, I contributed to the *Yearbook of Astronomy* an article on 'Amateur Opportunities in Contemporary Lunar Research' which attempted to identify a role for the amateur observer in the post-Orbiter and post-*Apollo* period. Then, as now, it was clear that the amateur who wished to undertake meaningful lunar study had to adopt new modes of thinking and follow a more systematic programme of observation, combining new approaches with the best of the old in order to yield results of value in areas where the space probe cannot yet compete. That is not to say that simply observing the Moon for its own sake is not an enjoyable and rewarding experience; indeed, it is one of the best ways of finding your way around the lunar surface. But eventually most regular observers will yearn for a more focused approach to their lunar studies, and the suggestions below are designed with that in mind.

NEW APPROACHES TO OLD PROBLEMS

The past decade or so has seen a revolution in the techniques available to the amateur astronomer. The Chinese economic miracle has seen large-aperture, good-quality telescopes become readily available at a fraction of their previous real prices. Whereas once the beginner might have started out with a 3-inch refractor or 4-inch reflector, nowadays they are just as likely to do so with the help of 8-inch reflectors or large Schmidt-Cassegrains. Moreover, modern CCD technology has transformed the world of lunar and planetary imaging, so that the present-day amateur is capable of taking extremely sharp and high-resolution pictures that lend themselves well to serious study. For the lunar observer this opens up immense possibilities, including that of using the technology of today in order to revisit the problems of the past. This is particularly so in those cases where past observers have suspected anomalies and changes on the lunar surface, and here the modern well-equipped amateur can shed much light. We need to be clear what we are talking about here: I am not suggesting that today's amateur should use his or her technological advantages in order to address the thorny question of whether or not significant changes have occurred on the lunar surface in the relatively recent past. It seems to me to be overwhelmingly likely that no large-scale change has occurred on the Moon for hundreds of millions of years, and certainly not during the comparatively brief period during which our satellite has been an object of telescopic scrutiny. And yet the history of lunar observation is littered with reports of changes and variations, often from experienced observers. The availability of high-resolution imaging allows us the opportunity to look anew at such instances, not in order to confirm or refute suspicions of change, but with a view to trying to establish what it was about the topography of the area in question that might have given rise to those suspicions. A couple of examples will serve to illustrate what I have in mind, but there are many more awaiting the diligent observer.

On 16 October 1866 the German J. F. J. Schmidt, one of the most dedicated and skilful of lunar observers, reported that the small crater Linné in the western half of the Mare Serenitatis had suddenly disappeared, or at least dramatically changed its nature. Previously recorded by Schmidt and others as a six-mile (8.5 km) crater, Linné now appeared as a nebulous white patch. Schmidt was convinced that a real change had occurred, and the 'Linné affair' reawakened interest in the Moon for generations to come. Even as late as the 1950s some observers were still arguing the case for a structural change possibly caused by some form of volcanic activity. Today Linné does appear in a large telescope as a whitish patch topped by a small crater pit, and spacecraft imagery has revealed it to be a relatively fresh impact crater surrounded by a blanket of bright ejecta. It is extremely unlikely that any change took place, but it would be good to understand exactly what it is about the surrounding terrain that might have encouraged the view that it had. What is needed is high-resolution imaging of Linné throughout the course of a lunar day, showing the feature under all angles of solar illumination. A similar instance is provided by the 58-km crater Eratosthenes, located at the tip of the Apennine mountain range. At the start of the twentieth century the American observer William H. Pickering described a network of apparently variable dark patches that he had seen in and around Eratosthenes, and which he ascribed initially to the diurnal development of primitive vegetation and then to the migration of huge swarms of insect life. This might sound ridiculous to us now, in the light of what we have subsequently learned about the Moon, but hindsight is a wonderful thing and such ideas, although eccentric, would not have seemed quite so outrageous in the age when Pickering worked. The possibility of some form of life on the Moon could not be discounted absolutely and it inhabited the popular imagination. The advanced Selenites described in H. G. Wells's *The First Men in the Moon* (1901) might well have been the stuff of dreams, but Wells's breathtaking description, in the same novel, of lunar vegetation awakening under the warmth of the Sun's morning rays was not so strange to the thinking of the time. The patches seen by Pickering undoubtedly exist, but they are simply part of the great ejecta blanket thrown out by the neighbouring crater Copernicus. Nevertheless, careful telescopic observation and imaging under all conditions of local lighting might help us to understand why Pickering thought he was witnessing changes on the Moon.

OBSERVATION OF SPECIFIC TYPES OF FEATURE

To revisit the problems posed by features such as Linné and Eratosthenes is more likely to shed light on the nature and history of lunar observation, rather than on the nature of the Moon itself. But the modern amateur is also capable of work that does the latter. There are many types of lunar formation, even familiar ones, which still have not been subjected to the sort of systematic study they deserve. Take, for example, the rilles (or *rimae*) that abound on the lunar surface and which look like cracks or channels in the Moon's crust. These have been known since the early years of telescopic observation and many hundreds have been discovered. However, they are still not properly understood or catalogued. What is needed is a systematic approach to the disposition and categorization of rilles, one that provides accurate measurements of their lengths and widths, as well as describing the relationship of such features to neighbouring topography. In particular, we should pay attention to the different types: perhaps the best-known rille is Rima Hadley (Figure 1), lying near the northern end of

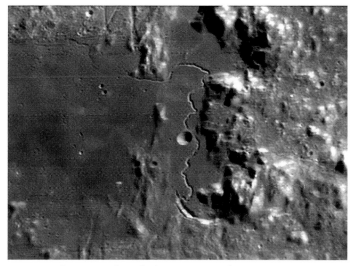

Figure 1. Hadley Rille. (Image courtesy of Damian Peach.)

the Apennine mountains, for it was here that *Apollo 15* landed. This is a fine example of a *sinuous rille*, and it looks for all the world like a dried-up river bed. In fact, it is a lava channel that once carried molten magma on to the mare surface. There are many more examples to be seen, and we should chart them carefully with a view to establishing what they have in common in terms of size and location. You will also find examples of rilles that follow a much straighter course than Rima Hadley: these are know as *linear rilles* and perhaps the finest example is Rima Ariadaeus, just south of the large ruined crater Julius Caesar. These features are similar to terrestrial *graben* and they were formed in the same way, by the slumping of terrain between parallel faults. At the edge of some of the lunar maria we find *arcuate rilles* that curve to follow the contours of the mare 'shore'. Particularly good examples are provided by the Liebig and Hippalus rilles on opposite sides of the Mare Humorum, and they mark the cracking of the lunar surface as the floor of the mare sank under the weight of the lavas that once flowed to fill it. However, Rima Hyginus (Figure 2), close to the Ariadaeus rille, provides a fine example of a rille formed by volcanic activity of a different kind: it is a *crateriform rille* and clearly consists of a chain of small craterlets that have formed along a surface fracture. These craterlets, along with Hyginus itself, are almost certainly volcanic calderas created when the Moon's interior was still molten and, unlike the vast majority of lunar craters, they are not primary impact features.

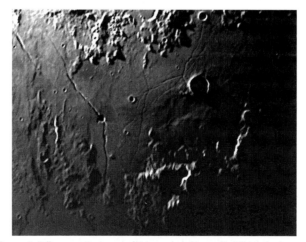

Figure 2. Rilles near Hyginus and Triesnecker. (Image by Bill Leatherbarrow.)

Careful scrutiny will reveal other categories of rille, including infrequent examples in highland areas (for example, in the crater Janssen, shown in Figure 3), as well as rilles that appear to turn into faults along their length (for example, Rima Liebig and the rille south-west of Bürg in the Lacus Mortis). Work on the charting and categorization of the

Figure 3. Rilles in Janssen. (Image by Bill Leatherbarrow.)

various types of rille is of undoubted value (and is long overdue), and it may be done either at the telescope or by making use of the vast amount of spacecraft imagery now available from the Internet. The latter will, of course, allow us to extend our studies to the far side of the Moon, which is never available for telescopic investigation.

Like rilles, bright rays are much observed, but our understanding of them is incomplete. The most obvious examples are to be found in the extensive system emanating from the relatively fresh crater Tycho, and this gives us the key to their origin. Rays represent bright ejecta material thrown out from the formation of impact craters, and they are to be found in association with comparatively young craters whose ejecta has not yet been overlaid by subsequent surface sculpture or darkened by the effects of space weathering. This much we know, but many questions remain to be answered by systematic study and classification. For example, we could learn much from careful measurement of lengths and analysis of shapes. Is the length of a ray always linked to the size of its parent crater? Are there clearly discernible patterns of ray formation? If so, how many, and how do they differ? Are rays interrupted along their length by other topographical features? If so, how? Are rays visible as soon as the Sun rises over their location, or do they appear at a later point in the local day? Can we use the eccentric disposition of ray material (for example, at the craters Messier and Messier A (Figure 4), where the ray pattern is not evenly distributed around the crater) to identify instances where craters have been formed by highly oblique impacts? This barrage of questions amply demonstrates that insufficient attention has been devoted to the analysis of rays, and this, too, is work that may be profitably pursued by the amateur using telescopes or photographic images.

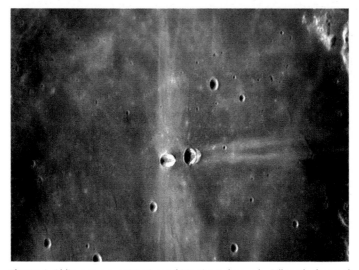

Figure 4. Oblique rays near Messier and Messier A. (Image by Bill Leatherbarrow.)

Even less is known about those enigmatic high-albedo swirls that may be found in certain areas of the lunar surface. The best known, and most photographed, is Reiner Gamma on the Oceanus Procellarum (Figure 5), but there are others and a list may be found on the Moon-Wiki website at: http://the-moon.wikispaces.com/swirl. This list is undoubtedly incomplete and there will certainly be more examples to be found, catalogued and categorized in a systematic way. One of the most intriguing things about bright swirls is that, although they appear to be mere albedo features, surface deposits without relief, most of the known examples appear to be associated with magnetic anomalies and several (but not Reiner Gamma) are antipodal to relatively recent

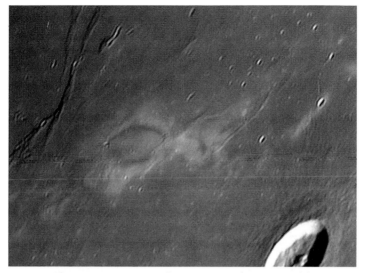

Figure 5. Reiner Gamma. (Image courtesy of Damian Peach.)

impact basins. They are amongst the most mysterious objects on the lunar surface and there is no certainty over how they were formed, although it has been suggested that they are the result of cometary impacts. More investigation is needed, and the amateur can contribute by seeking out new examples, studying the appearance of such swirls during the course of a lunation to see if that appearance changes, and perhaps observing and imaging them with different-coloured filters.

Many craters, not just on the Moon but also on other Solar System bodies, contain dusky radial bands on their inner walls. The brilliant crater Aristarchus (Figure 6), located prominently on the Oceanus Procellarum, is perhaps the best example, and its bands have been subjected to much study and comment since they were first reported in the nineteenth century. They have also been suspected of variability, but this is most unlikely. We do not really understand the nature of such bands. It has been suggested that banding is a natural by-product of the impact mechanism and is readily observable in most fresh craters, only to be covered up in older formations by the effects of space weathering. This may well be the case, but questions remain. We still have no complete listing or classification of banded craters, although attempts have been made in the past. Is banding indeed confined to smaller 'fresher'

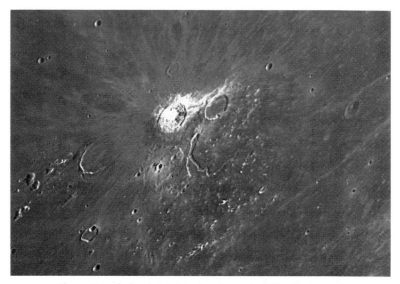

Figure 6. Dusky bands in Aristarchus. (Image by Bill Leatherbarrow.)

craters? Can we establish a relationship between crater size and banding? Moreover, there is evidence to suggest that some dusky bands are not merely surface deposits but are associated with radial dykes on the crater inner walls. This is true of Aristarchus, where even a small telescope will show the extension of that crater's inner shadow into the location of its bands, suggesting clearly that those bands lie in depressed dykes. We need further observation, using both the telescope and imagery, in order to see whether this is also true of other banded craters.

These are just a few examples of work waiting to be done, and space does not permit us to discuss at length the many other types of lunar features about which our knowledge is incomplete and which would benefit from the kind of sustained scrutiny, categorization and analysis that the amateur can carry out with the aid of telescope and/or photographs. Many of the most interesting craters contain dramatic fractures (Figure 7), probably caused by the uplift, slumping or tilting of their inner floors; there are domes (Figure 8) and other features that suggest areas of past volcanic activity; there are topographical alignments that might suggest the location of old impact basins; there are discrete areas of different-coloured regolith that might tell us much about the nature and history of that terrain; and there are craters that are surrounded by

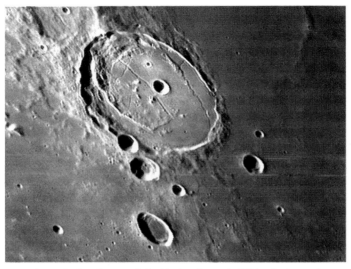

Figure 7. Floor fractures in Posidonius. (Image by Bill Leatherbarrow.)

dark haloes of material suggestive of volcanic vents, such as those to be found within Alphonsus. Many more examples could be cited and the Moon-Wiki website gives a full list of such special features (http:// the-moon.wikispaces.com/Spccial ι Fcatures+Lists). These will repay systematic study by Earth-based observers even in this age of ongoing spacecraft investigation of the Moon.

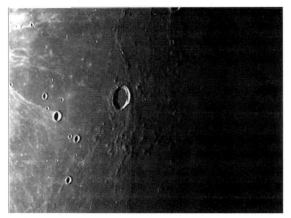

Figure 8. Domes near Marius. (Image by Bill Leatherbarrow.)

TRANSIENT LUNAR PHENOMENA

So far we have confined our discussion to observation of the Moon's more permanent surface features, but since observational records began there have been several thousand reports of suspected transient lunar phenomena, or TLPs, and these have been carefully logged. Such reports include sightings of temporary glows and obscurations, flashes of light on the Moon's darkened hemisphere, and observations of temporary colour. Some of these are undoubtedly real: flashes caused by meteoroid impacts have been widely observed and have been recorded by video cameras. We should not be surprised by this; meteors are a common phenomenon in the Earth's night sky, and there is no reason why they should not strike the Moon, where there is no atmosphere to dissipate the energy of impact, just as frequently. The observation of lunar meteoroid impacts is of real value and importance, and the amateur can play a full role. If astronauts ever return to the Moon it will be important to understand the nature and frequency of such impacts in order to minimize the danger to those exposed on the lunar surface. With this in mind, NASA has been running since 2005 an ongoing project of lunar impact observation based at the Marshall Space Flight Center, and this project includes provision for professional-amateur collaboration. Full details are available at http://www.nasa.gov/centers/marshall/news/lunar/, including guidance on how to participate in this work and what instrumentation is required. A telescope of 200–250 mm aperture will permit the detection of some impact flashes, while the use of video cameras suited to astronomical work (such as those manufactured by Watech and Mintron) allows the all-important recording and precise timing of such events. Moreover, software is now available that goes some way towards automating the detection of lunar impacts. Observation is best carried out when the moon is between 10 per cent and 50 per cent illuminated; at such phases the portion of the visible hemisphere that is in darkness is considerable and the glare from the illuminated surface is not so great that it might overwhelm any faint flashes occurring on the dark side. Meteoroid impacts are unpredictable, and may occur at any time, but the chances of seeing one are obviously greater when the Earth-Moon system encounters a known meteor stream.

Other examples of TLP are more contentious. Most have been reported by visual observers without the confirmation provided by recorded images. That is not to say that visual observations are inherently unreliable or that such reports should be discounted; but it is essential to treat them with due caution and to seek verification where possible. The human eye and brain, working at the limits of resolution, can sometimes play tricks and the observer of TLPs must be alert to the dangers and must take steps to minimize the possibility of being deceived by false appearances. For example, the use of achromatic (as opposed to apochromatic) refractors for TLP work is probably unwise, for they are susceptible to chromatic aberration and the production of images containing false colour. Valuable work can be done by attempting to confirm or refute past reports of TLPs. The Circular of the Lunar Section of the British Astronomical Association contains monthly predictions of repeat conditions, when both lighting and libration for a given area will match those of previously reported events. Careful visual observation or imaging at those times will allow us to discount reported TLPs that are simply attributable to tricks of the light. The professional community is divided over the reality or otherwise of transient phenomena, and this is another area where systematic and sustained amateur observation can do what orbiting spacecraft can't and thereby make a real contribution to our understanding of the Moon.

OTHER PROJECTS

Most of the observing projects outlined above may be carried out with modest telescopes. But – as has already been intimated – it is by no means essential to possess a telescope in order to make worthwhile contributions to lunar science. As well as the vast repository of spacecraft images, the amateur can also exploit the new possibilities opened up by the ready availability of powerful personal computers and the accessibility of the Internet. The sheer amount of data being returned from orbiting lunar craft and made available on the web is more than can be managed by professional lunar scientists. For example, the amateur is able to use data like the digital elevation models produced by probes such as Kaguya and the LRO, in conjunction with freely available software such as the Lunar Terminator Visualization Tool (LTVT), in order to produce virtual maps of the lunar surface that are capable of revealing much new information. Further details about LTVT, along with a download option, are available at: http://ltvt.wikispaces.com/LTVT.

Moreover, it is even possible to ask the LRO to take an image on your behalf! The NASA LROC website hosts a facility whereby members of the public can request a target, provided that they can make a basic scientific case for it. An on-line pro-forma for submitting a target request is available at: http://target.lroc.asu.edu/output/lroc/lroc_page.html, and this must surely be the ultimate in remote observing!

Finally, the amateur can participate in professional scientific work by contributing to the so-called 'citizen science' on-line projects that have sprung up in recent years. The best-known of these is *Galaxy Zoo*, a vast study of thousands of galactic images that has already yielded important results. The same team has recently launched a parallel project called *MoonZoo*. This seeks to provide, on the basis of widespread scrutiny by the public of LROC high-resolution imagery, detailed crater counts for as much of the lunar surface as possible. This is an approach that has been used by lunar scientists in the past to determine the age of parts of the lunar terrain – in general, older areas will show more evidence of impact cratering – but *MoonZoo* seeks to apply the technique on a level of resolution not previously available. Apart from crater counting, *MoonZoo* also hosts various spin-off projects, such as the identification of craters with local bouldering and of areas containing hardware from previous lunar missions. Further details, including a tutorial and advice on registration, are available at: http://www.moonzoo.org/.

So, there is much to be done, and the end of the great age of classical selenography does not mean that the amateur no longer has a part to play in lunar study. Certainly, the goalposts have shifted and whoever wants to undertake potentially useful work must be prepared to think differently from the way our predecessors did a generation or so ago. Some may regret the passing of what they see as the golden age of telescopic Moon mapping, but who is to say that the resources available to the amateur observer today do not hold out the promise of an even more golden future?

The Victorian Transits of Venus, 1874 and 1882, and Their Background

ALLAN CHAPMAN

On 5 or 6 June 2012, depending on your side of the International Date Line, Venus will form a line of sight between the Earth and the Sun, and over the course of some six and a half hours will appear to 'transit' the solar disk. Few people who saw the last transit, on 8 June 2004, are likely to forget that grand spectacle, and will, I am sure, do all they can to see it again in 2012. For those in Great Britain and in much of the Northern Hemisphere, however, only the last seventy-five minutes or so of the transit will be visible, the Sun rising with Venus already well into transit and approaching egress. But people in Japan, Eastern Australia and the Pacific Ocean will be able to witness the whole event, from ingress to egress, when Venus will slide off the Sun and into the blackness of space. And as a solar-transiting body, Venus will stay in that blackness for another 105 years, for the next pair of transits will not take place until 11 December 2117 and 8 December 2125.

THE DISCOVERY OF THE VENUS TRANSITS AND THEIR SIGNIFICANCE

At various periods during the Middle Ages, large sunspot groups visible to the naked eye were mistaken for Venus in transit, for even in the pre-Copernican, Earth-centred cosmology of the classical and medieval worlds, it was fully understood that Venus and Mercury came between the Earth and the Sun, and so, at least in theory, could pass in front of it. And at a time when no one knew how big (or small) the planets were in relation to the Sun, it was not unreasonable to attribute a large, dense sunspot group to Mercury or Venus. Yet modern astronomers, able to back-calculate the motions of the planets to a critical level of accuracy even for hundreds of years, realize that none of the ancient or medieval sightings of Sun-blotches could have been Venus, for on the dates in question Venus was at a quite different position in her orbit.

It was Johannes Kepler, the founding father of modern planetary dynamics calculation, who announced in print in 1629 that both Venus and Mercury would transit the Sun in 1631, on 7 November and 6 December respectively. But because the 6 December Venus transit would only begin around sunset, he did not believe that it would be practicable to observe it from Europe; and while Pierre Gassendi in Paris kept watch for it, he saw nothing. Kepler's calculations indicated that the next Venus transit would not take place until 1761. The 1631 Mercury transit, however, was successfully observed by Pierre Gassendi, who used his telescope to project an image of the solar disk: a technique, indeed, which had been employed to observe sunspots from the time of their discovery by Galileo in 1612.

Kepler's calculations had shown him that transits of Venus were very rare events, and only took place when the tilt between the Venusian and terrestrial orbits briefly produced a straight line. For although Venus rotates around the Sun in 225 days, and we could, in theory, have a transit every 8 months or so, the 3.5° tilt between the Earth's orbit and that of Venus led Kepler to conclude that a transit-creating alignment could only occur about every 130 years.

But it was a twenty-year-old Englishman, Jeremiah Horrocks, then living in the Lancashire village of Much Hoole, near Preston, who realized that Kepler had made an error regarding Venus. Yes, the transits did occur only every hundred years or so, but Horrocks came to believe, when comparing his own observations with the predictions of Kepler, that Venus would transit the Sun on Sunday 24 November 1639 (6 December in our calendar). Horrocks made his discovery too close to the event to spread the word far and wide, although he, and his friend William Crabtree, of Salford, Lancashire, both succeeded in securing accurate measurements of the position of the newly ingressed Venus on the disk of the setting Sun, using their telescopes to project the solar disk on to white paper. Tragically, Horrocks died suddenly just over two years after making the observation, but not before he had composed a ground-breakingly original treatise on the event, using his own observations, and drawing far-reaching conclusions. These included conclusions about the physical angular size of Venus (much smaller than everyone had previously believed), as well as the possible solar distance and other matters regarding the size of the Solar System. Horrocks's *Venus in Sole Visa* ('Venus in Transit upon the Sun') circulated in manuscript copies, before being sumptuously published in book form by Johannes Hevelius in Poland in 1662. This secured Horrocks's enduring posthumous fame, as one of the founders not only of British, but also of European observational and geometrical Copernican astronomy.

DR EDMOND HALLEY AND THE DISTANCE OF THE SUN

Venus, as we have seen, was not the only planet to occupy the attentions of Kepler, for he predicted the Mercury transit of 1631. A further Mercury transit followed in 1651, which was observed by the Englishman Jeremy Shakerley from Surat, India; and another in 1661, observed by Hevelius in Dantzig: an event that inspired the Polish astronomer to publish Horrocks's Venus transit work. Mercury transits, however, are much more frequent than those of Venus, for Mercury orbits the Sun in a mere eighty-eight days; and if the terrestrial orbit and that of Mercury had been in the same geometrical plane (which unfortunately they are not), then we would have four transits per year!

In 1677, the 23-year-old Edmond Halley was living on the island of St Helena, in the middle of the south Atlantic, and some 16° south of the Equator. He had gone there to make a new and much more accurate chart of the stars of the Southern Hemisphere, but on 7 November 1677, a transit of Mercury took place, which he witnessed, and of which he made meticulous observations and timings: noting the exact ingress and egress points of Mercury on the Sun's limb, and the precise duration of the planet's crossing of the solar disk. It then occurred to Halley, who was a geometer of genius, that the ingress and egress points which he had carefully noted on St Helena would *not* be the same for an observer in England, nor would the geometrical 'chord' line across the Sun's disk be the same length. This was caused, very straightforwardly, by the fact that the two observing stations, St Helena and London, were around 68° of latitude apart, and so must represent a base line or geometrical 'chord' through the Earth that was around 4,500 miles long.

Just as a pair of surveyors could map the lie of the land of an English county from two hilltops of a measured distance apart, why could not two sets of astronomers, thousands of miles apart, use Mercury in transit as a 'trig' point from which to measure the distance of the Sun? Of course, the displacement angles would be tiny, and the instruments – using the new telescopic sights and precision micrometers – would have to be critically accurate, but even by 1677 this was becoming technically possible.

The big problem, however, was the physical inappropriateness of Mercury for measuring what we now refer to as the 'Astronomical Unit': for the disk of Mercury on the Sun is tiny, making the limb contacts hard to measure with confidence, while the planet's rapid motion further complicated the issue. Venus, Halley realized, would be a much more suitable object to measure, with – as Horrocks had shown – its relatively larger disk (1' 12" arc) and much slower motion. The only difficulty lay in the fact that no one alive in 1677 would still be around to witness a transit of Venus, for the next pair would not take place until 1761 and 1769 – although Halley, dying at a ripe old age and with his faculties intact in 1742, was well on the way!

THE IMPORTANCE OF THE 'ASTRONOMICAL UNIT' AND THE 1761 AND 1769 TRANSITS

Even in the young Halley's day, the solar distance, or 'Astronomical Unit', was known pretty well. The Franco-Italian astronomer Giovanni Domenico Cassini, using a separate technique based on triangulating the opposition of Mars, had come up with a figure equivalent to 87 million miles, while even Horrocks by 1640 had extracted a solar distance of 15,000 Earth semi-diameters, or about 60 million miles, from his observations. But astronomy, the most advanced and accurate of all the sciences, did not deal in 'pretty wells'. For since the transformative work of Tycho Brahe in the 1580s, astronomers have striven to obtain the highest possible standards of geometrical accuracy in measuring celestial positions. And with the rise of gravitational astronomy, first with Hooke and Huygens, and, after 1687, with Newton, critical accuracy has become the order of the day. For how could we explain the various elliptical orbits of the planets, the seemingly bizarre behaviour of comets, and the long and short cycles that governed the lunar orbit, if we did not know the exact distance of the Sun? By the 1720s, astronomers realized that astronomy and physics were about proportions, masses and precise geometrical relationships, and if you could establish *one* of those relationships with absolute certainty, then the rest would automatically fall into place. And the key relationship for which they sought was the solar distance, or 'Astronomical Unit'.

In all these matters, Halley had led the way. Indeed, perceiving the usefulness of the 1761 and 1769 Venus transits for measuring the solar distance, Halley published some brilliant papers in the *Philosophical Transactions of the Royal Society*, which progressively refined his work on the Venusian orbit, and predicted what to expect in 1761. By the time of the appearance of his final Venus paper in 1716, Halley would be Savilian Professor of Geometry at Oxford, and within four years of becoming Astronomer Royal. As an indication of how important to astronomy the future transits were judged to be, the papers were written – unlike most of Halley's non-Venus publications – in Latin: the international learned language of Europe, so that a Frenchman, a Russian, or a Spaniard, could read and act upon them with no need for a translation.

It was the Venus transits of 1761 and 1769 that truly turned astronomy into a global science, for the more widely the observers were positioned across the globe, the longer the observing 'baselines' could be, and the more chances one had of gathering a rich harvest of positional data. It says something about the gentlemanly conventions of warfare in the eighteenth century, moreover, that although Great Britain and France were locked in the bloody 'Seven Years' War' in 1761 – being fought across the globe from India to Canada – the Royal Society and the French Académie issued government-backed certificates of safe conduct to protect the other nation's ships carrying transit expeditions. Alas, the certificates rarely worked, as English and French naval captains fought hammer and tongs in the Atlantic or Indian oceans, but the good intentions of the rival governments tell us much about the high esteem in which the international community held astronomical research.

In 1761, a dozen expeditions left mainland Europe to observe the transit, mostly British and French. And then, when Europe was at peace, twice that number were sent out, covering key locations from which it had been calculated that the transit would be observable. These included Canada, Russia, South Africa, India, Mauritius, China,

California and Tahiti. In both 1761 and 1769 the American colonists played a significant role in their own right, observing not only down the east coast of North America but also sending expeditions up into Canada. In total the 1761 transit was observed from 120 stations, and that of 1769 from 150. But it was to lead the 1769 expedition to Tahiti, on the longest possible baseline from Europe, that Lieutenant James Cook had been given command of HMS *Endeavour*. The spectacular success of this voyage would earn Cook a captaincy and a Fellowship of the Royal Society, and would pave his way to one of the most dazzling careers in global exploration.

To 'process' the observation results of all these returning expeditions, however, was a huge undertaking, for the amount of calculation required was enormous. And yet, there had been another entirely unforeseen problem, experienced by many observers, that muddied the hoped-for clarity of the observations: a thing which came to be referred to as the 'black drop'.

To obtain a truly useful observation of the transit, it was necessary to witness, measure and time with a good telescope and pendulum clock the very second when Venus's black limb touched and parted from the brilliant solar disk. Infuriatingly, however, instead of a clean break, many observers witnessed a dark, lenticular strand or blob, forming in the crucial seconds before and after the contacts: the 'black drop'. And the drop could make it impossible to time the contacts to an accuracy greater than several seconds, rather than the desired half a second or so. We now know that the drop was occasioned by direct sunlight being refracted through Venus's own dense atmosphere. Having learned from the black drop problems of 1761, the astronomers of 1769, whether they were observing from anywhere in mainland Europe or the American continent, or indeed anywhere else across the globe, were much better prepared. And when the observations were reduced, after the return home, the error spread of the 1769 observers fell across a much narrower band than did that of 1761. Some projections, indeed, came very close to the modern value of 8.798 arc seconds, though at the time no one knew which figures were the best.

Tragically, however, especially when one remembers that many men actually lost their lives on the 1761 and 1769 transit expeditions, no single accepted value for the solar parallax, and hence the distance of the Sun, emerged. Of course, the values were better than those available before 1761, but there was still no mathematical agreement. And while subsequent generations into the early nineteenth century attempted to re-analyse the 1761 and 1769 results, there was nothing definitive. And definitiveness was, above all, the thing that astronomers strove to achieve.

THE VENUS TRANSIT OF 1874

By the time that Venus next showed herself on the Sun, in the mid-Victorian age, the world had become a profoundly different and very much smaller place. The American continent, instead of having thirteen small colonies on its Atlantic seaboard and some Spanish mission stations in California, was now a booming, technological country in its own right. California no longer required months of dangerous travel of the kind that had claimed the life of the brave French astronomer Jean-Baptiste Chappe d'Auteroche and most of his colleagues in 1769, but was now a comfortable five-day train journey from New York. And the Australian continent, discovered by Captain Cook after observing the transit of 1769 from Tahiti, was a thriving country in daily contact with London by electric telegraph since 1872. Indeed, steamships, railways, electric telegraphs, newspapers, photographs, applied science and technology, and the beginnings of a global distribution system, had shrunk the world in a way which no one alive in 1769 could have imagined. America, its 'Frontier' a thing of the past by 1880, was now covered with new cities, universities, observatories, and science field stations.

Likewise, South Africa had a long established Royal Observatory in its own right by 1874, and there were fixed, well-equipped observatories, either private 'Grand Amateur' or publicly funded, in Australia, India, Canada, the Far East and elsewhere.

Not only was the globe remarkably well covered with observatories of varying quality by 1874, but getting access to far-flung islands that occupied key geometrical positions for securing observations was much easier than in 1769. For steamships, often running to scheduled timetables, the Suez Canal, and the transcontinental railroad across North America had shrunk global distances. Indeed, it had been the marvel of this rapidly shrinking world that had inspired Jules Verne's novel *Around the World in Eighty Days* (1873).

Telescopes were very much better by 1874 than they had been in the eighteenth century, as high-quality 6-, 8- and 10-inch-aperture achromatic refractors on finely engineered equatorial mounts were being routinely manufactured, especially in England, Ireland, Germany, France and the USA. This gave rise to the concept of attempting to observe the transit with instruments of a 'standard' quality and specification, which in turn should make the interpretation of the ensuing results much more uniform.

What is more, two new precision instruments were readily available in 1874 which had not existed in 1769, namely, the heliometer and the photographic plate. But more of these instruments and their significance later.

Central to the plans for 1874, especially for British and Imperial observers, was the work of Sir George Biddell Airy, Astronomer Royal. For as early as 1857, Airy had alerted the scientific world to the forthcoming transits, and to the need for meticulous long-term preparations if the solar parallax and distance were to be computed with accuracy. Indeed, Airy was assiduous in his attempt to impose a 'business efficiency' model on all the official British expeditions, running training courses for transit observers at the Royal Observatory. Amongst other things, Airy devised a transit 'simulator' in which a black disk was moved at the same speed as Venus in transit across a sharp white edge to simulate the ingress and egress. When viewed through a telescope at a distance of 400 feet, this gave an elegant representation of what the ingress and egress would look like. The hope was that after hours of practice with this machine, the critical measurement and timing of Venus on the Sun's disk on 8 or 9 December 1874 (depending on the longitude of your observing station east or west of Greenwich) would come as second nature. (American, French and German astronomers were also pre-trained in their own countries with other transit simulators.)

Yet in no way were the British preparations for observing the 1874 transit trouble-free. Richard Anthony Proctor, an eminent English private astronomer and influential writer and lecturer, openly criticized the Astronomer Royal on details of observing technique and the selection of foreign observing stations, backing up his criticisms in a book, *The Transits of Venus* (1874). The main bone of contention between Proctor and Airy was whether stations were to be selected in accordance with Halley's method of 1716, or with that of Joseph-Nicholas Delisle from the 1750s.

In Halley's method, each astronomer really needed to see the whole transit, observe and time the ingress and egress and all four Venus contacts with the solar limb, and fix the precise position of the chord line described by the moving Venus. This technique inevitably restricted observations to those locations, in either hemisphere, where the entire transit could be seen. In the Delisle method, however, an astronomer needed only to secure Venus's entry upon, or exit from, the solar disk – two contacts rather than four – provided he knew the exact longitude of his observing station. In this way, and only if the precise latitude and longitude of the respective stations were known, all the separate ingress and egress observations could be perfectly matched together back in Europe or America. The Delisle method was fine, but it did take for granted the critical geographical and astronomical accuracy of a single observation; whereas the Halley method, more limited in its geographical sweep of potential stations, had – weather permitting – the advantage of delivering a complete set of coordinates, and was less dependent upon exact knowledge of the longitude of the observing station.

THE HELIOMETER AND THE CAMERA

High hopes were placed on these two relatively new instruments, especially by German and French astronomers. The heliometer was a precision achromatic refracting telescope, the object glass of which was divided into a pair of perfectly matched 180° segments. Precision control rods from the eyepiece end of the heliometer moved micrometer screws that made the segments of the object glass slide laterally against each other. When the two lens segments formed a 360° circle, one saw a single image, as in a normal refracting telescope. But when they slid against each other, a split image was produced, in very much the same way as with a pre-digital camera rangefinder.

The heliometer was used to measure the distances between stars or other celestial objects to considerably less than a single arc second across angles that were bigger than those within the range of a conventional filar micrometer. For the transit of Venus, however, it was to be employed to measure the exact distance between Venus and the solar centre, and limb, and was naturally useful for observers following the Delisle method.

Photography was already an established astronomical technique by 1874, and in the late 1850s the Guernsey and London 'Grand Amateur'

Warren de la Rue had developed the first custom-built solar camera. It had a 3.5-inch-aperture, 50-inch-focal-length achromatic object glass, with an optical bias to the ultraviolet, and was set on an iron stand with an equatorial mount. It was used to take daily photographs of the Sun from the Kew Observatory, and took some classic Venus-in-transit plates in 1874. As with the heliometer, it was hoped that precisely timed photographs could be used to fix the position of Venus relative to the solar limb and centre, and that a parallax might be obtained.

But in addition to such a relatively straightforward camera, ingenious variants were produced. The Scottish 'Grand Amateur' Lord Lindsay, who observed the 1874 transit from the island of Mauritius, would subsequently come to devise (as did various American astronomers) a horizontal 'sideriostat' telescope of very long focal length, which would give a large prime focus image that would not require enlargement (and hence, risk distortion). Then the ingenious French pioneer of solar photography, Jules Janssen – ex-schoolteacher and pioneer astrophysicist – anticipated the movie-camera by twenty years. Realizing that things would happen rapidly, in an unfamiliar manner, and only once during the transit, he invented a camera with a circular glass plate capable of being rotated by a spring by a 'trigger' mechanism. This could get sixty separate photographs of Venus on the plate, at single-second intervals, and would, in effect, produce a 'movie' image of ingress or egress. If the exact second of the commencement of the sixty exposures was noted, then, at least in theory, every aspect of the transit could be recorded, automatically to the second and analysed at leisure later. A wonderful way, indeed, of obtaining data from which to calculate a parallax – provided everything went right!

THE 1874 TRANSIT EXPEDITIONS

In spite of the profusion of fixed observatories in 1874, an enormous number of expeditions were sent out by the principal scientific nations of the age. The Russians sent twenty-six (mostly within the Czar's vast dominions), the British twelve, America eight, France six, Germany six, Italy three, and Holland one. And in many ways, the more remote the location the better, for distant places, especially uninhabited rocky islands in the south seas, gave the widest triangulation points from the big European and North American observatories. The penguin-inhabited St Paul's and the Campbell Islands in the far Southern Hemisphere were used. The barren, windswept Kerguelen's Island – also known descriptively as 'Desolation Island' – at 49° south, in the middle of the Indian Ocean, only 1,000 miles north of Antarctica, was visited by no less than three expeditions. The one consolation, perhaps, was that as the transit fell on 8 December, the Southern Hemisphere was approaching mid-summer, which held out a good chance of their witnessing the entire event. Expeditions also went to bitterly cold Siberia to see what the short winter's day offered, the more welcoming Sandwich Islands in the Pacific were used for observations, while Jules Janssen was almost killed in a typhoon in the China Seas on his way to Nagasaki, Japan.

In this golden age of the British Grand Amateur, Lord James Ludovic Lindsay (who in 1880 would inherit his title as 26th Earl of Crawford) would lead his own privately funded expedition to Mauritius, in the Indian Ocean. Indeed, Lord Lindsay was a true research scientist by instinct, and on his estates at Dun Echt, Aberdeenshire, he ran a major astronomical observatory entirely out of his own pocket, paying particular attention to the new sciences of solar physics and spectroscopy. (Indeed, in 1894 he would give this observatory to the Scottish nation, to become the Royal Observatory, Edinburgh, or ROE.)

In 1873, Lord Lindsay ordered a state-of-the-art heliometer from the Repsold Brothers in Hamburg. He also suggested to his Observatory Director, Dr David (later, Sir David) Gill, that he should travel out to Pulkowa, near St Petersburg, Russia, to see heliometers and other instruments in the great research observatory funded by the Czar, then go down (on a 52-hour train journey) with its Director, Professor Otto Struve, to Hamburg, there to meet up with Lord Lindsay himself. They would then attend the Transit of Venus meeting of the Astronomische Gesellschaft, the German equivalent of the Royal Astronomical Society. Indeed, this suggestion in itself tells us not only how internationally based and co-operative astronomy had become by 1873, but also that Lord Lindsay treated the astronomers whom he employed as the scientific gentlemen they were, rather than as hired hands, as did some Grand Amateurs.

Lord Lindsay sent all the chronometers and equipment to observe the transit to Mauritius by steamer, through the new Suez Canal, on a six-week voyage from Southampton. But he himself, being a keen sailor, travelled with the new heliometer and other telescopes on his specially purchased 398-ton three-masted schooner, *Venus*, manned by 22 sailors and a ship's doctor. The *Venus* went the long way round, via the Cape of Good Hope, and arrived after 'a somewhat stormy and wearisome passage' of seventeen weeks.

Transit day, 9 December 1874, dawned cloudy, yet later cleared, allowing Lord Lindsay and his colleagues to obtain a full view of the transit. The beautiful Repsold heliometer enabled numerous measures of Venus and the Sun's centre to be made, and the astronomical camera produced many excellent photographs. Yet that accursed 'black drop' which had plagued the astronomers of 1761 and 1769 made its unwelcome appearance through Venus's atmosphere, and – in spite of the superb instruments – made it impossible to establish the contact times to anything better than a rough ten seconds of time.

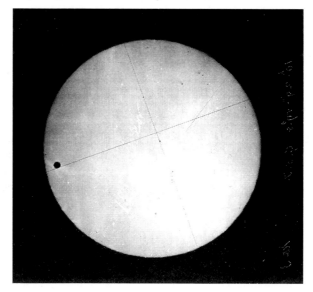

Figure 1. Venus is seen in transit across the face of the Sun in this image from a glass negative of the event taken in Luxor, Egypt, on 9 December 1874. (Image courtesy of the National Maritime Museum, Greenwich, London.)

And of the 'official' British expeditions of 1874, the most significant was that sent out to Honolulu, Hawaii, led by a Royal Marines scientific officer, Captain George Lyon Tupman. The seven astronomers in the

Hawaiian party left Liverpool with 100 tons of equipment in June 1873, aboard the commercial steamships the *Illumani* and *Britannia*. Sailing around Cape Horn, the astronomers arrived at Valparaiso, where they and their equipment were transhipped aboard HMS *Scout*, a British frigate stationed in Hawaiian waters, and they eventually arrived in Honolulu harbour on 9 September, where they took up a comfortable residence in the Hawaiian Hotel. Three British teams were stationed across the Hawaii archipelago, with the intention of obtaining three slightly different, yet congruent, observations of the start and early part of the transit.

And on transit day itself, 8 December, clear skies enabled a good sighting to be made, yet not without the troublesome 'black drop'. After packing up and dispatching the instruments back to Britain, the astronomers returned home separately. Fast steamers to San Francisco were followed by a few days on the transcontinental railroad to New York, and then another fast scheduled steamer for England. For as was indicated above, the world was already becoming a small place by 1874, and scheduled commercial steamers and trains had made global travel fast and remarkably comfortable, even if it was also a great adventure.

And great adventures inevitably abound with human stories, of which the British expedition to Hawaii had its share. One man, Charles Lambert, was tragically drowned, in spite of being a strong swimmer. He and some friends had gone down to the beach for a routine morning dip, but some undercurrent seems to have suddenly dragged him below to his death. Then there was the 'problem' man of the expedition, Henry Glanville Barnacle, a St John's, Cambridge, mathematics graduate who somehow could never quite get it right. He was sent home ahead of the rest of the party – via San Francisco and the railroad – in something resembling disgrace. But perhaps the most remarkable human story came from an unexpected Royal interest in the transit. For Hawaii was an independent monarchy in 1874, and recently ascended to the throne was King David Kalākaua, a cultured and educated gentleman, with a serious long-standing interest in astronomy. At His Majesty's request, the transit astronomers were formally presented to him, and, going by what the Hawaiian newspapers said, he freely offered whatever help and resources Queen Victoria's astronomers might require. And by Royal Order, silence had to be kept during the crucial parts of the transit, so as not to disturb the observers!

THE RESULTS OF THE 1874 TRANSIT

It had been generally hoped that the 'black drop', which had done so much to reduce the value of the data collected at the 1761 and 1769 transits, would be overcome by the much more sophisticated instruments and techniques used in 1874. Sadly, however, that hope was not fulfilled. As transit results began to fly in from across the globe by electric telegraph, two things became clear: first, that on the whole, weather conditions had been good, and most stations had obtained some results; second, that the 'black drop' had still raised its unwanted head at the astronomical banquet. Indeed, Captain Tupman had said that – even when 'personal equations' for individual nervous response times were factored in – two observers working side by side might differ in their timings for the Venus on solar limb contacts by as much as twenty or thirty seconds of time.

As a consequence, the calculated values for the solar parallax and distance were divergent to an almost unacceptable level. Much of the sheer hard work of reducing the British results descended on to the not unwilling shoulders of the Astronomer Royal, Airy, who was then in his late seventies. Indeed, this was a formidable undertaking for Airy, in addition to running the Royal Observatory. Airy extracted a solar parallax of 8.754 arc seconds, or to within 0.04 arc seconds of the modern value, which gave a Sun-distance of 93,375,000 miles. We know now that this was an excellent result, but at the time, it was clouded by others, such as Edmund Stone's independently computed parallax of 8.88 arc seconds, which placed the Sun at 92,000,000 miles.

Consequently, scientists and governments began to think long and hard about how much effort should be put into observing the 1882 transit. After all, as was pointed out by Sir David Gill, Johann Gottfried Galle, and others, there were other methods of measuring the solar parallax that had more potential and were less beset with problems, such as the black drop, than was the case with Venus transits. One such method was to measure the exact parallax angle of Mars at opposition: an improvement of the technique first used by Cassini's astronomers, triangulating between Paris and the South American island of Cayenne, in 1672. In a way, this was a bit like a planetary transit in reverse, in so far as it, like a transit, hinged upon the planet, the Earth and the Sun forming a straight line. Yet instead of demanding a wide global spread of observing stations, as did Venus transits, a Mars opposition could be observed by one man in one place. The movement was supplied by the Earth's own daily rotation, as the observer measured Mars against a background of dim stars, at night and morning, the Earth's own diameter providing a displacement of 8,000 miles over 12 hours. Sir David Gill used Lord Lindsay's Repsold heliometer to gain a set of observations from which he extracted a solar parallax of 8.78 arc seconds, and a corresponding solar distance of 93,080,000 miles.

Even better than Mars – which itself subtended a disk rather than a point – were the minor planets or asteroids, which only ever appeared as tiny points of light, making them ideal geometrically, and the orbits of which were well known. Indeed, even when still on Mauritius in 1874, Gill and Lord Lindsay had tried out this method on the minor planet Juno, and in later years Iris, Sappho and Victoria were successfully observed from solar parallax triangulation. After 1874, therefore, it looked increasingly as if the future of solar parallax determination was going to lie with more straightforward and less expensive methods than Venus transits. What is more, parallax determination of minor planets was best and most accurately performed from fixed observatories, with permanently mounted large transit circles and heliometers, rather than with the relatively portable instruments used on the 1874 expeditions. So did that mean that the forthcoming 1882 transit would be of no scientific value?

THE 1882 TRANSIT

The black drop and other problems of 1874 had made it clear to the international astronomical community that the significance of the 1882 transit would not be as great as had been hoped ten years before. After all, *three* atmospheres – terrestrial, Venusian, and solar – conspired against that clear and straightforward sighting that the solar parallax method demanded. The Russians refused to send out any expeditions, though France, Germany, Belgium, Britain and the USA sent out theirs, for after all, no one alive in 1882 would live to see the next transit in 2004. On the whole, the Germans and Belgians relied on heliometers to measure Venus's position on the Sun, while the French – influenced, no doubt, by Jules Janssen's brilliant solar photographic work – tended to favour the camera. But when the results came back home, and the mathematicians began calculating, they found once again that the spread of results was just too wide to allow a definitive solar parallax and distance to be computed. Indeed, as Herbert Hall Turner was to show in *Astronomical Discovery* (1904), p. 32, both transits displayed an error spread of 1¾ million miles for the solar distance, with British results reducing to 93¼ million miles (1874) and 92½ million (1882), and the French to 91½ million! Yet the 1882 transit was interesting for at least two reasons, both of which had to do with the ingenuity and achievements of two 'ordinary' self-taught astronomers.

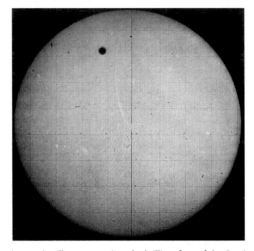

Figure 2. Venus is seen in silhouette against the brilliant face of the Sun in this image from a photographic plate showing the transit of Venus on 6 December 1882. (Image courtesy of US Naval Observatory Library.)

The first of these was Charles Grover, who was forty years old in 1882. He had begun life in a Buckinghamshire village, and at the age of twelve was apprenticed to a brush-maker. But astronomy soon became his passion, especially after viewing Donati's Comet in 1858. He began to read whatever he could find on astronomy, and was fortunate to be encouraged by several local gentlemen, including the famous 'Grand Amateur' Dr John Lee, of Hartwell House, Aylesbury. Moving to London, Grover began working for John Browning, whose company manufactured 'magic lanterns' and other optical instruments, where his job was not only that of a salesman and demonstrator, but also that of an expert who travelled around Britain on behalf of the company as a 'troubleshooter', to help wealthy clients who had bought Browning telescopes to use them correctly. It was through connection with this wealthy amateur astronomer world that Charles Grover entered the employ as a paid assistant of Cuthbert Peek, a wealthy astronomical gentleman whose family fortune derived from biscuit manufacture, whom he accompanied to Australia for the 1882 transit.

Charles Grover's 'Journal', much of which was published in 2005 (see bibliography), is a fascinating read, and full of both scientific and social detail. Grover seems to have been respected and well treated by Peek, for whom he worked until his boss's death in 1901, and then for the Peek family until his own death in 1921. His account of the voyage to Australia in the fast steamer *Liguria* was especially fascinating, along with his observations of the comet of 1882. Cuthbert Peek and his party arrived at Melbourne, and Grover visited and sketched the Melbourne Observatory with its several domes and weather station, as well as that at Sydney. Travelling 'up country' by train from Brisbane, they then took horses and wagons to their destination, Jimbour, a house on the Darling Downs. Having set up Cuthbert Peek's fine portable observatory with its large equatorial refractor, they had everything ready for the transit. But then, on transit day, 7 December 1882 (or 6 December in England), that curse of the astronomers' life set in – cloud and rain – so that nothing whatsoever was seen.

Modern-day serious amateur astronomers who travel the world to see eclipses and other phenomena all know the frustration of being clouded out at the critical moment. But imagine the frustration of having spent weeks traversing the globe by ship, train and wagon, not to mention the expenditure of a large sum of money, as Cuthbert Peek did, only to see clouds on the big day. Yet Charles Grover continued to work for Sir Cuthbert (after he inherited a baronetcy) as the director of his well-equipped astronomical and meteorological observatory on the family's estate at Rousden, Devonshire, his career transformed from that of brush-maker to astronomical gentleman. With a handsome house on the Peek estate, Charles and Elizabeth Grover became prominent, highly respected, and long-lived members of the local community. So without doubt, Charles Grover was one of the few outstanding success stories of the 1882 Venus transit.

The other success story in 1882 was that of Samuel Cooper, who sent articles on astronomy to the popular magazine *The English Mechanic* under the *nom de plume* of 'The Optical Bricklayer'.

Although Cooper was already the author of published pieces, however, it was the Welsh gentleman amateur astronomer G. Parry-Jenkins who told Cooper's story in 1911. For Samuel Cooper lived at Charminster, Dorset, not far from Bournemouth, where he earned his living as a bricklayer. Yet it was the 'marvels' of astronomy and optics that had 'appealed strongly to him from youth upwards'. Although he was entirely self-taught from books, Cooper's scientific skills 'succeeded so well that he became quite an expert on the construction of such instruments [reflecting telescopes]' and wrote popular articles on them.

When Parry-Jenkins had visited Cooper in the 1880s, he had been greatly impressed by his home-made 9-inch-aperture reflector. For it was with this instrument that, on 6 December 1882, Cooper secured three photographs of Venus ingressing into the solar limb: timed at 2.15, 2.30 and 2.45 p.m. Observing conditions had been poor on that December day, especially across England, but 'Samuel Cooper had the intense satisfaction of seeing and photographing the phenomenon which so many missed'. He gave the pictures to Parry-Jenkins, who published them.

Indeed, the more research I conduct into the Victorian amateur astronomical community, the more I am bedazzled by men like Cooper – working men who studied in their spare time at the end of the long working day, and who figured mirrors, built telescopes and became expert photographers. Yet Samuel Cooper is the only one I know who obtained three timed photographs of the elusive Venus transit of 1882.

THE TRANSITS OF 2004 AND 2012

Long before 2004, the astronomical unit had been measured to a high level of accuracy by methods that did not involve Venus. Where the transit was significant, however, was in its educative and synthesizing powers. It gave rise to conferences in which both professional and amateur astronomers took stock of where our knowledge stands today, and how research can best proceed in the future. The International Astronomical Union held its 196th meeting at the University of Central Lancashire, Preston, some eight miles from the village of Much Hoole, where Jeremiah Horrocks recorded the first observed transit in November 1639. I had the honour of delivering the keynote lecture on Horrocks and his work at the beginning of this conference, and I will

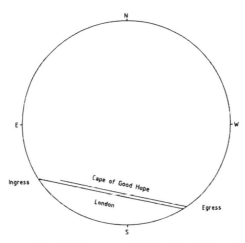

Figure 3. The apparent track of Venus across the Sun during the transit of 8 June 2004, as viewed from London, UK, and from the Cape of Good Hope, South Africa. Although the transit was visible in its entirety from London, the ingress was not visible from the Cape.

2004 and 2012 Transits of Venus

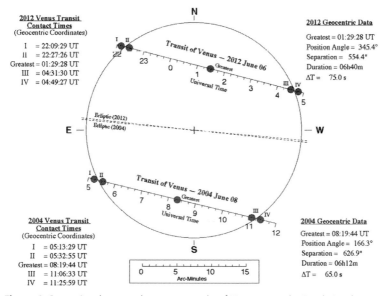

Figure 4. Comparison between the apparent paths of Venus across the Sun during the transits of 8 June 2004 and 5–6 June 2012. (Diagram courtesy of Fred Espenak, NASA/GSFC.)

never forget seeing the Sun rise on 8 June 2004 and, at 6.24 a.m., Venus making her ingress upon the solar disk. The I.A.U. astronomers observed the Sun rise and the ingress from the university's Wilfred Hall Observatory, eight miles outside Preston and on the foothills of the Pennines. After breakfast, we were all bussed to Much Hoole, about sixteen miles away, and then, from St Michael's Church yard (where Horrocks had been a bible clerk), we watched egress, at 12.20 p.m. The transit was quite unforgettable.

But across the country, and the world, telescopes were set up in public places, and countless people were able to watch the transit, and sense that fascination with astronomy which ran from Jeremiah Horrocks to Samuel Cooper, linking them, and us, across the centuries, with thousands, if not millions, of other people.

The next transit of Venus will be visible on 5–6 June 2012. Unlike Samuel Cooper, we in Britain will only catch the egress, visible shortly after sunrise; but if you want to travel to Australia, it will only take you as many hours to go there and back as it took Captain Cook and his colleagues *months* between 1768 and 1771.

Yet no matter where you choose to observe from in 2012, don't forget that you will have become part of a scientific tradition that has now spanned five centuries, from 1639 to 2012. And clear skies to you all!

FURTHER READING

Sir Robert Stawell Ball, *The Story of the Heavens* (London, 1893).

Hermann A. Brück, 'Lord Crawford's Observatory at Dun Echt, 1872–1892', *Vistas in Astronomy* 35 (Pergamon Press Ltd, 1922), 81–138.

Allan Chapman, 'Jeremiah Horrocks, William Crabtree and the Lancashire Observations of the Transit of Venus of 1639', in *Transits of Venus: New Views of the Solar System and Galaxy* (Proceedings of the 196th Colloquium of the International Astronomical Union, held at Preston, Lancashire, UK, 7–11 June 2004), ed. D. W. Kurtz (Cambridge University Press, 2005), 3–26.

Allan Chapman, *The Victorian Amateur Astronomer. Independent Astronomical Research in Britain 1820–1920* (Praxis-Wiley, Chichester, 1998).

Michael Chauvin, *Hōkūloa. The British 1874 Transit of Venus Expedition to Hawai'i* (Honolulu, Hawai'i, Bishop Museum Press, 2004).

Agnes M. Clerke, *A Popular History of Astronomy During the Nineteenth Century* (1893 edn.).

Alan Cook, *Edmond Halley: Charting the Heavens and the Seas* (Clarendon Press, Oxford, 1998).

George Forbes, *The Transits of Venus* (London, 1874).

David Gill, *A Determination of the Solar Parallax from Observations of Mars made at the Island of Ascension in 1877* (London, 1881).

Isobel Gill, *Six Months in Ascension* (London, 1878).

Robert Grant, *A History of Physical Astronomy from the Earliest Times to the Middle of the Nineteenth Century* (London, 1852).

Edmond Halley's papers on the Venus transits of 1761 and 1769 and the Mercury transit of 1677: letter, Halley to Hooke, in Robert Hooke's *Cometa* (1678); also 'De visibili conjunctione', *Philosophical Transactions of the Royal Society* 17, no. 194 (1691), 511–22; 'Methodus singularis quâ Solis parallaxis . . .', *Phil. Trans.* 29, no. 348 (1716), pp. 454–64.

Lord Lindsay, *Dun Echt Observatory Publications*, vol. 2 (1877).

Good on observing techniques and zones of visibility across the globe in 1874 and 1882: Simon Newcomb, *Popular Astronomy* (London, 1898), 187–95.

For an account of Samuel Cooper: G. Parry-Jenkins, 'A Plea for the Reflecting Telescope', *Journal of the R.A.S. of Canada*, V (1911), 59–75.

Barbara Slater, *The Astronomer of Rousdon, Charles Grover, 1842–1921* (Steam Mill Publishing, Norwich, 2005).

Herbert Hall Turner, *Astronomical Discovery* (London, 1904).

Harry Woolf, *The Transits of Venus: A Study of Eighteenth-Century Science* (Princeton University Press, 1959).

Part Three

Miscellaneous

Some Interesting Variable Stars

JOHN ISLES

All variable stars are of potential interest, and hundreds of them can be observed with the slightest optical aid – even with a pair of binoculars. The stars in the list that follows include many that are popular with amateur observers, as well as some less well-known objects that are, nevertheless, suitable for study visually. The periods and ranges of many variables are not constant from one cycle to another, and some are completely irregular.

Finder charts are given after the list for those stars marked with an asterisk. These charts are adapted with permission from those issued by the Variable Star Section of the British Astronomical Association. Apart from the eclipsing variables and others in which the light changes are purely a geometrical effect, variable stars can be divided broadly into two classes: the pulsating stars, and the eruptive or cataclysmic variables.

Mira (Omicron Ceti) is the best-known member of the long-period subclass of pulsating red-giant stars. The chart is suitable for use in estimating the magnitude of Mira when it reaches naked-eye brightness – typically from about a month before the predicted date of maximum until two or three months after maximum. Predictions for Mira and other stars of its class follow the section of finder charts.

The semi-regular variables are less predictable, and generally have smaller ranges. V Canum Venaticorum is one of the more reliable ones, with steady oscillations in a six-month cycle. Z Ursae Majoris, easily found with binoculars near Delta, has a large range, and often shows double maxima because of the presence of multiple periodicities in its light changes. The chart for Z is also suitable for observing another semi-regular star, RY Ursae Majoris. These semi-regular stars are mostly red giants or supergiants.

The RV Tauri stars are of earlier spectral class than the semi-regulars, and in a full cycle of variation they often show deep minima and double maxima that are separated by a secondary minimum. U Monocerotis is one of the brightest RV Tauri stars.

431

Among eruptive variable stars is the carbon-rich supergiant R Coronae Borealis. Its unpredictable eruptions cause it not to brighten, but to fade. This happens when one of the sooty clouds that the star throws out from time to time happens to come in our direction and blots out most of the star's light from our view. Much of the time R Coronae is bright enough to be seen in binoculars, and the chart can be used to estimate its magnitude. During the deepest minima, however, the star needs a telescope of 25-cm or larger aperture to be detected.

CH Cygni is a symbiotic star – that is, a close binary comprising a red giant and a hot dwarf star that interact physically, giving rise to outbursts. The system also shows semi-regular oscillations, and sudden fades and rises that may be connected with eclipses.

Observers can follow the changes of these variable stars by using the comparison stars whose magnitudes are given below each chart. Observations of variable stars by amateurs are of scientific value, provided they are collected and made available for analysis. This is done by several organizations, including the British Astronomical Association (see the list of astronomical societies beginning on p. 456), the American Association of Variable Star Observers (49 Bay State Road, Cambridge, Massachusetts 02138, USA), and the Royal Astronomical Society of New Zealand (PO Box 3181, Wellington, New Zealand).

Star	RA		Declination		Range	Type	Period	Spectrum
	h	m	°	′			(days)	
R Andromedae	00	24.0	+38	35	5.8–14.9	Mira	409	S
W Andromedae	02	17.6	+44	18	6.7–14.6	Mira	396	S
U Antliae	10	35.2	−39	34	5–6	Irregular	—	C
Theta Apodis	14	05.3	−76	48	5–7	Semi-regular	119	M
R Aquarii	23	43.8	−15	17	5.8–12.4	Symbiotic	387	M+Pec
T Aquarii	20	49.9	−05	09	7.2–14.2	Mira	202	M
R Aquilae	19	06.4	+08	14	5.5–12.0	Mira	284	M
V Aquilae	19	04.4	−05	41	6.6–8.4	Semi-regular	353	C
Eta Aquilae	19	52.5	+01	00	3.5–4.4	Cepheid	7.2	F–G
U Arae	17	53.6	−51	41	7.7–14.1	Mira	225	M
R Arietis	02	16.1	+25	03	7.4–13.7	Mira	187	M
U Arietis	03	11.0	+14	48	7.2–15.2	Mira	371	M
R Aurigae	05	17.3	+53	35	6.7–13.9	Mira	458	M
Epsilon Aurigae	05	02.0	+43	49	2.9–3.8	Algol	9892	F+B
R Boötis	14	37.2	+26	44	6.2–13.1	Mira	223	M
X Camelopardalis	04	45.7	+75	06	7.4–14.2	Mira	144	K–M

Some Interesting Variable Stars

Star	RA h	m	Declination °	′	Range	Type	Period (days)	Spectrum
R Cancri	08	16.6	+11	44	6.1–11.8	Mira	362	M
X Cancri	08	55.4	+17	14	5.6–7.5	Semi-regular	195?	C
R Canis Majoris	07	19.5	−16	24	5.7–6.3	Algol	1.1	F
VY Canis Majoris	07	23.0	−25	46	6.5–9.6	Unique	—	M
S Canis Minoris	07	32.7	+08	19	6.6–13.2	Mira	333	M
R Canum Ven.	13	49.0	+39	33	6.5–12.9	Mira	329	M
*V Canum Ven.	13	19.5	+45	32	6.5–8.6	Semi-regular	192	M
R Carinae	09	32.2	−62	47	3.9–10.5	Mira	309	M
S Carinae	10	09.4	−61	33	4.5–9.9	Mira	149	K–M
l Carinae	09	45.2	−62	30	3.3–4.2	Cepheid	35.5	F–K
Eta Carinae	10	45.1	−59	41	−0.8–7.9	Irregular	—	Pec
R Cassiopeiae	23	58.4	+51	24	4.7–13.5	Mira	430	M
S Cassiopeiae	01	19.7	+72	37	7.9–16.1	Mira	612	S
W Cassiopeiae	00	54.9	+58	34	7.8–12.5	Mira	406	C
Gamma Cas.	00	56.7	+60	43	1.6–3.0	Gamma Cas.	—	B
Rho Cassiopeiae	23	54.4	+57	30	4.1–6.2	Semi-regular	—	F–K
R Centauri	14	16.6	−59	55	5.3–11.8	Mira	546	M
S Centauri	12	24.6	−49	26	7–8	Semi-regular	65	C
T Centauri	13	41.8	−33	36	5.5–9.0	Semi-regular	90	K–M
S Cephei	21	35.2	+78	37	7.4–12.9	Mira	487	C
T Cephei	21	09.5	+68	29	5.2–11.3	Mira	388	M
Delta Cephei	22	29.2	+58	25	3.5–4.4	Cepheid	5.4	F–G
Mu Cephei	21	43.5	+58	47	3.4–5.1	Semi-regular	730	M
U Ceti	02	33.7	−13	09	6.8–13.4	Mira	235	M
W Ceti	00	02.1	−14	41	7.1–14.8	Mira	351	S
*Omicron Ceti	02	19.3	−02	59	2.0–10.1	Mira	332	M
R Chamaeleontis	08	21.8	−76	21	7.5–14.2	Mira	335	M
T Columbae	05	19.3	−33	42	6.6–12.7	Mira	226	M
R Comae Ber.	12	04.3	+18	47	7.1–14.6	Mira	363	M
*R Coronae Bor.	15	48.6	+28	09	5.7–14.8	R Coronae Bor.	—	C
S Coronae Bor.	15	21.4	+31	22	5.8–14.1	Mira	360	M
T Coronae Bor.	15	59.6	+25	55	2.0–10.8	Recurrent nova	—	M+Pec
V Coronae Bor.	15	49.5	+39	34	6.9–12.6	Mira	358	C
W Coronae Bor.	16	15.4	+37	48	7.8–14.3	Mira	238	M
R Corvi	12	19.6	−19	15	6.7–14.4	Mira	317	M
R Crucis	12	23.6	−61	38	6.4–7.2	Cepheid	5.8	F–G
R Cygni	19	36.8	+50	12	6.1–14.4	Mira	426	S
U Cygni	20	19.6	+47	54	5.9–12.1	Mira	463	C
W Cygni	21	36.0	+45	22	5.0–7.6	Semi-regular	131	M
RT Cygni	19	43.6	+48	47	6.0–13.1	Mira	190	M

Star	RA		Declination		Range	Type	Period	Spectrum
	h	m	°	′			(days)	
SS Cygni	21	42.7	+43	35	7.7−12.4	Dwarf nova	50±	K+Pec
*CH Cygni	19	24.5	+50	14	5.6−9.0	Symbiotic	—	M+B
Chi Cygni	19	50.6	+32	55	3.3−14.2	Mira	408	S
R Delphini	20	14.9	+09	05	7.6−13.8	Mira	285	M
U Delphini	20	45.5	+18	05	5.6−7.5	Semi-regular	110?	M
EU Delphini	20	37.9	+18	16	5.8−6.9	Semi-regular	60	M
Beta Doradûs	05	33.6	−62	29	3.5−4.1	Cepheid	9.8	F−G
R Draconis	16	32.7	+66	45	6.7−13.2	Mira	246	M
T Eridani	03	55.2	−24	02	7.2−13.2	Mira	252	M
R Fornacis	02	29.3	−26	06	7.5−13.0	Mira	389	C
R Geminorum	07	07.4	+22	42	6.0−14.0	Mira	370	S
U Geminorum	07	55.1	+22	00	8.2−14.9	Dwarf nova	105±	Pec+M
Zeta Geminorum	07	04.1	+20	34	3.6−4.2	Cepheid	10.2	F−G
Eta Geminorum	06	14.9	+22	30	3.2−3.9	Semi-regular	233	M
S Gruis	22	26.1	−48	26	6.0−15.0	Mira	402	M
S Herculis	16	51.9	+14	56	6.4−13.8	Mira	307	M
U Herculis	16	25.8	+18	54	6.4−13.4	Mira	406	M
Alpha Herculis	17	14.6	+14	23	2.7−4.0	Semi-regular	—	M
68, u Herculis	17	17.3	+33	06	4.7−5.4	Algol	2.1	B+B
R Horologii	02	53.9	−49	53	4.7−14.3	Mira	408	M
U Horologii	03	52.8	−45	50	6−14	Mira	348	M
R Hydrae	13	29.7	−23	17	3.5−10.9	Mira	389	M
U Hydrae	10	37.6	−13	23	4.3−6.5	Semi-regular	450?	C
VW Hydri	04	09.1	−71	18	8.4−14.4	Dwarf nova	27±	Pec
R Leonis	09	47.6	+11	26	4.4−11.3	Mira	310	M
R Leonis Minoris	09	45.6	+34	31	6.3−13.2	Mira	372	M
R Leporis	04	59.6	−14	48	5.5−11.7	Mira	427	C
Y Librae	15	11.7	−06	01	7.6−14.7	Mira	276	M
RS Librae	15	24.3	−22	55	7.0−13.0	Mira	218	M
Delta Librae	15	01.0	−08	31	4.9−5.9	Algol	2.3	A
R Lyncis	07	01.3	+55	20	7.2−14.3	Mira	379	S
R Lyrae	18	55.3	+43	57	3.9−5.0	Semi-regular	46?	M
RR Lyrae	19	25.5	+42	47	7.1−8.1	RR Lyrae	0.6	A−F
Beta Lyrae	18	50.1	+33	22	3.3−4.4	Eclipsing	12.9	B
U Microscopii	20	29.2	−40	25	7.0−14.4	Mira	334	M
*U Monocerotis	07	30.8	−09	47	5.9−7.8	RV Tauri	91	F−K
V Monocerotis	06	22.7	−02	12	6.0−13.9	Mira	340	M
R Normae	15	36.0	−49	30	6.5−13.9	Mira	508	M
T Normae	15	44.1	−54	59	6.2−13.6	Mira	241	M
R Octantis	05	26.1	−86	23	6.3−13.2	Mira	405	M

Some Interesting Variable Stars

Star	RA h	RA m	Declination °	Declination ′	Range	Type	Period (days)	Spectrum
S Octantis	18	08.7	−86	48	7.2–14.0	Mira	259	M
V Ophiuchi	16	26.7	−12	26	7.3–11.6	Mira	297	C
X Ophiuchi	18	38.3	+08	50	5.9–9.2	Mira	329	M
RS Ophiuchi	17	50.2	−06	43	4.3–12.5	Recurrent nova	—	OB+M
U Orionis	05	55.8	+20	10	4.8–13.0	Mira	368	M
W Orionis	05	05.4	+01	11	5.9–7.7	Semi-regular	212	C
Alpha Orionis	05	55.2	+07	24	0.0–1.3	Semi-regular	2335	M
S Pavonis	19	55.2	−59	12	6.6–10.4	Semi-regular	381	M
Kappa Pavonis	18	56.9	−67	14	3.9–4.8	W Virginis	9.1	G
R Pegasi	23	06.8	+10	33	6.9–13.8	Mira	378	M
X Persei	03	55.4	+31	03	6.0–7.0	Gamma Cas.	—	O9.5
Beta Persei	03	08.2	+40	57	2.1–3.4	Algol	2.9	B
Zeta Phoenicis	01	08.4	−55	15	3.9–4.4	Algol	1.7	B+B
R Pictoris	04	46.2	−49	15	6.4–10.1	Semi-regular	171	M
RS Puppis	08	13.1	−34	35	6.5–7.7	Cepheid	41.4	F–G
L² Puppis	07	13.5	−44	39	2.6–6.2	Semi-regular	141	M
T Pyxidis	09	04.7	−32	23	6.5–15.3	Recurrent nova	7000±	Pec
U Sagittae	19	18.8	+19	37	6.5–9.3	Algol	3.4	B+G
WZ Sagittae	20	07.6	+17	42	7.0–15.5	Dwarf nova	1900±	A
R Sagittarii	19	16.7	−19	18	6.7–12.8	Mira	270	M
RR Sagittarii	19	55.9	−29	11	5.4–14.0	Mira	336	M
RT Sagittarii	20	17.7	−39	07	6.0–14.1	Mira	306	M
RU Sagittarii	19	58.7	−41	51	6.0–13.8	Mira	240	M
RY Sagittarii	19	16.5	−33	31	5.8–14.0	R Coronae Bor.	—	G
RR Scorpii	16	56.6	−30	35	5.0–12.4	Mira	281	M
RS Scorpii	16	55.6	−45	06	6.2–13.0	Mira	320	M
RT Scorpii	17	03.5	−36	55	7.0–15.2	Mira	449	S
Delta Scorpii	16	00.3	−22	37	1.6–2.3	Irregular	—	B
S Sculptoris	00	15.4	−32	03	5.5–13.6	Mira	363	M
R Scuti	18	47.5	−05	42	4.2–8.6	RV Tauri	146	G–K
R Serpentis	15	50.7	+15	08	5.2–14.4	Mira	356	M
S Serpentis	15	21.7	+14	19	7.0–14.1	Mira	372	M
T Tauri	04	22.0	+19	32	9.3–13.5	T Tauri	—	F–K
SU Tauri	05	49.1	+19	04	9.1–16.9	R Coronae Bor.	—	G
Lambda Tauri	04	00.7	+12	29	3.4–3.9	Algol	4.0	B+A
R Trianguli	02	37.0	+34	16	5.4–12.6	Mira	267	M
R Ursae Majoris	10	44.6	+68	47	6.5–13.7	Mira	302	M
T Ursae Majoris	12	36.4	+59	29	6.6–13.5	Mira	257	M
*Z Ursae Majoris	11	56.5	+57	52	6.2–9.4	Semi-regular	196	M
*RY Ursae Majoris	12	20.5	+61	19	6.7–8.3	Semi-regular	310?	M

Star	RA		Declination		Range	Type	Period	Spectrum
	h	m	°	′			(days)	
U Ursae Minoris	14	17.3	+66	48	7.1–13.0	Mira	331	M
R Virginis	12	38.5	+06	59	6.1–12.1	Mira	146	M
S Virginis	13	33.0	−07	12	6.3–13.2	Mira	375	M
SS Virginis	12	25.3	+00	48	6.0–9.6	Semi-regular	364	C
R Vulpeculae	21	04.4	+23	49	7.0–14.3	Mira	137	M
Z Vulpeculae	19	21.7	+25	34	7.3–8.9	Algol	2.5	B+A

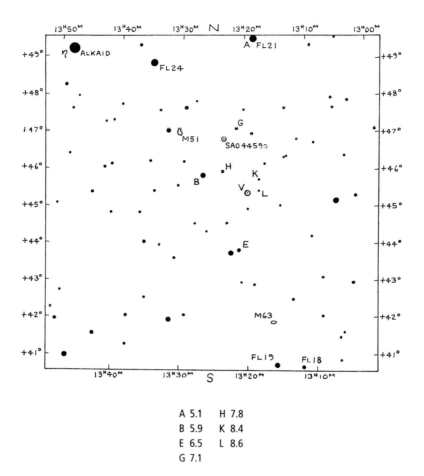

V CANUM VENATICORUM   13h 19.5m +45° 32′ (2000)

A 5.1	H 7.8
B 5.9	K 8.4
E 6.5	L 8.6
G 7.1	

o (MIRA) CETI   02h 19.3m −02° 59′ (2000)

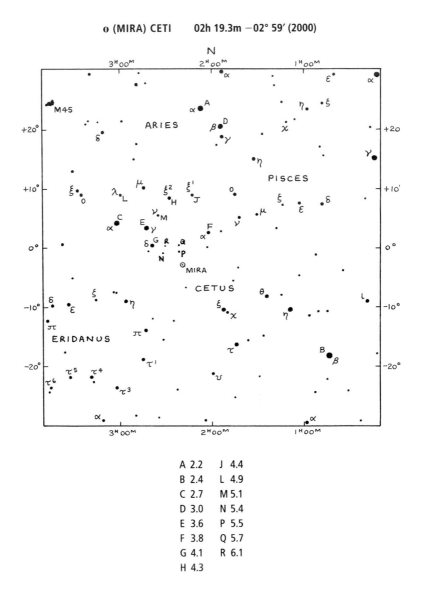

A 2.2	J 4.4
B 2.4	L 4.9
C 2.7	M 5.1
D 3.0	N 5.4
E 3.6	P 5.5
F 3.8	Q 5.7
G 4.1	R 6.1
H 4.3	

R CORONAE BOREALIS   15h 48.6m +28° 09' (2000)

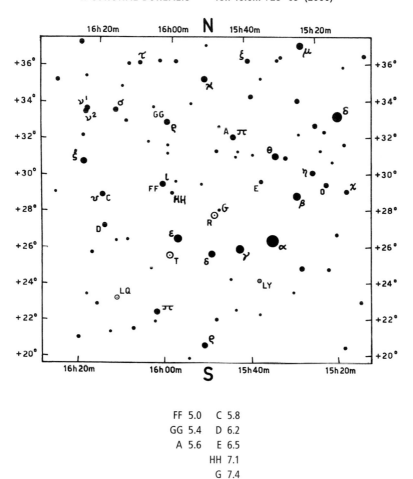

FF 5.0	C 5.8
GG 5.4	D 6.2
A 5.6	E 6.5
	HH 7.1
	G 7.4

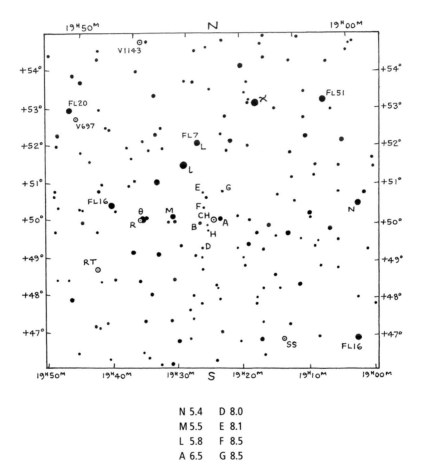

CH CYGNI   19h 24.5m +50° 14' (2000)

N 5.4	D 8.0
M 5.5	E 8.1
L 5.8	F 8.5
A 6.5	G 8.5
B 7.4	H 9.2

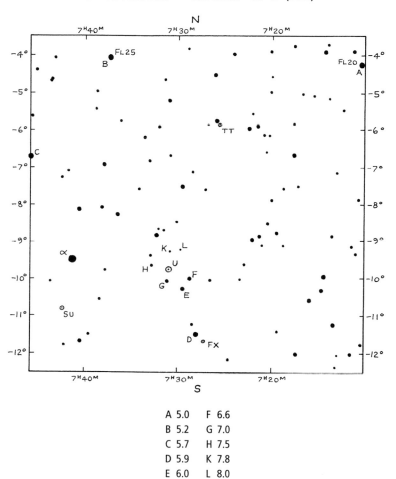

U MONOCEROTIS   07h 30.8m −09° 47′ (2000)

A 5.0	F 6.6
B 5.2	G 7.0
C 5.7	H 7.5
D 5.9	K 7.8
E 6.0	L 8.0

RY URSAE MAJORIS   12h 20.5m +61° 19′ (2000)
Z URSAE MAJORIS   11h 56.5m +57° 52′ (2000)

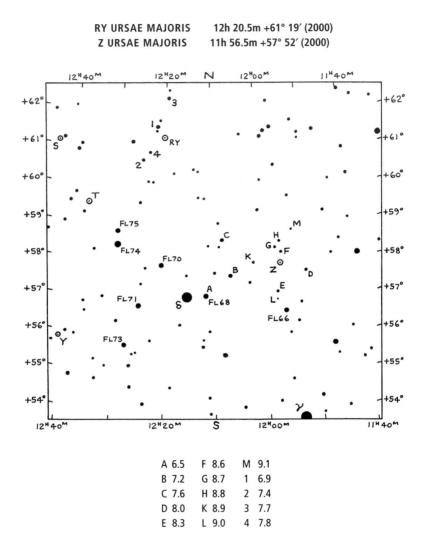

A	6.5	F	8.6	M	9.1
B	7.2	G	8.7	1	6.9
C	7.6	H	8.8	2	7.4
D	8.0	K	8.9	3	7.7
E	8.3	L	9.0	4	7.8

Mira Stars: Maxima, 2012

JOHN ISLES

Below are the predicted dates of maxima for Mira stars that reach magnitude 7.5 or brighter at an average maximum. Individual maxima can in some cases be brighter or fainter than average by a magnitude or more, and all dates are only approximate. The positions, extreme ranges and mean periods of these stars can be found in the preceding list of interesting variable stars.

Star	Mean magnitude at maximum	Dates of maxima
W Andromedae	7.4	14 Oct
R Aquarii	6.5	17 Feb
R Aquilae	6.1	23 Sep
R Bootis	7.2	8 Aug
R Cancri	6.8	8 Aug
S Canis Minoris	7.5	6 Sep
R Carinae	4.6	27 Jul
S Carinae	5.7	18 Apr, 14 Sep
R Cassiopeiae	7.0	17 Jul
R Centauri	5.8	31 Jul
T Cephei	6.0	14 Mar
U Ceti	7.5	27 Feb, 18 Oct
Omicron Ceti	3.4	13 Aug
T Columbae	7.5	27 Mar, 8 Nov
S Coronae Borealis	7.3	14 Sep
V Coronae Borealis	7.5	2 Dec
R Corvi	7.5	23 Jul
R Cygni	7.5	17 Jun
U Cygni	7.2	21 Aug
RT Cygni	7.3	3 Mar, 9 Sep
Chi Cygni	5.2	23 Mar
R Geminorum	7.1	4 Jan

Star	Mean magnitude at maximum	Dates of maxima
U Herculis	7.5	18 Jan
R Horologii	6.0	2 May
R Hydrae	4.5	13 Dec
R Leonis	5.8	27 Mar
R Leonis Minoris	7.1	11 Feb
R Leporis	6.8	14 Jun
RS Librae	7.5	28 Jun
V Monocerotis	7.0	10 Feb
T Normae	7.4	6 Apr, 3 Dec
V Ophiuchi	7.5	26 May
X Ophiuchi	6.8	4 Sep
U Orionis	6.3	12 Mar
R Sagittarii	7.3	11 Feb, 6 Nov
RR Sagittarii	6.8	12 Aug
RT Sagittarii	7.0	12 Jan, 13 Nov
RU Sagittarii	7.2	26 May
RR Scorpii	5.9	10 Apr
RS Scorpii	7.0	26 Sep
S Sculptoris	6.7	19 Dec
R Serpentis	6.9	1 Sep
R Trianguli	6.2	18 Jul
R Ursae Majoris	7.5	1 Jul
R Virginis	6.9	6 May, 29 Sep
S Virginis	7.0	25 Feb

Some Interesting Double Stars

BOB ARGYLE

The positions, angles and separations given below correspond to epoch 2012.0.

No.	RA	Declin-ation	Star	Magni-tudes	Separa-tion	PA	Cata-logue	Comments
	h m	° ′			arcsec	°		
1	00 31.5	−62 58	β Tuc	4.4,4.8	27.1	169	LCL 119	Both stars again difficult doubles.
2	00 49.1	+57 49	η Cas	3.4,7.5	13.3	322	Σ60	Easy. Creamy, bluish. P = 480 years.
3	00 55.0	+23 38	36 And	6.0,6.4	1.1	325	Σ73	P = 168 years. Both yellow. Slowly opening.
4	01 13.7	+07 35	ζ Psc	5.6,6.5	23.1	63	Σ100	Yellow, reddish-white
5	01 39.8	−56 12	p Eri	5.8,5.8	11.7	188	Δ5	Period = 483 years.
6	01 53.5	+19 18	γ Ari	4.8,4.8	7.5	1	Σ180	Very easy. Both white.
7	02 02.0	+02 46	α Psc	4.2,5.1	1.8	263	Σ202	Binary, period = 933 years.
8	02 03.9	+42 20	γ¹ And	2.3,5.0	9.6	63	Σ205	Yellow, blue. Relatively fixed.
			γ² And	5.1,6.3	0.2	96	OΣ38	BC. now beyond range of amateur instruments.
9	02 29.1	+67 24	ι Cas AB	4.9,6.9	2.6	229	Σ262	AB is long-period binary. P = 620 years.
			ι Cas AC	4.9,8.4	7.2	118		
10	02 33.8	−28 14	ω For	5.0,7.7	10.8	245	HJ 3506	Common proper motion.
11	02 43.3	+03 14	γ Cet	3.5,7.3	2.3	298	Σ299	Not too easy.
12	02 58.3	−40 18	θ Eri	3.4,4.5	8.3	90	PZ 2	Both white.

No.	RA	Declin- ation	Star	Magni- tudes	Separa- tion	PA	Cata- logue	Comments
	h m	° ′			arcsec	°		
13	02 59.2	+21 20	ε Ari	5.2,5.5	1.4	209	Σ333	Closing slowly. P=350 years? Both white.
14	03 00.9	+52 21	Σ331 Per	5.3,6.7	12.0	85	–	Fixed.
15	03 12.1	−28 59	α For	4.0,7.0	5.3	300	HJ 3555	P = 269 years. B variable?
16	03 48.6	−37 37	f Eri	4.8,5.3	8.2	215	Δ16	Pale yellow. Fixed.
17	03 54.3	−02 57	32 Eri	4.8,6.1	6.9	348	Σ470	Fixed.
18	04 32.0	+53 55	1 Cam	5.7,6.8	10.3	308	Σ550	Fixed.
19	04 50.9	−53 28	ι Pic	5.6,6.4	12.4	58	Δ18	Good object for small apertures. Fixed.
20	05 13.2	−12 56	κ Lep	4.5,7.4	2.0	357	Σ661	Visible in 7.5 cm. Slowly closing.
21	05 14.5	−08 12	β Ori	0.1,6.8	9.5	204	Σ668	Companion once thought to be close double.
22	05 21.8	−24 46	41 Lep	5.4,6.6	3.4	93	HJ 3752	Deep yellow pair in a rich field.
23	05 24.5	−02 24	η Ori	3.8,4.8	1.7	78	DA 5	Slow-moving binary.
24	05 35.1	+09 56	λ Ori	3.6,5.5	4.3	44	Σ738	Fixed.
25	05 35.3	−05 23	θ Ori AB	6.7,7.9	8.6	32	Σ748	Trapezium in M42.
			θ Ori CD	5.1,6.7	13.4	61		
26	05 40.7	−01 57	ζ Ori	1.9,4.0	2.6	166	Σ774	Can be split in 7.5 cm. Long-period binary.
27	06 14.9	+22 30	η Gem	var,6.5	1.6	253	β1008	Well seen with 20 cm. Primary orange.
28	06 46.2	+59 27	12 Lyn AB	5.4,6.0,	1.9	68	Σ948	AB is binary, P = 706 years.
			12 Lyn AC	5.4,7.3	8.7	309	–	
29	07 08.7	−70 30	γ Vol	3.9,5.8	14.1	298	Δ42	Very slow binary.
30	07 16.6	−23 19	h3945 cMa	4.8,6.8	26.8	51	–	Contrasting colours. Yellow and blue.
31	07 20.1	+21 59	δ Gem	3.5,8.2	5.6	228	Σ1066	Not too easy. Yellow, pale blue.
32	07 34.6	+31 53	α Gem	1.9,2.9	4.8	56	Σ1110	Widening. Easy with 7.5 cm.

Some Interesting Double Stars

No.	RA h m	Declin- ation ° ′	Star	Magni- tudes	Separa- tion arcsec	PA °	Cata- logue	Comments
33	07 38.8	−26 48	κ Pup	4.5,4.7	9.8	318	H III 27	Both white.
34	08 12.2	+17 39	ζ Cnc AB	5.6,6.0	1.1	31	Σ1196	Period (AB)= 60 years. Near maximum separation.
			ζ Cnc AB-C	5.0,6.2	5.9	68	Σ1196	Period (AB−C = 1150 years.
35	08 46.8	+06 25	ε Hyd	3.3,6.8	2.9	306	Σ1273	PA slowly increasing. A is a very close pair.
36	09 18.8	+36 48	38 Lyn	3.9,6.6	2.6	226	Σ1334	Almost fixed.
37	09 47.1	−65 04	υ Car	3.1,6.1	5.0	129	RMK 11	Fixed. Fine in small telescopes.
38	10 20.0	+19 50	γ Leo	2.2,3.5	4.6	126	Σ1424	Binary, period = 510 years. Both orange.
39	10 32.0	−45 04	s Vel	6.2,6.5	13.5	218	PZ 3	Fixed.
40	10 46.8	−49 26	μ Vel	2.7,6.4	2.6	56	R 155	P = 138 years. Near widest separation.
41	10 55.6	+24 45	54 Leo	4.5,6.3	6.6	111	Σ1487	Slowly widening. Pale yellow and white.
42	11 18.2	+31 32	ξ UMa	4.3,4.8	1.6	197	Σ1523	Binary, 60 years. Needs 7.5 cm.
43	11 23.9	+10 32	ι Leo	4.0,6.7	2.0	99	Σ1536	Binary, period = 186 years.
44	11 32.3	−29 16	N Hya	5.8,5.9	9.4	210	H III 96	Both yellow. Long period binary.
45	12 14.0	−45 43	D Cen	5.6,6.8	2.8	243	RMK 14	Orange and white. Closing.
46	12 26.6	−63 06	α Cru	1.4,1.9	4.0	114	Δ252	Glorious pair. Third star in a low power field.
47	12 41.5	−48 58	γ Cen	2.9,2.9	0.2	300	HJ 4539	Period = 84 years. Closing. Both yellow.
48	12 41.7	−01 27	γ Vir	3.5,3.5	1.8	14	Σ1670	Now widening quickly. Beautiful pair for 10 cm.

No.	RA h m	Declin-ation ° '	Star	Magni-tudes	Separa-tion arcsec	PA °	Cata-logue	Comments
49	12 46.3	−68 06	β Mus	3.7,4.0	0.9	55	R 207	Both white. Closing slowly. P = 194 years.
50	12 54.6	−57 11	μ Cru	4.3,5.3	34.9	17	Δ126	Fixed. Both white.
51	12 56.0	+38 19	α CVn	2.9,5.5	19.3	229	Σ1692	Easy. Yellow, bluish.
52	13 22.6	−60 59	J Cen	4.6,6.5	60.0	343	Δ133	Fixed. A is a close pair.
53	13 24.0	+54 56	ζ UMa	2.3,4.0	14.4	152	Σ1744	Very easy. Naked eye pair with Alcor.
54	13 51.8	−33 00	3 Cen	4.5,6.0	7.7	102	H III 101	Both white. Closing slowly.
55	14 39.6	−60 50	α Cen	0.0,1.2	5.4	257	RHD 1	Finest pair in the sky. P = 80 years. Closing.
56	14 41.1	+13 44	ζ Boo	4.5,4.6	0.5	293	Σ1865	Both white. Closing – highly inclined orbit.
57	14 45.0	+27 04	ε Boo	2.5,4.9	2.9	344	Σ1877	Yellow, blue. Fine pair.
58	14 46.0	−25 27	54 Hya	5.1,7.1	8.3	122	H III 97	Closing slowly.
59	14 49.3	−14 09	μ Lib	5.8,6.7	1.8	6	β106	Becoming wider. Fine in 7.5 cm.
60	14 51.4	+19 06	ξ Boo	4.7,7.0	5.9	306	Σ1888	Fine contrast. Easy.
61	15 03.8	+47 39	44 Boo	5.3,6.2	1.4	62	Σ1909	Period = 206 years. Beginning to close.
62	15 05.1	−47 03	π Lup	4.6,4.7	1.7	66	HJ 4728	Widening.
63	15 18.5	−47 53	μ Lup AB	5.1,5.2	1.1	300	HJ 4753	AB closing. Under observed.
			μ Lup AC	4.4,7.2	22.7	127	Δ180	AC almost fixed
64	15 23.4	−59 19	γ Cir	5.1,5.5	0.8	359	HJ 4757	Closing. Needs 20 cm. Long-period binary.
65	15 34.8	+10 33	δ Ser	4.2,5.2	4.0	172	Σ1954	Long-period binary.
66	15 35.1	−41 10	γ Lup	3.5,3.6	0.8	277	HJ 4786	Binary. Period = 190 years. Needs 20 cm.
67	15 56.9	−33 58	γ Lup	5.3,5.8	10.2	49	PZ 4	Fixed.
68	16 14.7	+33 52	σ CrB	5.6,6.6	7.1	238	Σ2032	Long-period binary. Both white.

Some Interesting Double Stars

No.	RA h m	Declination ° ′	Star	Magnitudes	Separation arcsec	PA °	Catalogue	Comments
69	16 29.4	−26 26	α Sco	1.2,5.4	2.6	277	GNT 1	Red, green. Difficult from mid-northern latitudes.
70	16 30.9	+01 59	λ Oph	4.2,5.2	1.4	39	Σ2055	P = 129 years. Fairly difficult in small apertures.
71	16 41.3	+31 36	ζ Her	2.9,5.5	1.2	161	Σ2084	Period 34 years. Now widening. Needs 20 cm.
72	17 05.3	+54 28	μ Dra	5.7,5.7	2.4	6	Σ2130	Period 672 years.
73	17 14.6	+14 24	α Her	var,5.4	4.6	104	Σ2140	Red, green. Long-period binary.
74	17 15.3	−26 35	36 Oph	5.1,5.1	5.0	142	SHJ 243	Period = 471 years.
75	17 23.7	+37 08	ρ Her	4.6,5.6	4.1	319	Σ2161	Slowly widening.
76	17 26.9	−45 51	HJ 4949 AB	5.6,6.5	2.1	251	HJ 4949	Beautiful coarse triple. All white.
			Δ 216 AC	7.1	105.0	310		
77	18 01.5	+21 36	95 Her	5.0,5.1	6.5	257	Σ2264	Colours thought variable in C19.
78	18 05.5	+02 30	70 Oph	4.2,6.0	5.9	129	Σ2272	Opening. Easy in 7.5 cm.
79	18 06.8	−43 25	HJ 5014 CrA	5.7,5.7	1.7	1	–	Period = 450 years. Needs 10 cm.
80	18 25.4	−20 33	21 Sgr	5.0,7.4	1.7	279	JC 6	Slowly closing binary, orange and green.
81	18 35.9	+16 58	OΣ358 Her	6.8,7.0	1.5	148	–	Period = 380 years.
82	18 44.3	+39 40	ε1 Lyr	5.0,6.1	2.4	347	Σ2382	Quadruple system with epsilon2. Both pairs
83	18 44.3	+39 40	ε2 Lyr	5.2,5.5	2.4	77	Σ2383	visible in 7.5 cm.
84	18 56.2	+04 12	θ Ser	4.5,5.4	22.4	104	Σ2417	Fixed. Very easy
85	19 06.4	−37 04	γ CrA	4.8,5.1	1.4	3	HJ 5084	Beautiful pair. Period = 122 years.
86	19 30.7	+27 58	β Cyg AB	3.1,5.1	34.3	54	Σ I 43	Glorious. Yellow, blue-greenish.

No.	RA	Declin-ation	Star	Magni-tudes	Separa-tion	PA	Cata-logue	Comments
	h m	° '			arcsec	°		
			β Cyg Aa	3.1,5.2	0.4	92	MCA 55	Aa. Very difficult. Period = 214 years
87	19 45.0	+45 08	δ Cyg	2.9,6.3	2.7	219	Σ2579	Slowly widening. Period = 780 years.
88	19 48.2	+70 16	ε Dra	3.8,7.4	3.3	20	Σ2603	Slow binary.
89	19 54.6	−08 14	57 Aql	5.7,6.4	36.0	170	Σ2594	Easy pair. Contrasting colours.
90	20 46.7	+16 07	γ Del	4.5,5.5	9.0	265	Σ2727	Easy. Yellowish. Long period binary.
91	20 59.1	+04 18	ε Equ AB	6.0,6.3	0.4	283	Σ2737	Fine triple. AB a test for 30 cm. P = 101.5 years
			ε Equ AC	6.0,7.1	10.3	66		
92	21 06.9	+38 45	61 Cyg	5.2,6.0	31.4	151	Σ2758	Nearby binary. Both orange. Period = 659 years.
93	21 19.9	−53 27	θ Ind	4.5,7.0	7.0	271	HJ 5258	Pale yellow and reddish. Long-period binary.
94	21 44.1	+28 45	μ Cyg	4.8,6.1	1.6	318	Σ2822	Period = 789 years.
95	22 03.8	+64 37	ξ Cep	4.4,6.5	8.4	274	Σ2863	White and blue. Long-period binary.
96	22 14.3	−21 04	41 Aqr	5.6,6.7	5.1	113	H N 56	Yellowish and purple?
97	22 26.6	−16 45	53 Aqr	6.4,6.6	1.3	47	SHJ 345	Long-period binary; periastron in 2023.
98	22 28.8	−00 01	ζ Aqr	4.3,4.5	2.3	169	Σ2909	Period = 487 years. Slowly widening.
99	23 19.1	−13 28	94 Aqr	5.3,7.0	12.3	351	Σ2988	Yellow and orange. Probable binary.
100	23 59.5	+33 43	Σ3050 And	6.6,6.6	2.1	338	–	Period = 320 years. Visible in 7.5 cm.

Some Interesting Nebulae, Clusters and Galaxies

Object	RA		Declina-tion		Remarks
	h	m	°	′	
M31 Andromedae	00	40.7	+41	05	Andromeda Galaxy, visible to naked eye.
H VIII 78 Cassiopeiae	00	41.3	+61	36	Fine cluster, between Gamma and Kappa Cassiopeiae.
M33 Trianguli	01	31.8	+30	28	Spiral. Difficult with small apertures.
H VI 33–4 Persei, C14	02	18.3	+56	59	Double cluster; Sword-handle.
Δ142 Doradûs	05	39.1	−69	09	Looped nebula round 30 Doradus. Naked eye. In Large Magellanic Cloud.
M1 Tauri	05	32.3	+22	00	Crab Nebula, near Zeta Tauri.
M42 Orionis	05	33.4	−05	24	Orion Nebula. Contains the famous Trapezium, Theta Orionis.
M35 Geminorum	06	06.5	+24	21	Open cluster near Eta Geminorum.
H VII 2 Monocerotis, C50	06	30.7	+04	53	Open cluster, just visible to naked eye.
M41 Canis Majoris	06	45.5	−20	42	Open cluster, just visible to naked eye.
M47 Puppis	07	34.3	−14	22	Mag. 5.2. Loose cluster.
H IV 64 Puppis	07	39.6	−18	05	Bright planetary in rich neighbourhood.
M46 Puppis	07	39.5	−14	42	Open cluster.
M44 Cancri	08	38	+20	07	Praesepe. Open cluster near Delta Cancri. Visible to naked eye.
M97 Ursae Majoris	11	12.6	+55	13	Owl Nebula, diameter 3′. Planetary.
Kappa Crucis, C94	12	50.7	−60	05	'Jewel Box'; open cluster, with stars of contrasting colours.
M3 Can. Ven.	13	40.6	+28	34	Bright globular.
Omega Centauri, C80	13	23.7	−47	03	Finest of all globulars. Easy with naked eye.
M80 Scorpii	16	14.9	−22	53	Globular, between Antares and Beta Scorpii.
M4 Scorpii	16	21.5	−26	26	Open cluster close to Antares.

Object	RA		Declina-tion		Remarks
	h	m	°	′	
M13 Herculis	16	40	+36	31	Globular. Just visible to naked eye.
M92 Herculis	16	16.1	+43	11	Globular. Between Iota and Eta Herculis.
M6 Scorpii	17	36.8	−32	11	Open cluster; naked eye.
M7 Scorpii	17	50.6	−34	48	Very bright open cluster; naked eye.
M23 Sagittarii	17	54.8	−19	01	Open cluster nearly 50′ in diameter.
H IV 37 Draconis, C6	17	58.6	+66	38	Bright planetary.
M8 Sagittarii	18	01.4	−24	23	Lagoon Nebula. Gaseous. Just visible with naked eye.
NGC 6572 Ophiuchi	18	10.9	+06	50	Bright planetary, between Beta Ophiuchi and Zeta Aquilae.
M17 Sagittarii	18	18.8	−16	12	Omega Nebula. Gaseous. Large and bright.
M11 Scuti	18	49.0	−06	19	Wild Duck. Bright open cluster.
M57 Lyrae	18	52.6	+32	59	Ring Nebula. Brightest of planetaries.
M27 Vulpeculae	19	58.1	+22	37	Dumb-bell Nebula, near Gamma Sagittae.
H IV 1 Aquarii, C55	21	02.1	−11	31	Bright planetary, near Nu Aquarii.
M15 Pegasi	21	28.3	+12	01	Bright globular, near Epsilon Pegasi.
M39 Cygni	21	31.0	+48	17	Open cluster between Deneb and Alpha Lacertae. Well seen with low powers.

(M = Messier number; NGC = New General Catalogue number; C = Caldwell number.)

Our Contributors

Howard G. Miles MBE is a former lecturer in Mathematics at the Lanchester Polytechnic in Coventry (which subsequently became Coventry Polytechnic and is now Coventry University). He has a long-standing interest in artificial satellites and in fireballs and meteorites. He was founder director of the Artificial Satellite Section of the British Astronomical Association (BAA) and was also director of the British Fireball Survey, set up by the BAA to investigate major fireball events and possible meteorite falls. He also served as President of the BAA from 1974 to 1976.

Dr David Allen's extraordinary life had an inauspicious start due to a misdiagnosed hip complaint. He spent seven years of his childhood with his body and legs encased in plaster, always claiming to have been 'born at the age of nine'. Educated in Cambridge where he began his pioneering work on infrared astronomy, David soon gravitated to the Anglo–Australian Observatory. There, his outstanding scientific research was matched by a prolific output in popular astronomy, including many articles for the *Yearbook of Astronomy*. David died in July 1994 at the age of forty-seven.

Dr Steven Bell is Head of HM Nautical Almanac Office (HMNAO) based at the UK Hydrographic Office in Taunton, Somerset. HMNAO publishes a number of almanacs jointly with the US Naval Observatory in Washington, including *The Astronomical Almanac,* and makes a variety of astronomical data available through its website http://astro.ukho.gov.uk.

Professor Anthony Fairall spent most of his career at the University of Cape Town, where he had studied as an undergraduate before moving briefly to Texas to undertake his PhD degree. He was an international authority on large-scale structure in the universe, publishing more than two hundred research papers and several books. He was also a prominent and much-loved science communicator. When he died in a tragic accident in 2008, shortly before retirement, his loss was felt deeply throughout southern Africa and the international astronomical community.

Iain Nicolson was formerly Principal Lecturer in Astronomy at the University of Hertfordshire, and is a writer and lecturer in the fields of astronomy and space science. A Contributing Consultant to the magazine *Astronomy Now*, he has been a frequent contributor to BBC Television's *The Sky at Night*. He is author or co-author of more than twenty books, the most recent of which, *Dark Side of the Universe*, was published in April 2007 by Canopus Publishing Limited.

Martin Mobberley is one of the UK's most active imagers of comets, planets, asteroids, variable stars, novae and supernovae and served as President of the British Astronomical Association from 1997 to 1999. In 2000 he was awarded the Association's Walter Goodacre Award. He is the author of eight popular astronomy books published by Springer as well as three children's 'Space Exploration' books published by Top That Publishing. In addition, he has authored hundreds of articles in *Astronomy Now* and numerous other astronomical publications.

Professor Fred Watson is Astronomer-in-Charge of the Australian Astronomical Observatory at Coonabarabran in north-western New South Wales, and one of Australia's best-known science communicators. He is a regular contributor to the *Yearbook of Astronomy*, and his recent books include *Universe* (for which he was Chief Consultant), *Stargazer: The Life and Times of the Telescope* and *Why is Uranus Upside Down? And Other Questions About the Universe*. In 2006 he was awarded the Australian Government Eureka Prize for Promoting Understanding of Science. In 2010 he was appointed a Member of the Order of Australia. Visit Fred's website at http://fredwatson.com.au/.

Bill Leatherbarrow is Emeritus Professor of Russian Studies at the University of Sheffield and a lifelong amateur astronomer. He is an active observer of the Moon and planets and has previously contributed to the *1971 Yearbook of Astronomy*. He is currently President of the British Astronomical Association (BAA) and Director of the BAA Lunar Section.

Dr Allan Chapman, of Wadham College, Oxford, is probably Britain's leading authority on the history of astronomy. He has published many research papers and several books, as well as numerous popular accounts. He is a frequent and welcome contributor to the *Yearbook*.

Dr David M. Harland gained his BSc in astronomy in 1977 and a doctorate in computational science. Subsequently, he has taught computer science, worked in industry and managed academic research. In 1995 he 'retired' and has since published many books on space themes.

Paul G. Abel is a co-presenter on the BBC *Sky at Night* and works in the Department of Physics & Astronomy at the University of Leicester. Initially he trained as a mathematician and works in the area of Theoretical Physics concerned with Hawking Radiation; the radiation emitted by black holes. His other research interest is concerned with Quantum Field theories in curved spacetimes. He also teaches mathematics in the Physics Department and is an active visual amateur astronomer; his primary interest being the Moon and planets. He is assistant director of the BAA Saturn Section and a Fellow of the Royal Astronomical Society. He frequently writes for various popular astronomy magazines.

Ninian Boyle is an experienced astronomer, teacher, lecturer and a contributor to both the BBC's *Sky at Night* and *Astronomy Now* magazines. He has appeared on BBC Television's *The Sky at Night* programme on several occasions and on the Sky Information Channel. His special interest is the Sun and solar observing. He is a member of the British Astronomical Association and a Fellow of the Royal Astronomical Society.

Richard Myer Baum is a former Director of the Mercury and Venus Section of the British Astronomical Association, an amateur astronomer and an independent scholar. He is author of *The Planets: Some Myths and Realities* (1973, with W. Sheehan), *In Search of Planet Vulcan: The Ghost in Newton's Clockwork Universe* (1997) and *The Haunted Observatory* (2007). He has contributed to the *Journal of the British Astronomical Association*, *Journal for the History of Astronomy*, *Sky and Telescope* and many other publications, including *The Dictionary of Nineteenth-Century British Scientists* (2004) and *The Biographical Encyclopedia of Astronomers* (2007).

Astronomical Societies in the British Isles

Association for Astronomy Education
Secretary: Teresa Grafton, The Association for Astronomy Education, c/o The Royal Astronomical Society, Burlington House, Piccadilly, London W1V 0NL.

Astronomical Society of Edinburgh
Secretary: Graham Rule, 105/19 Causewayside, Edinburgh EH9 1QG.
Website: www.roe.ac.uk/asewww/; *Email:* asewww@roe.ac.uk
Meetings: City Observatory, Calton Hill, Edinburgh. 1st Friday each month, 8 p.m.

Astronomical Society of Glasgow
Secretary: Mr David Degan, 5 Hillside Avenue, Alexandria, Dunbartonshire G83 0BB.
Website: www.astronomicalsocietyofglasgow.org.uk
Meetings: Royal College, University of Strathclyde, Montrose Street, Glasgow. 3rd Thursday each month, Sept.–Apr., 7.30 p.m.

Astronomical Society of Haringey
Secretary: Jerry Workman, 91 Greenslade Road, Barking, Essex IG11 9XF.
Meetings: Palm Court, Alexandra Palace, 3rd Wednesday each month, 8 p.m.

Astronomy Ireland
Secretary: Tony Ryan, PO Box 2888, Dublin 1, Eire.
Website: www.astronomy.ie; *Email:* info@astronomy.ie
Meetings: 2nd Monday of each month. Telescope meetings every clear Saturday.

British Astronomical Association
Assistant Secretary: Burlington House, Piccadilly, London W1V 9AG.
Meetings: Lecture Hall of Scientific Societies, Civil Service Commission Building, 23 Savile Row, London W1. Last Wednesday each month (Oct.–June), 5 p.m. and some Saturday afternoons.

Federation of Astronomical Societies
Secretary: Clive Down, 10 Glan-y-Llyn, North Cornelly, Bridgend, County Borough CF33 4EF.
Email: clivedown@btinternet.com

Junior Astronomical Society of Ireland
Secretary: K. Nolan, 5 St Patrick's Crescent, Rathcoole, Co. Dublin.
Meetings: The Royal Dublin Society, Ballsbridge, Dublin 4. Monthly.

Society for Popular Astronomy
Secretary: Guy Fennimore, 36 Fairway, Keyworth, Nottingham NG12 5DU.
Website: www.popastro.com; *Email:* SPAstronomy@aol.com
Meetings: Last Saturday in Jan., Apr., July, Oct., 2.30 p.m. in London.

Webb Deep-Sky Society
Membership Secretary/Treasurer: Steve Rayner, 11 Four Acres, Weston, Portland, Dorset DT5 2JG.

Email: stephen.rayner@tesco,net
Website: www.webbdeepsky.com
Aberdeen and District Astronomical Society
Secretary: Ian C. Giddings, 95 Brentfield Circle, Ellon, Aberdeenshire AB41 9DB.
Meetings: Robert Gordon's Institute of Technology, St Andrew's Street, Aberdeen.
Fridays, 7.30 p.m.
Abingdon Astronomical Society (was **Fitzharry's Astronomical Society**)
Secretary: Chris Holt, 9 Rutherford Close, Abingdon, Oxon OX14 2AT.
Website: www.abingdonastro.org.uk; *Email:* info@abingdonastro.co.uk
Meetings: All Saints' Methodist Church Hall, Dorchester Crescent, Abingdon, Oxon.
2nd Monday Sept.–June, 8 p.m. and additional beginners' meetings and observing
evenings as advertised.
Altrincham and District Astronomical Society
Secretary: Derek McComiskey, 33 Tottenham Drive, Manchester M23 9WH.
Meetings: Timperley Village Club. 1st Friday Sept.–June, 8 p.m.
Andover Astronomical Society
Secretary: Mrs S. Fisher, Staddlestones, Aughton, Kingston, Marlborough, Wiltshire
SN8 3SA.
Meetings: Grately Village Hall. 3rd Thursday each month, 7.30 p.m.
Astra Astronomy Section
Secretary: c/o Duncan Lunan, Flat 65, Dalraida House, 56 Blythswood Court,
Anderston, Glasgow G2 7PE.
Meetings: Airdrie Arts Centre, Anderson Street, Airdrie. Weekly.
Astrodome Mobile School Planetarium
Contact: Peter J. Golding, 53 City Way, Rochester, Kent ME1 2AX.
Website: www.astrodome.clara.co.uk; *Email:* astrodome@clara.co.uk
Aylesbury Astronomical Society
Secretary: Alan Smith, 182 Marley Fields, Leighton Buzzard, Bedfordshire LU7 8WN.
Meetings: 1st Monday in month at 8 p.m., venue in Aylesbury area. Details from
Secretary.
Bassetlaw Astronomical Society
Secretary: Andrew Patton, 58 Holding, Worksop, Notts S81 0TD.
Meetings: Rhodesia Village Hall, Rhodesia, Worksop, Notts. 2nd and 4th Tuesdays of
month at 7.45 p.m.
Batley & Spenborough Astronomical Society
Secretary: Robert Morton, 22 Links Avenue, Cleckheaton, West Yorks BD19 4EG.
Meetings: Milner K. Ford Observatory, Wilton Park, Batley. Every Thursday, 8 p.m.
Bedford Astronomical Society
Secretary: Mrs L. Harrington, 24 Swallowfield, Wyboston, Bedfordshire MK44 3AE.
Website: www.observer1.freeserve.co.uk/bashome.html
Meetings: Bedford School, Burnaby Rd, Bedford. Last Wednesday each month.
Bingham & Brooks Space Organization
Secretary: N. Bingham, 15 Hickmore's Lane, Lindfield, West Sussex.
Birmingham Astronomical Society
Contact: P. Bolas, 4 Moat Bank, Bretby, Burton-on-Trent DE15 0QJ.
Website: www.birmingham-astronomical.co.uk; *Email:* pbolas@aol.com
Meetings: Room 146, Aston University. Last Tuesday of month. Sept.–June (except
Dec., moved to 1st week in Jan.).

Blackburn Leisure Astronomy Section
Secretary: Mr H. Murphy, 20 Princess Way, Beverley, East Yorkshire HU17 8PD.
Meetings: Blackburn Leisure Welfare. Mondays, 8 p.m.

Blackpool & District Astronomical Society
Secretary: Terry Devon, 30 Victory Road, Blackpool, Lancashire FY1 3JT.
Website: www.blackpoolastronomy.org.uk; *Email:* info@blackpoolastronomy.org.uk
Meetings: St Kentigern's Social Centre, Blackpool. 1st Wednesday of the month,
7.45 p.m.

Bolton Astronomical Society
Secretary: Peter Miskiw, 9 Hedley Street, Bolton, Lancashire BL1 3LE.
Meetings: Ladybridge Community Centre, Bolton. 1st and 3rd Tuesdays Sept.–May,
7.30 p.m.

Border Astronomy Society
Secretary: David Pettitt, 14 Sharp Grove, Carlisle, Cumbria CA2 5QR.
Website: www.members.aol.com/P3pub/page8.html
Email: davidpettitt@supanet.com
Meetings: The Observatory, Trinity School, Carlisle. Alternate Thursdays, 7.30 p.m.,
Sept.–May.

Boston Astronomers
Secretary: Mrs Lorraine Money, 18 College Park, Horncastle, Lincolnshire LN9 6RE.
Meetings: Blackfriars Arts Centre, Boston. 2nd Monday each month, 7.30 p.m.

Bradford Astronomical Society
Contact: Mrs J. Hilary Knaggs, 6 Meadow View, Wyke, Bradford BD12 9LA.
Website: www.bradford-astro.freeserve.co.uk/index.htm
Meetings: Eccleshill Library, Bradford. Alternate Mondays, 7.30 p.m.

Braintree, Halstead & District Astronomical Society
Secretary: Mr J. R. Green, 70 Dorothy Sayers Drive, Witham, Essex CM8 2LU.
Meetings: BT Social Club Hall, Witham Telephone Exchange. 3rd Thursday each
month, 8 p.m.

Breckland Astronomical Society (was **Great Ellingham and District Astronomy Club**)
Contact: Martin Wolton, Willowbeck House, Pulham St Mary, Norfolk IP21 4QS.
Meetings: Great Ellingham Recreation Centre, Watton Road (B1077), Great
Ellingham, 2nd Friday each month, 7.15 p.m.

Bridgend Astronomical Society
Secretary: Clive Down, 10 Glan-y-Llyn, Broadlands, North Cornelly, Bridgend
County CF33 4EF.
Email: clivedown@btinternet.com
Meetings: Bridgend Bowls Centre, Bridgend. 2nd Friday, monthly, 7.30 p.m.

Bridgwater Astronomical Society
Secretary: Mr G. MacKenzie, Watergore Cottage, Watergore, South Petherton,
Somerset TA13 5JQ.
Website: www.ourworld.compuserve.com/hompages/dbown/Bwastro.htm
Meetings: Room D10, Bridgwater College, Bath Road Centre, Bridgwater. 2nd
Wednesday each month, Sept.–June.

Bridport Astronomical Society
Secretary: Mr G.J. Lodder, 3 The Green, Walditch, Bridport, Dorset DT6 4LB.
Meetings: Walditch Village Hall, Bridport. 1st Sunday each month, 7.30 p.m.

Brighton Astronomical and Scientific Society
Secretary: Ms T. Fearn, 38 Woodlands Close, Peacehaven, East Sussex BN10 7SF.
Meetings: St John's Church Hall, Hove. 1st Tuesday each month, 7.30 p.m.

Bristol Astronomical Society
Secretary: Dr John Pickard, 'Fielding', Easter Compton, Bristol BS35 5SJ.
Meetings: Frank Lecture Theatre, University of Bristol Physics Dept., alternate Fridays in term time, and Westbury Park Methodist Church Rooms, North View, other Fridays.

Callington Community Astronomy Group
Secretary: Beccy Watson. *Tel:* 07891 573786
Email: enquiries@callington-astro.org.uk
Website: www.callington-astro.org.uk
Meetings: Callington Space Centre, Callington Community College, Launceston Road, Callington, Cornwall PL17 7DR. 1st Friday of each month, 7.30 p.m., Sept.–June.

Cambridge Astronomical Society
Secretary: Brian Lister, 80 Ramsden Square, Cambridge CB4 2BL.
Meetings: Institute of Astronomy, Madingley Road. 3rd Friday each month.

Cardiff Astronomical Society
Secretary: D.W.S. Powell, 1 Tal-y-Bont Road, Ely, Cardiff CF5 5EU.
Meetings: Dept. of Physics and Astronomy, University of Wales, Newport Road, Cardiff. Alternate Thursdays, 8 p.m.

Castle Point Astronomy Club
Secretary: Andrew Turner, 3 Canewdon Hall Close, Canewdon, Rochford, Essex SS4 3PY.
Meetings: St Michael's Church Hall, Daws Heath. Wednesdays, 8 p.m.

Chelmsford Astronomers
Secretary: Brendan Clark, 5 Borda Close, Chelmsford, Essex.
Meetings: Once a month.

Chester Astronomical Society
Secretary: John Gilmour, 2 Thomas Brassey Close, Chester CH2 3AE.
Tel.: 07974 948278
Email: john_gilmour@ouvip.com
Website: www.manastro.co.uk/nwgas/chester/
Meetings: Burley Memorial Hall, Waverton, near Chester. Last Wednesday of each month except August and December at 7.30 p.m.

Chester Society of Natural Science, Literature and Art
Secretary: Paul Braid, 'White Wing', 38 Bryn Avenue, Old Colwyn, Colwyn Bay LL29 8AH.
Email: p.braid@virgin.net
Meetings: Once a month.

Chesterfield Astronomical Society
President: Mr D. Blackburn, 71 Middlecroft Road, Stavely, Chesterfield, Derbyshire S41 3XG. Tel: 07909 570754.
Website: www.chesterfield-as.org.uk
Meetings: Barnet Observatory, Newbold, each Friday.

Clacton & District Astronomical Society
Secretary: C. L. Haskell, 105 London Road, Clacton-on-Sea, Essex.

Cleethorpes & District Astronomical Society
Secretary: C. Illingworth, 38 Shaw Drive, Grimsby, South Humberside.
Meetings: Beacon Hill Observatory, Cleethorpes. 1st Wednesday each month.

Cleveland & Darlington Astronomical Society
Contact: Dr John McCue, 40 Bradbury Rd., Stockton-on-Tees, Cleveland TS20 1LE.
Meetings: Grindon Parish Hall, Thorpe Thewles, near Stockton-on-Tees. 2nd Friday, monthly.

Cork Astronomy Club
Website: www.corkastronomyclub.com
Email: astronomycork@yahoo.ie
Meetings: UCC, Civil Engineering Building, 2nd Monday each month, Sept.–May, 8.00 p.m.

Cornwall Astronomical Society
Secretary: J.M. Harvey, 1 Tregunna Close, Porthleven, Cornwall TR13 9LW.
Meetings: Godolphin Club, Wendron Street, Helston, Cornwall. 2nd and 4th Thursday of each month, 7.30 for 8 p.m.

Cotswold Astronomical Society
Secretary: Rod Salisbury, Grove House, Christchurch Road, Cheltenham, Gloucestershire GL50 2PN.
Website: www.members.nbci.com/CotswoldAS
Meetings: Shurdington Church Hall, School Lane, Shurdington, Cheltenham. 2nd Saturday each month, 8 p.m.

Coventry & Warwickshire Astronomical Society
Secretary: Steve Payne, 68 Stonebury Avenue, Eastern Green, Coventry CV5 7FW.
Website: www.cawas.freeserve.co.uk; *Email:* sjp2000@thefarside57.freeserve.co.uk
Meetings: The Earlsdon Church Hall, Albany Road, Earlsdon, Coventry. 2nd Friday, monthly, Sept.–June.

Crawley Astronomical Society
Secretary: Ron Gamer, 1 Pevensey Close, Pound Hill, Crawley, West Sussex RH10 7BL.
Meetings: Ifield Community Centre, Ifield Road, Crawley. 3rd Friday each month, 7.30 p.m.

Crayford Manor House Astronomical Society
Secretary: Roger Pickard, 28 Appletons, Hadlow, Kent TM1 0DT.
Meetings: Manor House Centre, Crayford. Monthly during term time.

Crewkerne and District Astronomical Society (CADAS)
Chairman: Kevin Dodgson, 46 Hermitage Street, Crewkerne, Somerset TA18 8ET.
Email: crewastra@aol.com

Croydon Astronomical Society
Secretary: John Murrell, 17 Dalmeny Road, Carshalton, Surrey.
Meetings: Lecture Theatre, Royal Russell School, Combe Lane, South Croydon. Alternate Fridays, 7.45 p.m.

Derby & District Astronomical Society
Secretary: Ian Bennett, Freers Cottage, Sutton Lane, Etwall.
Website: www.derby-astro-soc.fsnet/index.html
Email: bennett.lovatt@btinternet.com
Meetings: Friends Meeting House, Derby. 1st Friday each month, 7.30 p.m.

Doncaster Astronomical Society
Secretary: A. Anson, 15 Cusworth House, St James Street, Doncaster DN1 3AY
Website: www.donastro.freeserve.co.uk; *Email:* space@donastro.freeserve.co.uk

Meetings: St George's Church House, St George's Church, Church Way, Doncaster. 2nd and 4th Thursday of each month, commencing at 7.30 p.m.

Dumfries Astronomical Society
Secretary: Klaus Schiller, lesley.burrell@btinternet.com.
Website: www.astronomers.ukscientists.com
Meetings: George St Church Hall, George St, Dumfries. 2nd Tuesday of each month, Sept.–May.

Dundee Astronomical Society
Secretary: G. Young, 37 Polepark Road, Dundee, Tayside DD1 5QT.
Meetings: Mills Observatory, Balgay Park, Dundee. 1st Friday each month, 7.30 p.m. Sept.–Apr.

Easington and District Astronomical Society
Secretary: T. Bradley, 52 Jameson Road, Hartlepool, Co. Durham.
Meetings: Easington Comprehensive School, Easington Colliery. Every 3rd Thursday throughout the year, 7.30 p.m.

East Antrim Astronomical Society
Secretary: Stephen Beasant
Website: www.eaas.co.uk
Meetings: Ballyclare High School, Ballyclare, County Antrim. First Monday each month.

Eastbourne Astronomical Society
Secretary: Peter Gill, 18 Selwyn House, Selwyn Road, Eastbourne, East Sussex BN21 2LF.
Meetings: Willingdon Memorial Hall, Church Street, Willingdon. One Saturday per month, Sept –July, 7.30 p.m.

East Riding Astronomers
Secretary: Tony Scaife, 15 Beech Road, Elloughton, Brough, North Humberside HU15 1JX.
Meetings: As arranged.

East Sussex Astronomical Society
Secretary: Marcus Croft, 12 St Mary's Cottages, Ninfield Road, Bexhill-on-Sea, East Sussex.
Website: www.esas.org.uk
Meetings: St Mary's School, Wrestwood Road, Bexhill. 1st Thursday of each month, 8 p.m.

Edinburgh University Astronomical Society
Secretary: c/o Dept. of Astronomy, Royal Observatory, Blackford Hill, Edinburgh.

Ewell Astronomical Society
Secretary: Richard Gledhill, 80 Abinger Avenue, Cheam SM2 7LW.
Website: www.ewell-as.co.uk
Meetings: St Mary's Church Hall, London Road, Ewell. 2nd Friday of each month except August, 7.45 p.m.

Exeter Astronomical Society
Secretary: Tim Sedgwick, Old Dower House, Half Moon, Newton St Cyres, Exeter, Devon EX5 5AE.
Meetings: The Meeting Room, Wynards, Magdalen Street, Exeter. 1st Thursday of month.

Farnham Astronomical Society
Secretary: Laurence Anslow, 'Asterion', 18 Wellington Lane, Farnham, Surrey
GU9 9BA.
Meetings: Central Club, South Street, Farnham. 2nd Thursday each month, 8 p.m.

Foredown Tower Astronomy Group
Secretary: M. Feist, Foredown Tower Camera Obscura, Foredown Road, Portslade,
East Sussex BN41 2EW.
Meetings: At the above address, 3rd Tuesday each month. 7 p.m. (winter), 8 p.m.
(summer).

Greenock Astronomical Society
Secretary: Carl Hempsey, 49 Brisbane Street, Greenock.
Meetings: Greenock Arts Guild, 3 Campbell Street, Greenock.

Grimsby Astronomical Society
Secretary: R. Williams, 14 Richmond Close, Grimsby, South Humberside.
Meetings: Secretary's home. 2nd Thursday each month, 7.30 p.m.

Guernsey: La Société Guernesiasie Astronomy Section
Secretary: Debby Quertier, Lamorna, Route Charles, St Peter Port, Guernsey GY1 1QS.
and Jessica Harris, Keanda, Les Sauvagees, St Sampson's, Guernsey GY2 4XT.
Meetings: Observatory, Rue du Lorier, St Peter's. Tuesdays, 8 p.m.

Guildford Astronomical Society
Secretary: A. Langmaid, 22 West Mount, The Mount, Guildford, Surrey GU2 5HL.
Meetings: Guildford Institute, Ward Street, Guildford. 1st Thursday each month
except Aug., 7.30 p.m.

Gwynedd Astronomical Society
Secretary: Mr Ernie Greenwood, 18 Twrcelyn Street, Llanerchymedd, Anglesey
LL74 8TL.
Meetings: Dept. of Electronic Engineering, Bangor University. 1st Thursday each
month except Aug., 7.30 p.m.

The Hampshire Astronomical Group
Secretary: Geoff Mann, 10 Marie Court, 348 London Road, Waterlooville,
Hampshire PO7 7SR.
Website: www.hantsastro.demon.co.uk; *Email:* Geoff.Mann@hazleton97.fsnet.co.uk
Meetings: 2nd Friday, Clanfield Memorial Hall, all other Fridays Clanfield
Observatory.

Hanney & District Astronomical Society
Secretary: Bob Church, 47 Upthorpe Drive, Wantage, Oxfordshire OX12 7DG.
Meetings: Last Thursday each month, 8 p.m.

Harrogate Astronomical Society
Secretary: Brian Bonser, 114 Main Street, Little Ouseburn TO5 9TG.
Meetings: National Power HQ, Beckwith Knowle, Harrogate. Last Friday each
month.

Havering Astronomical Society
Secretary: Frances Ridgley, 133 Severn Drive, Upminster, Essex RM14 1PP.
Meetings: Cranham Community Centre, Marlborough Gardens, Upminster, Essex.
3rd Wednesday each month except July and Aug., 7.30 p.m.

Heart of England Astronomical Society
Secretary: John Williams, 100 Stanway Road, Shirley, Solihull B90 3JG.
Website: www.members.aol.com/hoeas/home.html; *Email:* hoeas@aol.com

Meetings: Furnace End Village, over Whitacre, Warwickshire. Last Thursday each month, except June, July & Aug., 8 p.m.

Hebden Bridge Literary & Scientific Society, Astronomical Section
Secretary: Peter Jackson, 44 Gilstead Lane, Bingley, West Yorkshire BD16 3NP.
Meetings: Hebden Bridge Information Centre. Last Wednesday, Sept.–May.

Herefordshire Astronomical Society
Secretary: Paul Olver, The Buttridge, Wellington Lane, Canon Pyon, Hereford HR4 8NL.
Email: info@hsastro.org.uk
Meetings: The Kindle Centre, ASDA Supermarket, Hereford. 1st Thursday of every month (except August) 7 p.m.

Herschel Astronomy Society
Secretary: Kevin Bishop, 106 Holmsdale, Crown Wood, Bracknell, Berkshire RG12 3TB.
Meetings: Eton College. 2nd Friday each month, 7.30 p.m.

Highlands Astronomical Society
Secretary: Richard Green, 11 Drumossie Avenue, Culcabock, Inverness IV2 3SJ.
Meetings: The Spectrum Centre, Inverness. 1st Tuesday each month, 7.30 p.m.

Hinckley & District Astronomical Society
Secretary: Mr S. Albrighton, 4 Walnut Close, The Bridleways, Hartshill, Nuneaton, Warwickshire CV10 0XH.
Meetings: Burbage Common Visitors Centre, Hinckley. 1st Tuesday Sept.–May, 7.30 p.m.

Horsham Astronomy Group (was **Forest Astronomical Society**)
Secretary: Dan White, 32 Burns Close, Horsham, West Sussex RH12 5PF.
Email: secretary@horshamastronomy.com
Meetings: 1st Wednesday each month.

Howards Astronomy Club
Secretary: H. Ilett, 22 St George's Avenue, Warblington, Havant, Hampshire.
Meetings: To be notified.

Huddersfield Astronomical and Philosophical Society
Secretary: Lisa B. Jeffries, 58 Beaumont Street, Netherton, Huddersfield, West Yorkshire IID4 7IIE.
Email: l.b.jeffries@hud.ac.uk
Meetings: 4a Railway Street, Huddersfield. Every Wednesday and Friday, 7.30 p.m.

Hull and East Riding Astronomical Society
President: Sharon E. Long
Email: charon@charon.karoo.co.uk
Website: http://www.heras.org.uk
Meetings: The Wilberforce Building, Room S25, University of Hull, Cottingham Road, Hull. 2nd Monday each month, Sept.–May, 7.30–9.30 p.m.

Ilkeston & District Astronomical Society
Secretary: Mark Thomas, 2 Elm Avenue, Sandiacre, Nottingham NG10 5EJ.
Meetings: The Function Room, Erewash Museum, Anchor Row, Ilkeston. 2nd Tuesday monthly, 7.30 p.m.

Ipswich, Orwell Astronomical Society
Secretary: R. Gooding, 168 Ashcroft Road, Ipswich.
Meetings: Orwell Park Observatory, Nacton, Ipswich. Wednesdays, 8 p.m.

Irish Astronomical Association
President: Terry Moseley, 31 Sunderland Road, Belfast BT6 9LY, Northern Ireland.
Email: terrymosel@aol.com
Meetings: Ashby Building, Stranmillis Road, Belfast. Alternate Wednesdays,
7.30 p.m.

Irish Astronomical Society
Secretary: James O'Connor, PO Box 2547, Dublin 15, Eire.
Meetings: Ely House, 8 Ely Place, Dublin 2. 1st and 3rd Monday each month.

Isle of Man Astronomical Society
Secretary: James Martin, Ballaterson Farm, Peel, Isle of Man IM5 3AB.
Email: ballaterson@manx.net
Meetings: Isle of Man Observatory, Foxdale. 1st Thursday of each month, 8 p.m.

Isle of Wight Astronomical Society
Secretary: J. W. Feakins, 1 Hilltop Cottages, High Street, Freshwater, Isle of Wight.
Meetings: Unitarian Church Hall, Newport, Isle of Wight. Monthly.

Jersey Astronomy Club
Secretary: Jodie Masterman, *Tel:* 07797 813681
Email: jodiemasterman@yahoo.co.uk
Chairman: Martin Ahier, *Tel:* (01534) 732157
Meetings: Sir Patrick Moore Astronomy Centre, Lex Creux Country Park, St Brelade,
Jersey, 2nd Monday of every month (except August), 8.00 p.m.

Keele Astronomical Society
Secretary: Natalie Webb, Department of Physics, University of Keele, Keele,
Staffordshire ST5 5BG.
Meetings: As arranged during term time.

Kettering and District Astronomical Society
Asst. Secretary: Steve Williams, 120 Brickhill Road, Wellingborough,
Northamptonshire.
Meetings: Quaker Meeting Hall, Northall Street, Kettering, Northamptonshire.
1st Tuesday each month, 7.45 p.m.

King's Lynn Amateur Astronomical Association
Secretary: P. Twynman, 17 Poplar Avenue, RAF Marham, King's Lynn.
Meetings: As arranged.

Lancaster and Morecambe Astronomical Society
Secretary: Mrs E. Robinson, 4 Bedford Place, Lancaster LA1 4EB.
Email: ehelenerob@btinternet.com
Meetings: Church of the Ascension, Torrisholme. 1st Wednesday each month except
July and Aug.

Knowle Astronomical Society
Secretary: Nigel Foster, 21 Speedwell Drive, Balsall Common, Coventry,
West Midlands CV7 7AU.
Meetings: St George & St Theresa's Parish Centre, 337 Station Road, Dorridge,
Solihull, West Midlands B93 8TZ. 1st Monday of each month (+/– 1 week for Bank
Holidays) except August.

Lancaster University Astronomical Society
Secretary: c/o Students' Union, Alexandra Square, University of Lancaster.
Meetings: As arranged.

Astronomical Societies in the British Isles

Layman's Astronomical Society
Secretary: John Evans, 10 Arkwright Walk, The Meadows, Nottingham.
Meetings: The Popular, Bath Street, Ilkeston, Derbyshire. Monthly.

Leeds Astronomical Society
Secretary: Mark A. Simpson, 37 Roper Avenue, Gledhow, Leeds LS8 1LG.
Meetings: Centenary House, North Street. 2nd Wednesday each month, 7.30 p.m.

Leicester Astronomical Society
Secretary: Dr P. J. Scott, 21 Rembridge Close, Leicester LE3 9AP.
Meetings: Judgemeadow Community College, Marydene Drive, Evington, Leicester. 2nd and 4th Tuesdays each month, 7.30 p.m.

Letchworth and District Astronomical Society
Secretary: Eric Hutton, 14 Folly Close, Hitchin, Hertfordshire.
Meetings: As arranged.

Lewes Amateur Astronomers
Secretary: Christa Sutton, 8 Tower Road, Lancing, West Sussex BN15 9HT.
Meetings: The Bakehouse Studio, Lewes. Last Wednesday each month.

Limerick Astronomy Club
Secretary: Tony O'Hanlon, 26 Ballycannon Heights, Meelick, Co. Clare, Eire.
Meetings: Limerick Senior College, Limerick. Monthly (except June and Aug.), 8 p.m.

Lincoln Astronomical Society
Secretary: David Swaey, 'Everglades', 13 Beaufort Close, Lincoln LN2 4SF.
Meetings: The Lecture Hall, off Westcliffe Street, Lincoln. 1st Tuesday each month.

Liverpool Astronomical Society
Secretary: Mr K. Clark, 31 Sandymount Drive, Wallasey, Merseyside L45 0LJ.
Meetings: Lecture Theatre, Liverpool Museum. 3rd Friday each month, 7 p.m.

Norman Lockyer Observatory Society
Secretary: G. E. White, PO Box 9, Sidmouth EX10 0YQ.
Website: www.cx.ac.uk/nlo/; *Email:* g.e.white@ex.ac.uk
Meetings: Norman Lockyer Observatory, Sidmouth. Fridays and 2nd Monday each month, 7.30 p.m.

Loughton Astronomical Society
Secretary: Charles Munton, 14a Manor Road, Wood Green, London N22 4YJ.
Meetings: 1st Theydon Bois Scout Hall, Loughton Lane, Theydon Bois. Weekly.

Lowestoft and Great Yarmouth Regional Astronomers (LYRA) Society
Secretary: Simon Briggs, 28 Sussex Road, Lowestoft, Suffolk.
Meetings: Community Wing, Kirkley High School, Kirkley Run, Lowestoft. 3rd Thursday each month, 7.30 p.m.

Luton Astronomical Society
Secretary: Mr G. Mitchell, Putteridge Bury, University of Luton, Hitchin Road, Luton.
Website: www.lutonastrosoc.org.uk; *Email:* user998491@aol.com
Meetings: Univ. of Luton, Putteridge Bury (except June, July and August), or Someries Junior School, Wigmore Lane, Luton (July and August only), last Thursday each month, 7.30–9.00 p.m.

Lytham St Anne's Astronomical Association
Secretary: K. J. Porter, 141 Blackpool Road, Ansdell, Lytham St Anne's, Lancashire.
Meetings: College of Further Education, Clifton Drive South, Lytham St Anne's. 2nd Wednesday monthly Oct.–June.

Macclesfield Astronomical Society
Secretary: Mr John H. Thomson, 27 Woodbourne Road, Sale, Cheshire M33 3SY
Website: www.maccastro.com; *Email:* jhandlc@yahoo.com
Meetings: Jodrell Bank Science Centre, Goostrey, Cheshire. 1st Tuesday of every month, 7 p.m.

Maidenhead Astronomical Society
Secretary: Tim Haymes, Hill Rise, Knowl Hill Common, Knowl Hill, Reading RG10 9YD.
Meetings: Stubbings Church Hall, near Maidenhead. 1st Friday Sept.–June.

Maidstone Astronomical Society
Secretary: Stephen James, 4 The Cherry Orchard, Haddow, Tonbridge, Kent.
Meetings: Nettlestead Village Hall. 1st Tuesday in the month except July and Aug., 7.30 p.m.

Manchester Astronomical Society
Secretary: Mr Kevin J. Kilburn FRAS, Godlee Observatory, UMIST, Sackville Street, Manchester M60 1QD.
Website: www.u-net.com/ph/mas/; *Email:* kkilburn@globalnet.co.uk
Meetings: At the Godlee Observatory. Thursdays, 7 p.m., except below.
Free Public Lectures: Renold Building UMIST, third Thursday Sept.–Mar., 7.30 p.m.

Mansfield and Sutton Astronomical Society
Secretary: Angus Wright, Sherwood Observatory, Coxmoor Road, Sutton-in-Ashfield, Nottinghamshire NG17 5LF.
Meetings: Sherwood Observatory, Coxmoor Road. Last Tuesday each month, 7.30 p.m.

Mexborough and Swinton Astronomical Society
Secretary: Mark R. Benton, 14 Sandalwood Rise, Swinton, Mexborough, South Yorkshire S64 8PN.
Website: www.msas.org.uk; *Email:* mark@masas.f9.co.uk
Meetings: Swinton WMC. Thursdays, 7.30 p.m.

Mid-Kent Astronomical Society
Secretary: Peter Parish, 30 Wooldeys Road, Rainham, Kent ME8 7NU.
Meetings: Bredhurst Village Hall, Hurstwood Road, Bredhurst, Kent. 2nd and last Fridays each month except August, 7.45 p.m.
Website: www.mkas-site.co.uk

Milton Keynes Astronomical Society
Secretary: Mike Leggett, 19 Matilda Gardens, Shenley Church End, Milton Keynes MK5 6HT.
Website: www.mkas.org.uk; *Email:* mike-pat-leggett@shenley9.fsnet.co.uk
Meetings: Rectory Cottage, Bletchley. Alternate Fridays.

Moray Astronomical Society
Secretary: Richard Pearce, 1 Forsyth Street, Hopeman, Elgin, Moray, Scotland.
Meetings: Village Hall Close, Co. Elgin.

Newbury Amateur Astronomical Society (NAAS)
Secretary: Mrs Monica Balstone, 37 Mount Pleasant, Tadley RG26 4BG.
Meetings: United Reformed Church Hall, Cromwell Place, Newbury. 1st Friday of month, Sept.–June.

Newcastle-on-Tyne Astronomical Society
Secretary: C. E. Willits, 24 Acomb Avenue, Seaton Delaval, Tyne and Wear.
Meetings: Zoology Lecture Theatre, Newcastle University. Monthly.

North Aston Space & Astronomical Club
Secretary: W. R. Chadburn, 14 Oakdale Road, North Aston, Sheffield.
Meetings: To be notified.

Northamptonshire Natural History Society (Astronomy Section)
Secretary: R. A. Marriott, 24 Thirlestane Road, Northampton NN4 8HD.
Email: ram@hamal.demon.co.uk
Meetings: Humfrey Rooms, Castilian Terrace, Northampton. 2nd and last Mondays, most months, 7.30 p.m.

Northants Amateur Astronomers
Secretary: Mervyn Lloyd, 76 Havelock Street, Kettering, Northamptonshire.
Meetings: 1st and 3rd Tuesdays each month, 7.30 p.m.

North Devon Astronomical Society
Secretary: P. G. Vickery, 12 Broad Park Crescent, Ilfracombe, Devon EX34 8DX.
Meetings: Methodist Hall, Rhododendron Avenue, Sticklepath, Barnstaple. 1st Wednesday each month, 7.15 p.m.

North Dorset Astronomical Society
Secretary: J. E. M. Coward, The Pharmacy, Stalbridge, Dorset.
Meetings: Charterhay, Stourton, Caundle, Dorset. 2nd Wednesday each month.

North Downs Astronomical Society
Secretary: Martin Akers, 36 Timber Tops, Lordswood, Chatham, Kent ME5 8XQ.
Meetings: Vigo Village Hall. 3rd Thursday each month. 7.30 p.m.

North-East London Astronomical Society
Secretary: Mr B. Beeston, 38 Abbey Road, Bush Hill Park, Enfield EN1 2QN.
Meetings: Wanstead House, The Green, Wanstead. 3rd Sunday each month (except Aug.), 3 p.m.

North Gwent and District Astronomical Society
Secretary: Jonathan Powell, 14 Lancaster Drive, Gilwern, nr Abergavenny, Monmouthshire NP7 0AA.
Meetings: Gilwern Community Centre. 15th of each month, 7.30 p.m.

North Staffordshire Astronomical Society
Secretary: Duncan Richardson, Halmerend Hall Farm, Halmerend, Stoke-on-Trent, Staffordshire ST7 8AW.
Email: dwr@enterprise.net
Meetings: 21st Hartstill Scout Group HQ, Mount Pleasant, Newcastle-under-Lyme ST5 1DR. 1st Tuesday each month (except July and Aug.), 7–9.30 p.m.

Northumberland Astronomical Society
Contact: Dr Adrian Jametta, 1 Lake Road, Hadston, Morpeth, Northumberland NE65 9TF.
Email: adrian@themoon.co.uk
Website: www.nastro.org.uk
Meetings: Hauxley Nature Reserve (near Amble). Last Thursday of every month (except December), 7.30 pm. Additional meetings and observing sessions listed on website.
Tel: 07984 154904

North Western Association of Variable Star Observers
Secretary: Jeremy Bullivant, 2 Beaminster Road, Heaton Mersey, Stockport, Cheshire.
Meetings: Four annually.

Yearbook of Astronomy 2012

Norwich Astronomical Society
Secretary: Dave Balcombe, 52 Folly Road, Wymondham, Norfolk NR18 0QR.
Website: www.norwich.astronomical.society.org.uk
Meetings: Seething Observatory, Toad Lane, Thwaite St Mary, Norfolk. Every Friday, 7.30 p.m.

Nottingham Astronomical Society
Secretary: C. Brennan, 40 Swindon Close, The Vale, Giltbrook, Nottingham NG16 2WD.
Meetings: Djanogly City Technology College, Sherwood Rise (B682). 1st and 3rd Thursdays each month, 7.30 p.m.

Oldham Astronomical Society
Secretary: P. J. Collins, 25 Park Crescent, Chadderton, Oldham.
Meetings: Werneth Park Study Centre, Frederick Street, Oldham. Fortnightly, Friday.

Open University Astronomical Society
Secretary: Dr Andrew Norton, Department of Physics and Astronomy, The Open University, Walton Hall, Milton Keynes MK7 6AA.
Website: www.physics.open.ac.uk/research/astro/a_club.html
Meetings: Open University, Milton Keynes. 1st Tuesday of every month, 7.30 p.m.

Orpington Astronomical Society
Secretary: Dr Ian Carstairs, 38 Brabourne Rise, Beckenham, Kent BR3 2SG.
Meetings: High Elms Nature Centre, High Elms Country Park, High Elms Road, Farnborough, Kent. 4th Thursday each month, Sept.–July, 7.30 p.m.

Papworth Astronomy Club
Contact: Keith Tritton, Magpie Cottage, Fox Street, Great Gransden, Sandy, Bedfordshire SG19 3AA.
Email: kpt2@tutor.open.ac.uk
Meetings: Bradbury Progression Centre, Church Lane, Papworth Everard, nr Huntingdon. 1st Wednesday each month, 7 p.m.

Peterborough Astronomical Society
Secretary: Sheila Thorpe, 6 Cypress Close, Longthorpe, Peterborough.
Meetings: 1st Thursday every month, 7.30 p.m.

Plymouth Astronomical Society
Secretary: Alan G. Penman, 12 St Maurice View, Plympton, Plymouth, Devon PL7 1FQ.
Email: oakmount12@aol.com
Meetings: Glynis Kingham Centre, YMCA Annex, Lockyer Street, Plymouth. 2nd Friday each month, 7.30 p.m.

PONLAF
Secretary: Matthew Hepburn, 6 Court Road, Caterham, Surrey CR3 5RD.
Meetings: Room 5, 6th floor, Tower Block, University of North London. Last Friday each month during term time, 6.30 p.m.

Port Talbot Astronomical Society (formerly **Astronomical Society of Wales**)
Secretary: Mr J. Hawes, 15 Lodge Drive, Baglan, Port Talbot, West Glamorgan SA12 8UD.
Meetings: Port Talbot Arts Centre. 1st Tuesday each month, 7.15 p.m.

Preston & District Astronomical Society
Secretary: P. Sloane, 77 Ribby Road, Wrea Green, Kirkham, Preston, Lancashire.
Meetings: Moor Park (Jeremiah Horrocks) Observatory, Preston. 2nd Wednesday, last Friday each month, 7.30 p.m.

Reading Astronomical Society
Secretary: Mrs Ruth Sumner, 22 Anson Crescent, Shinfield, Reading RG2 8JT.
Meetings: St Peter's Church Hall, Church Road, Earley. 3rd Friday each month, 7 p.m.

Renfrewshire Astronomical Society
Secretary: Ian Martin, 10 Aitken Road, Hamilton, South Lanarkshire ML3 7YA.
Website: www.renfrewshire-as.co.uk; *Email:* RenfrewAS@aol.com
Meetings: Coats Observatory, Oakshaw Street, Paisley. Fridays, 7.30 p.m.

Rower Astronomical Society
Secretary: Mary Kelly, Knockatore, The Rower, Thomastown, Co. Kilkenny, Eire.

St Helens Amateur Astronomical Society
Secretary: Carl Dingsdale, 125 Canberra Avenue, Thatto Heath, St Helens, Merseyside WA9 5RT.
Meetings: As arranged.

Salford Astronomical Society
Secretary: Mrs Kath Redford, 2 Albermarle Road, Swinton, Manchester M27 5ST.
Meetings: The Observatory, Chaseley Road, Salford. Wednesdays.

Salisbury Astronomical Society
Secretary: Mrs R. Collins, 3 Fairview Road, Salisbury, Wiltshire SP1 1JX.
Meetings: Glebe Hall, Winterbourne Earls, Salisbury. 1st Tuesday each month.

Sandbach Astronomical Society
Secretary: Phil Benson, 8 Gawsworth Drive, Sandbach, Cheshire.
Meetings: Sandbach School, as arranged.

Sawtry & District Astronomical Society
Secretary: Brooke Norton, 2 Newton Road, Sawtry, Huntingdon, Cambridgeshire PE17 5UT.
Meetings: Greenfields Cricket Pavilion, Sawtry Fen. Last Friday each month.

Scarborough & District Astronomical Society
Secretary: Mrs S. Anderson, Basin House Farm, Sawdon, Scarborough, North Yorkshire.
Meetings: Scarborough Public Library. Last Saturday each month, 7–9 p.m.

Scottish Astronomers Group
Secretary: Dr Ken Mackay, Hayford House, Cambusbarron, Stirling FK7 9PR.
Meetings: North of Hadrian's Wall, twice yearly.

Sheffield Astronomical Society
Secretary: Darren Swindels, 102 Sheffield Road, Woodhouse, Sheffield, South Yorkshire S13 7EU.
Website: www.sheffieldastro.org.uk; *Email:* info@sheffieldastro.org.uk
Meetings: Twice monthly at Mayfield Environmental Education Centre, David Lane, Fulwood, Sheffield S10, 7.30–10 p.m.

Shetland Astronomical Society
Secretary: Peter Kelly, The Glebe, Fetlar, Shetland ZE2 9DJ.
Email: theglebe@zetnet.co.uk
Meetings: Fetlar, Fridays, Oct.–Mar.

Shropshire Astronomical Society
Contact: Mr David Woodward, 20 Station Road, Condover, Shrewsbury, Shropshire SY5 7BQ.
Website: http://www.shropshire-astro.com; *Email:* jacquidodds@ntlworld.com
Meetings: Quarterly talks at the Gateway Arts and Education Centre, Chester Street, Shrewsbury and monthly observing meetings at Rodington Village Hall.

Sidmouth and District Astronomical Society
Secretary: M. Grant, Salters Meadow, Sidmouth, Devon.
Meetings: Norman Lockyer Observatory, Salcombe Hill. 1st Monday in each month.

Solent Amateur Astronomers
Secretary: Ken Medway, 443 Burgess Road, Swaythling, Southampton SO16 3BL.
Website: www.delscope.demon.co.uk;
Email: ken@medway1875.freeserve.co.uk
Meetings: Communications Room 2 Oasis Academy, Fairisle Road, Lordshill, Southampton, SO16 8BY. 3rd Tuesday each month, 7.30 p.m.

Southampton Astronomical Society
Secretary: John Thompson, 4 Heathfield, Hythe, Southampton SO45 5BJ.
Website: www.home.clara.net/lmhobbs/sas.html;
Email: John.G.Thompson@Tesco.net
Meetings: Conference Room 3, The Civic Centre, Southampton. 2nd Thursday each month (except Aug.), 7.30 p.m.

South Downs Astronomical Society
Secretary: J. Green, 46 Central Avenue, Bognor Regis, West Sussex PO21 5HH.
Website: www.southdowns.org.uk
Meetings: Chichester High School for Boys. 1st Friday in each month (except Aug.).

South-East Essex Astronomical Society
Secretary: C. P. Jones, 29 Buller Road, Laindon, Essex.
Website: www.seeas.dabsol.co.uk/; *Email:* cpj@cix.co.uk
Meetings: Lecture Theatre, Central Library, Victoria Avenue, Southend-on-Sea. Generally 1st Thursday in month, Sept.–May, 7.30 p.m.

South-East Kent Astronomical Society
Secretary: Andrew McCarthy, 25 St Paul's Way, Sandgate, near Folkestone, Kent CT20 3NT.
Meetings: Monthly.

South Lincolnshire Astronomical & Geophysical Society
Secretary: Ian Farley, 12 West Road, Bourne, Lincolnshire PE10 9PS.
Meetings: Adult Education Study Centre, Pinchbeck. 3rd Wednesday each month, 7.30 p.m.

Southport Astronomical Society
Secretary: Patrick Brannon, Willow Cottage, 90 Jacksmere Lane, Scarisbrick, Ormskirk, Lancashire L40 9RS.
Meetings: Monthly Sept.–May, plus observing sessions.

Southport, Ormskirk and District Astronomical Society
Secretary: J. T. Harrison, 92 Cottage Lane, Ormskirk, Lancashire L39 3NJ.
Meetings: Saturday evenings, monthly, as arranged.

South Shields Astronomical Society
Secretary: c/o South Tyneside College, St George's Avenue, South Shields.
Meetings: Marine and Technical College. Each Thursday, 7.30 p.m.

South Somerset Astronomical Society
Secretary: G. McNelly, 11 Laxton Close, Taunton, Somerset.
Meetings: Victoria Inn, Skittle Alley, East Reach, Taunton, Somerset. Last Saturday each month, 7.30 p.m.

Astronomical Societies in the British Isles

South-West Hertfordshire Astronomical Society
Secretary: Tom Walsh, 'Finches', Coleshill Lane, Winchmore Hill, Amersham, Buckinghamshire HP7 0NP.
Meetings: Rickmansworth. Last Friday each month, Sept.–May.

Stafford and District Astronomical Society
Secretary: Miss L. Hodkinson, 6 Elm Walk, Penkridge, Staffordshire ST19 5NL.
Meetings: Weston Road High School, Stafford. Every 3rd Thursday, Sept.–May, 7.15 p.m.

Stirling Astronomical Society
Secretary: Hamish MacPhee, 10 Causewayhead Road, Stirling FK9 5ER.
Meetings: Smith Museum & Art Gallery, Dumbarton Road, Stirling. 2nd Friday each month, 7.30 p.m.

Stoke-on-Trent Astronomical Society
Secretary: M. Pace, Sundale, Dunnockstold, Alsager, Stoke-on-Trent.
Meetings: Cartwright House, Broad Street, Hanley. Monthly.

Stratford-upon-Avon Astronomical Society
Secretary: Robin Swinbourne, 18 Old Milverton, Leamington Spa, Warwickshire CV32 6SA.
Meetings: Tiddington Home Guard Club. 4th Tuesday each month, 7.30 p.m.

Sunderland Astronomical Society
Contact: Don Simpson, 78 Stratford Avenue, Grangetown, Sunderland SR2 8RZ.
Meetings: Friends Meeting House, Roker. 1st, 2nd and 3rd Sundays each month.

Sussex Astronomical Society
Secretary: Mrs C. G. Sutton, 75 Vale Road, Portslade, Sussex.
Meetings: English Language Centre, Third Avenue, Hove. Every Wednesday, 7.30–9.30 p.m., Sept.–May.

Swansea Astronomical Society
Secretary: Dr Michael Morales, 238 Heol Dulais, Birch Grove, Swansea SA7 9LH.
Website: www.crysania.co.uk/sas/astro/star
Meetings: Lecture Room C, Science Tower, University of Swansea. 2nd and 4th Thursday each month from Sept.–June, 7 p.m.

Tavistock Astronomical Society
Secretary: Mrs Ellie Coombes, Rosemount, Under Road, Gunnislake, Cornwall PL18 9JL.
Meetings: Science Laboratory, Kelly College, Tavistock. 1st Wednesday each month, 7.30 p.m.

Thames Valley Astronomical Group
Secretary: K. J. Pallet, 82a Tennyson Street, South Lambeth, London SW8 3TH.
Meetings: As arranged.

Thanet Amateur Astronomical Society
Secretary: P. F. Jordan, 85 Crescent Road, Ramsgate.
Meetings: Hilderstone House, Broadstairs, Kent. Monthly.

Torbay Astronomical Society
Secretary: Tim Moffat, 31 Netley Road, Newton Abbot, Devon TQ12 2LL.
Meetings: Torquay Boys' Grammar School, 1st Thursday in month; and Town Hall, Torquay, 3rd Thursday in month, Oct.–May, 7.30 p.m.

Tullamore Astronomical Society
Secretary: Tom Walsh, 25 Harbour Walk, Tullamore, Co. Offaly, Eire.
Website: www.iol.ie/seanmck/tas.htm; *Email:* tcwalsh25@yahoo.co.uk
Meetings: Order of Malta Lecture Hall, Tanyard, Tullamore, Co. Offaly, Eire.
Mondays at 8 p.m., every fortnight.

Tyrone Astronomical Society
Secretary: John Ryan, 105 Coolnafranky Park, Cookstown, Co. Tyrone, Northern Ireland.
Meetings: Contact Secretary.

Usk Astronomical Society
Secretary: Bob Wright, 'Llwyn Celyn', 75 Woodland Road, Croesyceiliog, Cwmbran NP44 2OX.
Meetings: Usk Community Education Centre, Maryport Street, Usk. Every Thursday during school term, 7 p.m.

Vectis Astronomical Society
Secretary: Rosemary Pears, 1 Rockmount Cottages, Undercliff Drive, St Lawrence, Ventnor, Isle of Wight PO38 1XG.
Website: www.wightskies.fsnet.co.uk/main.html;
Email: may@tatemma.freeserve.co.uk
Meetings: Lord Louis Library Meeting Room, Newport. 4th Friday each month except Dec., 7.30 p.m.

Vigo Astronomical Society
Secretary: Robert Wilson, 43 Admers Wood, Vigo Village, Meopham, Kent DA13 0SP.
Meetings: Vigo Village Hall. As arranged.

Walsall Astronomical Society
Secretary: Bob Cleverley, 40 Mayfield Road, Sutton Coldfield B74 3PZ.
Meetings: Freetrade Inn, Wood Lane, Pelsall North Common. Every Thursday.

Wealden Astronomical Society
Secretary: K.A. Woodcock, 24 Emmanuel Road, Hastings, East Sussex TN34 3LB.
Email: wealdenas@hotmail.co.uk
Meetings: Herstmonceux Science Centre. Dates, as arranged.

Wellingborough District Astronomical Society
Secretary: S. M. Williams, 120 Brickhill Road, Wellingborough, Northamptonshire.
Meetings: Gloucester Hall, Church Street, Wellingborough. 2nd Wednesday each month, 7.30 p.m.

Wessex Astronomical Society
Secretary: Leslie Fry, 14 Hanhum Road, Corfe Mullen, Dorset.
Meetings: Allendale Centre, Wimborne, Dorset. 1st Tuesday of each month.

West Cornwall Astronomical Society
Secretary: Dr R. Waddling, The Pines, Pennance Road, Falmouth, Cornwall TR11 4ED.
Meetings: Helston Football Club, 3rd Thursday each month, and St Michall's Hotel, 1st Wednesday each month, 7.30 p.m.

West of London Astronomical Society
Secretary: Duncan Radbourne, 28 Tavistock Road, Edgware, Middlesex HA8 6DA.
Website: www.wocas.org.uk
Meetings: Monthly, alternately in Uxbridge and North Harrow. 2nd Monday in month, except Aug.

Astronomical Societies in the British Isles

West Midlands Astronomical Association
Secretary: Miss S. Bundy, 93 Greenridge Road, Handsworth Wood, Birmingham.
Meetings: Dr Johnson House, Bull Street, Birmingham. As arranged.

West Yorkshire Astronomical Society
Secretary: Pete Lunn, 21 Crawford Drive, Wakefield, West Yorkshire.
Meetings: Rosse Observatory, Carleton Community Centre, Carleton Road,
Pontefract. Each Tuesday, 7.15 p.m.

Whitby and District Astronomical Society
Secretary: Rosemary Bowman, The Cottage, Larpool Drive, Whitby, North Yorkshire
YO22 4ND.
Meetings: Whitby Mission, Seafarers' Centre, Haggersgate, Whitby. 1st Tuesday of
the month, 7.30 p.m.

Whittington Astronomical Society
Secretary: Peter Williamson, The Observatory, Top Street, Whittington, Shropshire.
Meetings: The Observatory. Every month.

Wiltshire Astronomical Society
Chair: Mr Andrew J. Burns, The Knoll, Lowden Hill, Chippenham, SN15 2BT;
01249 654541
Website: www.wasnet.co.uk; *Email:* angleburns@hotmail.com
Secretary: Simon Barnes, 25 Woodcombe, Melksham, Wiltshire SN12 6HA.
Meetings: The Field Pavilion, Rusty Lane, Seend, Nr Devizes, Wiltshire. 1st Tuesday
each month, Sept.–June. Viewing evenings 4th Friday plus special events, Lacock
Playing Fields, Lacock, Wilsthire.

Wolverhampton Astronomical Society
Secretary: Mr M. Bryce, Iona, 16 Yellowhammer Court, Kidderminster,
Worcestershire DY10 4RR.
Website: www.wolvas.org.uk; *Email:* michaelbryce@wolvas.org.uk
Meetings: Beckminster Methodist Church Hall, Birches Barn Road, Wolverhampton.
Alternate Mondays, Sept.–Apr., extra dates in summer, 7.30 p.m.

Worcester Astronomical Society
Secretary: Mr S. Bateman, 12 Bozward Street, Worcester WR2 5DE.
Meetings: Room 117, Worcester College of Higher Education, Henwick Grove,
Worcester. 2nd Thursday each month, 8 p.m.

Worthing Astronomical Society
Contact: G. Boots, 101 Ardingly Drive, Worthing, West Sussex BN12 4TW.
Website: www.worthingastro.freeserve.co.uk;
Email: gboots@observatory99.freeserve.co.uk
Meetings: Heene Church Rooms, Heene Road, Worthing. 1st Wednesday each
month (except Aug.), 7.30 p.m.

Wycombe Astronomical Society
Secretary: Mr P. Treherne, 34 Honeysuckle Road, Widmer End, High Wycombe,
Buckinghamshire HP15 6BW.
Meetings: Woodrow High House, Amersham. 3rd Wednesday each month, 7.45 p.m.

The York Astronomical Society
Contact: Hazel Collett, Public Relations Officer
Tel: 07944 751277
Website: www.yorkastro.freeserve.co.uk; *Email:* info@yorkastro.co.uk
Meetings: The Knavesmire Room, York Priory Street Centre, Priory Street, York.
1st and 3rd Friday of each month (except Aug.), 8 p.m.

Any society wishing to be included in this list of local societies or to update details, including any website addresses, is invited to write to the Editor (c/o Pan Macmillan, 20 New Wharf Road, London N1 9RR or astronomy@macmillan.co.uk), so that the relevant information may be included in the next edition of the Yearbook.

The William Herschel Society maintains the museum established at 19 New King Street, Bath BA1 2BL – the only surviving Herschel House. It also undertakes activities of various kinds. New members would be welcome; those interested are asked to contact the Membership Secretary at the museum.

The South Downs Planetarium (Kingsham Farm, Kingsham Road, Chichester, West Sussex PO19 8RP) is now fully operational. For further information, visit www.southdowns.org.uk/sdpt or telephone (01243) 774400